Structure & Property Relationships for Interfaces

Edited by

John L. Walter
General Electric Company

Alexander H. King
State University of New York—Stony Brook

Kris Tangri
University of Manitoba

ASM INTERNATIONAL®
The Materials Information Society

First printing, October 1991
Second printing, January 1993

Library of Congress Catalog Card No.: 91-73858
ISBN: 0-87170-426-9
SAN: 204-7586

PRINTED IN THE UNITED STATES OF AMERICA

Preface

It is becoming increasingly clear that interfaces play an important and often controlling role in determining the chemical, mechanical, and electronic properties of materials. In view of this and the many new studies of interfaces in ceramics, composites, and ceramic superconductors that are in progress, it seemed pertinent to gather these together so that each area of investigation might learn from the other. Thus, a symposium on this subject was arranged by the Committee on Surfaces and Interfaces for the 1990 Fall Meeting of ASM International. The symposium that evolved included 24 invited speakers over four sessions. ASM's technical book publication division agreed to publish this book from manuscripts provided by the speakers. We hope that this is a timely and worthy successor to the 1979 symposium book, "Grain Boundary Structure and Kinetics," edited by R.W. Balluffi. The scope and content of this book are different, reflecting the considerable changes of emphasis in interfacial studies that have occurred over a ten-year period. Most notably different is the shift away from emphasis on pure metals and a deliberate attempt to set the stage for an era of "interfacial engineering."

The book has six divisions: "Interfaces in Metals," "Interfaces in Ordered Alloys," "Interfaces in Oxides and Ceramics," "Mechanical Properties of Interfaces," "Migration and Transformation of Interfaces," and "Electrical Properties of Interfaces." The first chapter introduces the concept of grain-boundary design and control to improve properties of polycrystalline materials. The understanding of grain-boundary structure and the effect of structure on chemical and mechanical properties are discussed in succeeding chapters.

The "weak-link" behavior of some grain boundaries in ceramic superconductors in terms of their morphology, structure, and composition is discussed in two chapters. Electrical properties of grain boundaries and the effect of current-induced mass transport along grain boundaries are discussed in three chapters. Of particular interest is the chapter on adhesion and short-range forces of mica, which, when applied to metals, might provide new insight into surface characteristics of metals.

The need to keep the symposium and the book a reasonable size belies the wealth of recent publications and indicates that the study of grain boundaries and interfaces in materials is flourishing. This book includes but a small fraction of the published information on interfaces and the control of their properties. It is, however, a beginning.

J.L. Walter
A.H. King
K. Tangri

Contents

Section I:

Interfaces in Metals

1

Interface Control in Metals

K.T. Aust and G. Palumbo
Department of Metallurgy and Materials Science
University of Toronto, Toronto, Ontario
Canada M5S 1A4

Interface control in polycrystalline metals has previously involved the control of grain size in the μm range and the formation of low-angle boundaries, for example, by obtaining a strong preferred grain orientation. As a result of studies of the structures and properties of interfaces, it is now possible to control the structure and chemistry of certain large-angle grain boundaries, control the structure and quantity of triple-line defects (i.e., the intersection line of three or more crystals), and produce single-phase and multiple-phase materials having a grain size of a few nanometers. The methods used to achieve these interface characteristics and their effects on several properties of materials are presented in this chapter.

Grain Boundaries: Control of Structure and Chemistry

Watanabe (Ref 1) introduced the concept of grain-boundary design and control to improve various properties in polycrystalline materials. This concept is based on the fact that grain boundaries having certain coincidence site lattice (CSL) orientations with relatively low Σ values have been found to exert beneficial effects on various properties in materials. The coincidence site lattice corresponds to a special rotation of one lattice relative to the other to form a three-dimensional atomic pattern in which a certain fraction of lattice points coincide (Ref 2). The Σ value is defined as the reciprocal of this fraction.

Boundaries deviating from CSL relationships by angles (θ) greater than that defined by Brandon's criterion (Ref 3), i.e., $\Delta\theta = 15°\Sigma^{-1/2}$, and those with $\Sigma > 49$, are classified as random here. The complete geometric characterization of interfaces also involves the determination of boundary plane indexes, referenced to both adjoining lattices. However, the effect of grain-boundary plane on interface properties has generally been observed to be greatest when the grain boundary is already in a low-Σ CSL orientation relationship (for example, coherent and incoherent segments of a $\Sigma3$ twin boundary) (Ref 4).

Low-Σ grain boundaries generally possess well-ordered structures and display special properties. For example, based on the published literature in this field, the present authors concluded (Ref 4) that low-Σ grain boundaries, i.e., $\Sigma \leq 29$ (and $\Delta\theta \leq 15°\Sigma^{-5/6}$), have the following properties, in contrast to higher-Σ and random boundaries:

- Less susceptibility to solute segregation
- Lower energy in *pure* metals
- Greater mobility with specific solutes in a certain concentration range
- Smaller diffusivity
- Greater resistance to grain-boundary sliding, fracture, and cavitation
- Lower intrinsic electrical resistivity
- Greater resistance to the initiation of localized corrosion

Consequently, the control of the character and density of low-Σ grain boundaries may result in the generation of new and improved properties that are quite different from those of the usual polycrystal produced without any control of the grain boundaries. Warrington and Boon have found that only 8.5% of the grain boundaries in a random distribution (i.e., no grain-boundary control) would be expected to be of the CSL type with $\Sigma \leq 25$ (excluding low-angle boundaries) ($\Sigma = 1$) and twins ($\Sigma = 3$) (Ref 5).

Solute and Boundary Mobility. One of the first experiments to indicate that a large-angle-type grain boundary may be controlled was conducted by Aust and Rutter (Ref 6). New grains, introduced by recrystallization, were allowed to grow competitively (at 300 °C) at the expense of the substructure of a melt-grown crystal. The orientations of the largest recrystallized grains were determined at 300 °C for zone-refined lead with varying tin (solute) additions. Generally, large-angle random boundaries were obtained with less than 5 ppm by weight of tin and with 50 to 1000 ppm tin. However, with additions of 5 to 40 ppm tin, a preferred orientation was obtained, corresponding to "special" boundaries with $\Sigma = 7$, 13, 17, and 19. These results were due to the different solute (Sn) interactions with the various grain boundaries, producing differences in grain-boundary migration velocities. The frequency of Σ grain boundaries in the recrystallization structure of zone-refined Al was also strongly affected by solute atoms. As seen in Fig. 1 (Ref 7), the frequency of coincidence boundaries, up to Σ17, increased from about 20 to 50% by a 10 ppm addition of Sn, while a smaller effect was observed for Cu addition, and little effect was seen for Ti.

The effect of solute concentration on the frequency of CSL-related grain boundaries is also demonstrated in Fig. 2 (Ref 8). This figure shows the influence of sulfur concentrations (0.3 to 10 ppm) on the distribution of grain-boundary types in high-purity nickel (99.999%) following recrystallization and grain growth at 1450 K (average grain size = 750 μm). More than 1100

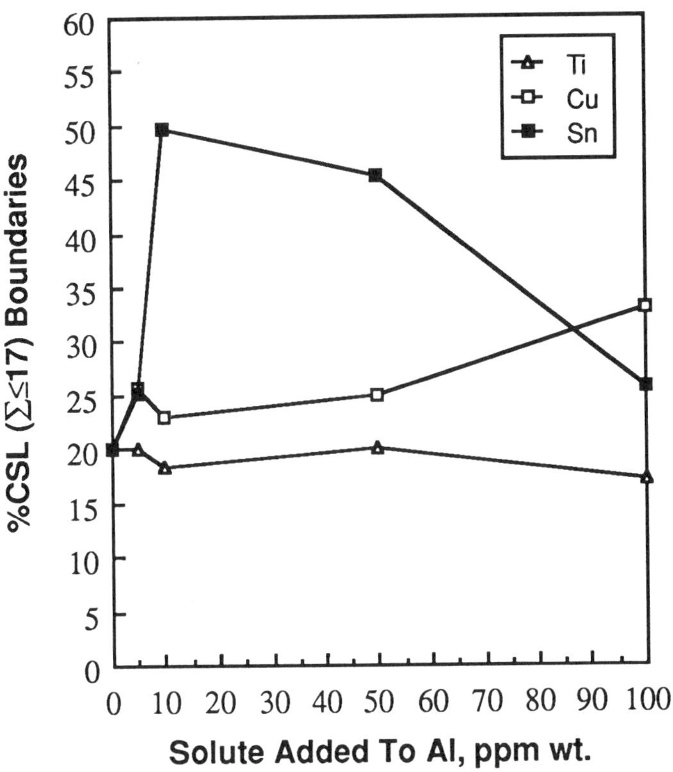

Fig. 1 The effect of Sn, Ti, and Cu additions on the frequency of coincidence boundaries (Σ ≤ 17) in zone-refined Al (Ref 7)

individual grain boundaries were characterized using the selected area channeling method and classified in the range $\Sigma = 3$ to 49. The most pronounced effect of sulfur content is observed with the frequency of CSL boundaries in the range $\Sigma = 5$ to 25, which increases from 18% with Ni containing 0.3 ppm S to 28% with Ni-3 ppm S. With a further increase in S content to 10 ppm, the frequency decreases to 21%.

Twinning and Boundary Energy. The energy, as well as the mobility, of grain boundaries must be considered. Energy arises in the formation of annealing twins, which can occur when grain boundaries migrate. It has been observed that in zone-refined metals such as Pb (Ref 9), Al (Ref 10), and Cu (Ref 11, 12) the frequency of low-Σ boundaries can be increased by twinning. In materials with a low twin boundary energy, such as Cu and Pb, repeated twinning can occur, which produces lower-Σ grain boundaries for each twinning event. The driving force for the twinning is the lower energies of the Σ grain boundaries in pure metals.

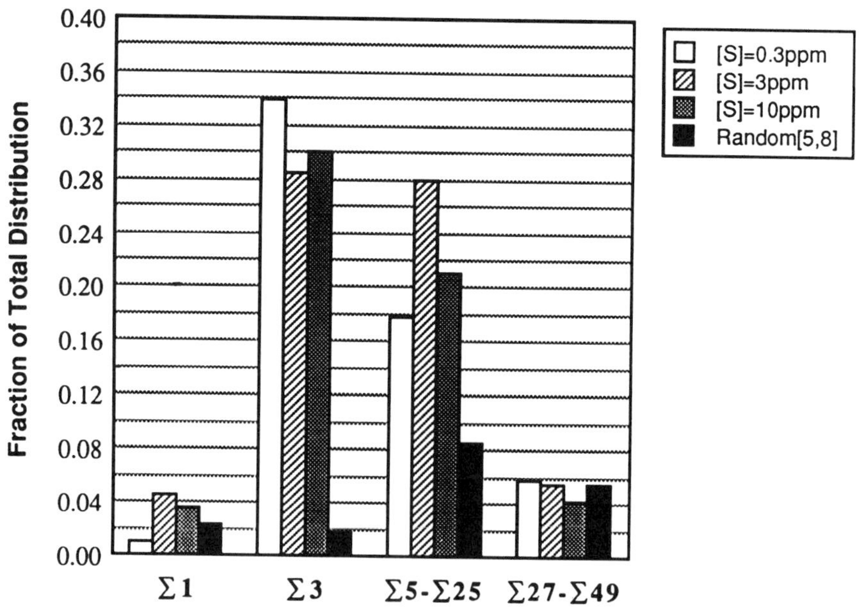

Fig. 2 The effect of sulfur concentration on the grain-boundary character distribution in strain-annealed 99.999% Ni (Ref 8)

The role of twinning in grain-boundary character distributions (GBCDs) is demonstrated in Fig. 3, which shows the effect of S content on the fraction of CSL-related grain boundaries (Σ5 to Σ49) in 99.999% Ni that are associated with twin boundaries (i.e., Σ3) at triple junctions (Ref 8). This fraction is a measure of the contribution of twinning in generating CSL-related grain boundaries, and it is observed to increase with increasing specimen purity (i.e., decreasing S content), achieving a value of 92% with Ni-0.3 ppm S. The solute effects can be rationalized in terms of selective impurity segregation, whereby increasing impurity content results in diminished energy differences between CSL- and non-CSL-related interfaces.

Summary of Solute Effects. The frequency of low-Σ grain boundaries as a function of solute concentration can be summarized in terms of energetic, kinetic, and geometric influences (Ref 8), as indicated schematically in Fig. 4. Geometric contributions to the GBCD can be considered to be independent of solute content. These geometric effects involve the generation of interfaces through the sole interaction of Σ3 boundaries, i.e., $\Sigma\ 3^n$ (for example, Σ = 9, 27, 81), referred to as twin-related variants by Randle and Brown (Ref 13). Energetic influences are primarily manifested through twinning events and are expected to be significant only at very low impurity levels. With increasing solute concentration, energetic contributions diminish. When materials are of

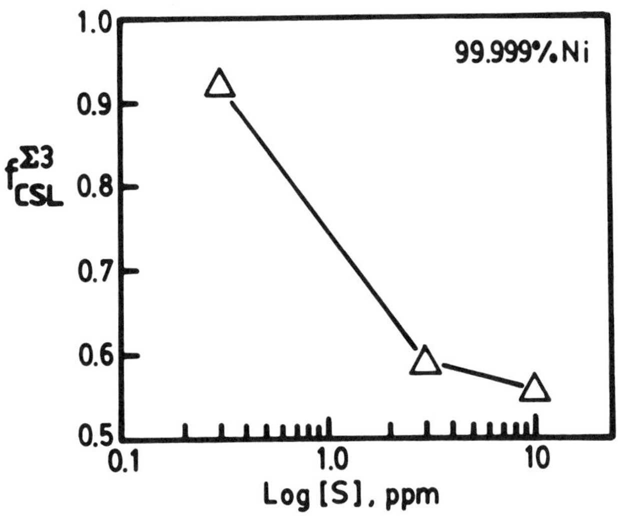

Fig. 3 The effect of sulfur concentration on the fraction of the total number of CSLs (Σ5 to Σ49) associated with twin boundaries (Σ3) at triple junctions (Ref 8)

Table 1 Effect of heat treatment on the grain-boundary character distribution for Nimonic PE 16

Heat treatment	No. of grain boundaries analyzed	CSL% ($\Sigma \leq 49$)
Unaged	43	18
Aged	74	27
Overaged	77	47

Source: Ref 14, 15

high purity, kinetic influences corresponding to differences in mobility between coincidence and noncoincidence boundaries are small. Maximum kinetic contributions to the GBCD occur at intermediate solute contents, where CSL interfaces have higher mobility than non-CSL boundaries. With further increases in solute concentration, this influence is diminished. At high solute concentrations, both energetic and kinetic factors may be considered negligible, and the GBCD should approach that of a random distribution.

Precipitation. It should be noted that Fig. 4 refers only to solute effects; it does not include the possible role of alloying elements in the form of precipitates. Randle *et al.* (Ref 14, 15) measured the orientation relationships of about 200 boundaries in a Ni-based superalloy (Nimonic PE 16) using an electron backscattering technique in conjunction with scanning electron microscopy (SEM). As shown in Table 1, the frequency of Σ boundaries increases with aging treatment. Almost half the boundaries are of the CSL type ($\Sigma \leq 49$) in the overaged samples. This overaging treatment reduces the amount of solute in solid solution and is expected to lead to reduced solute-boundary effects. This may be analogous to the Pb-Sn, Al-Sn, and Ni-S results previously discussed, in which too much solute leads to random boundaries, while a suitable solute concentration favors coincidence boundaries.

Process Control. Watanabe (Ref 16, 17), using the electron channeling pattern (ECP) method for orientation determination, found that the frequency of Σ boundaries depends on the original orientation and the amount of prestrain in Al polycrystals produced from single crystals. For example, the frequency of coincidence boundaries ($\leq \Sigma 29$) increased from 20 to 30% when compressive prestrain was increased from 60 to 80%. In addition, there is some indication that the frequency of coincidence boundaries increases with decreasing grain size (~1000 to 2 μm), based on studies by Watanabe (Ref 16) and Grabski (Ref 18) for recrystallized polycrystals of Fe alloys and Al. Aust and Rutter (Ref 19) found that the frequency of low-Σ boundaries in a pure metal (Pb) increased with decreased annealing temperature; these results are largely due to the observed greater mobility of low-Σ boundaries over random boundaries at the lower temperature.

Haasen and coworkers studied the initial state of recrystallization in deformed single crystals of Cu and Al using high-voltage electron microscopy (HVEM) and x-ray texture measurements (Ref 20-22). Nearly all observed orientations of the recrystallized grains can be described by coincidence relations to the deformed material. Also, twinning was found to lead to a lower Σ value, or to a better approximation (i.e., smaller $\Delta\theta$) to the same Σ value than before twinning. Typical Σ values observed were near Σ = 7, 11, and 13. Harase *et al.* (Ref 23) showed that the presence of low-Σ boundaries after primary recrystallization was required to supply potential nuclei for secondary recrystallization in Fe-3% Si and thereby obtain the Goss orientation texture. The latter texture is used commercially to produce the desired magnetic properties in Fe-Si transformer materials. Makita *et al.* (Ref 24) found, in addition, that the potential nuclei of secondary recrystallization in Cu already exist in the primary cube texture grains of Cu (Ref 24). These potential nuclei possess CSL orientations corresponding to Σ values of 7, 13, 21, 25, and 37. These preferred

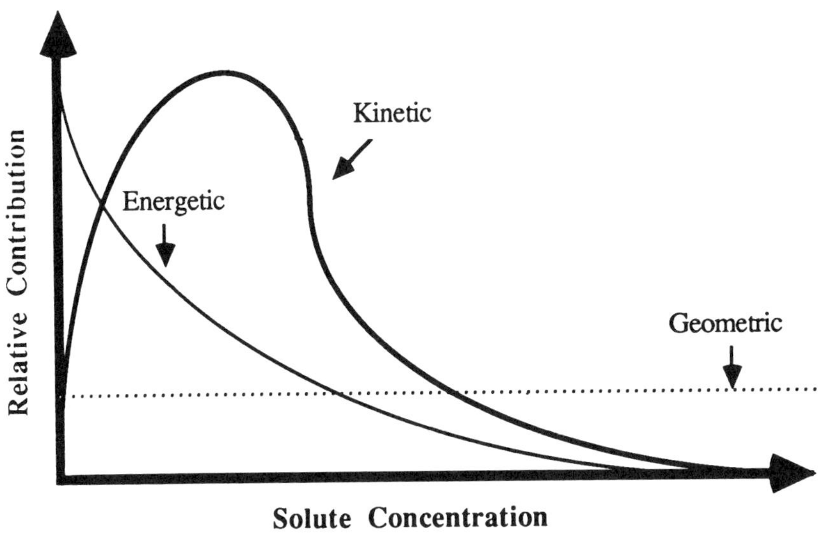

Fig. 4 The solute dependence of the primary influences governing the frequency of CSL-related boundaries in polycrystalline materials (Ref 8)

orientations of secondary grains in Cu are similar to those previously reported by Kronberg and Wilson (Ref 2) and by Sharp and Dunn (Ref 25).

Grain-Boundary Chemistry. Hondros (Ref 26) has discussed an important aspect of grain-boundary design deriving from grain-boundary microchemistry. He has considered various approaches to the problem of control of bulk material properties by changing the grain-boundary chemistry. One approach involves the use of rapid solidification techniques, producing microcrystalline alloys yielding the following advantages: the dilution of the harmful solute species by spreading the small quantity of solute over a very large grain-boundary surface; and the increase of the solid solubility of the damaging species, thereby reducing the grain-boundary concentration of the harmful solute. The production of nanocrystalline materials, to be discussed later, is a similar approach to this problem.

Another approach is that of "selective segregation," in which a harmless surface-active element having a strong repulsive interaction with a harmful segregant may expel the latter from the grain boundary. Hondros has discussed several observations dealing with embrittlement that support this idea (Ref 26). In two cases it was shown that boron is enriched at the grain boundary

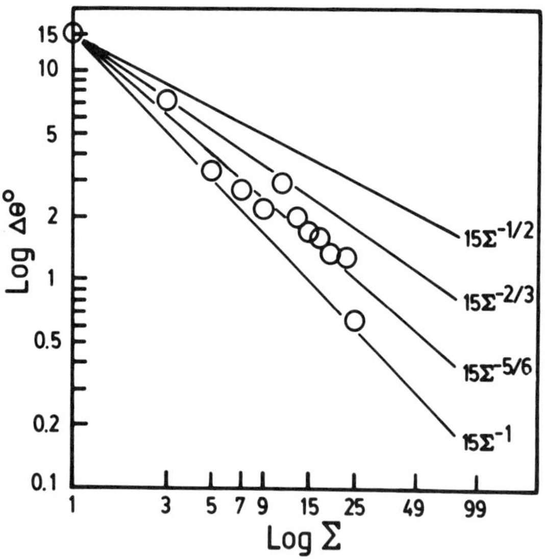

Fig. 5 Maximum deviation angle (Δθ) from any CSL (Σ) at which selective immunity to intergranular corrosion was noted following potentiostatic and coulostatic etching of polycrystalline 99.999% Ni in 2N H_2SO_4 (Ref 31)

and that this combats grain-boundary decohesion, with a probable mechanism being the expulsion of the embrittling trace element. The chemical stabilization of the harmful impurity elements through the formation of stable microprecipitates has also shown some promise (Ref 26). A somewhat different concept is that of "tying up" a harmful grain-boundary segregant by means of separate second-phase particles at the grain boundaries (Ref 27). This was shown to be useful in the case of intergranular corrosion in a highly oxidizing environment (Ref 28) and in environmentally induced intergranular cracking (Ref 29) of austenitic stainless steels.

It has been shown that susceptibility to the initiation of localized corrosion at grain boundaries in high-purity (99.999%) Ni is strongly dependent on the nature of the boundaries (i.e., CSL) (Ref 30, 31). Figure 5 summarizes the structure dependence of intergranular corrosion observed in 2N H_2SO_4. Grain boundaries close to low-Σ CSL relationships display a high resistance to the initiation of localized corrosion as manifested in selective immunity to corrosion. In this study, special properties were generally observed with grain boundaries deviating by less than $\Delta\theta = 15°\Sigma^{-5/6}$ from CSLs having $\Sigma \leq 25$. The intergranular corrosion observed here cannot be attributed entirely to structural effects since it has been shown that structure-sensitive impurity

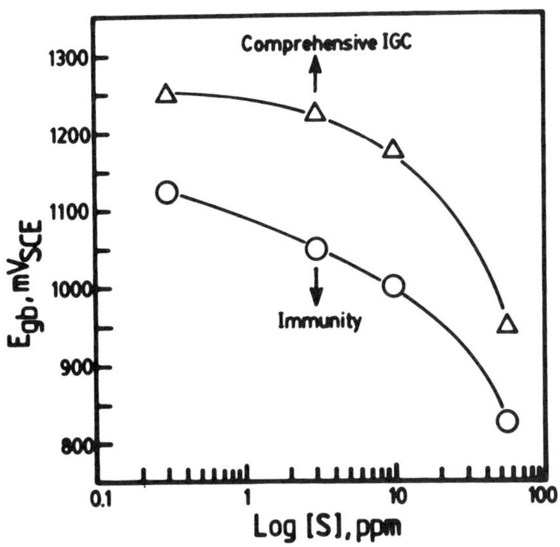

Fig. 6 Effect of sulfur concentration on the potential limits for intergranular corrosion of polycrystalline 99.999% Ni in 2N H_2SO_4 (Ref 31)

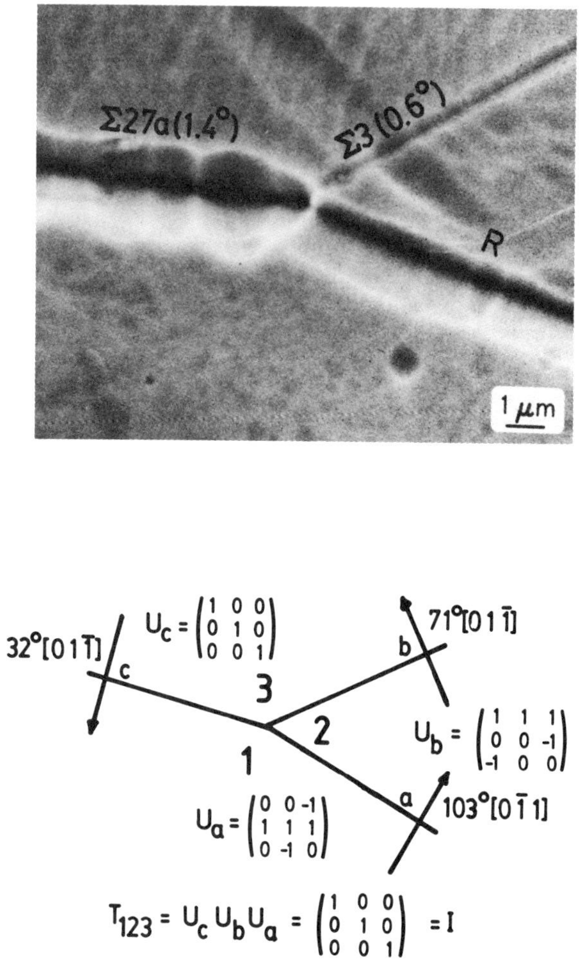

a

b

Fig. 7 (a) Scanning electron micrograph of the corrosion morphology evident at the intersection of a Σ3 boundary, a Σ27a boundary, and a non-CSL (R) interface in 99.999% Ni following potentiostatic etching at 1200 mV_{SCE} in 2N H_2SO_4. (Values in parentheses following Σ values represent angular deviations from the exact CSL.) (b) Corresponding orientation relationships and U-matrices yielding nearest-neighbor relationships (i.e., most widely spaced grain-boundary dislocation arrays) for adjoining grain boundaries in the O-lattice framework (Ref 36-38). Counterclockwise sequence of U-matrices yields the characteristic tensor of the triple junction, T_{123}, which here is the identity matrix, I (Bollmann's I-line) (Ref 35).

Table 2 Triple-line character in random and textured (fiber) crystal orientation distribution obtained by computer simulation

	Random	⟨100⟩		⟨110⟩		⟨111⟩	
		$\Delta\theta \leq 20^\circ$	$\Delta\theta \leq 10^\circ$	$\Delta\theta \leq 20^\circ$	$\Delta\theta \leq 10^\circ$	$\Delta\theta \leq 20^\circ$	$\Delta\theta \leq 10^\circ$
No. of triple junctions	**99**	**12**	**4**	**39**	**10**	**31**	**8**
O-lattice							
I-lines, %	4.04	8.33	25.00	10.26	20.00	6.45	12.50
U-lines, %	95.96	91.6	75.00	89.74	80.00	93.55	87.50
CSL/DSC ($\Sigma \leq 51$)							
CSL, %	24.24	33.33	50.00	35.90	40.00	29.03	50.00
CAD							
$\Pi \leq 8$, %	11.11	33.32	75.00	20.51	50.00	25.80	62.50
$\Pi \leq 24$, %	33.32	49.98	100.00	35.90	70.00	38.71	75.00

segregation can influence corrosion properties even in high-purity materials (Ref 32). Figure 6 shows the effect of sulfur content on the electrochemical potential limits for intergranular corrosion of 99.999% Ni in 2N H_2SO_4. Increasing the S content (0.3 to 50 ppm) results in reduced overpotential requirements for the initiation of grain-boundary corrosion. The region bounded by the upper and lower curves in the figure (i.e., comprehensive attack and immunity to localized corrosion) represents a potential range in which grain-boundary selectivity in the initiation of localized corrosion is observed. The magnitude of this range is thus a measure of the maximum difference in corrosion susceptibility of the grain boundaries (i.e., from the most susceptible to least susceptible). This difference is observed to be greatest (i.e., 175 mV) at intermediate sulfur levels of 3 and 10 ppm. This potential range decreases to 125 mV with 50 and 0.3 ppm S. These effects can be attributed to selective S segregation at different grain boundaries at the intermediate impurity levels.

Triple Lines

Susceptibility to intergranular corrosion can also be modified by the interaction of grain boundaries at triple junctions. It has been demonstrated (Ref 33-35) that susceptibility to corrosion at triple junctions is dependent on the specific crystallographic arrangement of crystals at the junction, in accordance with Bollmann's criteria (Ref 36-38) for a nodal balance of intersecting grain-boundary dislocation arrays. For example, Fig. 7 and 8 illustrate the differences in corrosion morphology observed at the grain-boundary intersections in high-purity Ni (0.3 wt ppm S) at E_b = 1200 mV in 2N H_2SO_4. In Fig. 7 the extent of corrosion at the triple junction is consistent with that observed along the adjoining grain boundaries, i.e., no preferential corrosion at the triple line. In

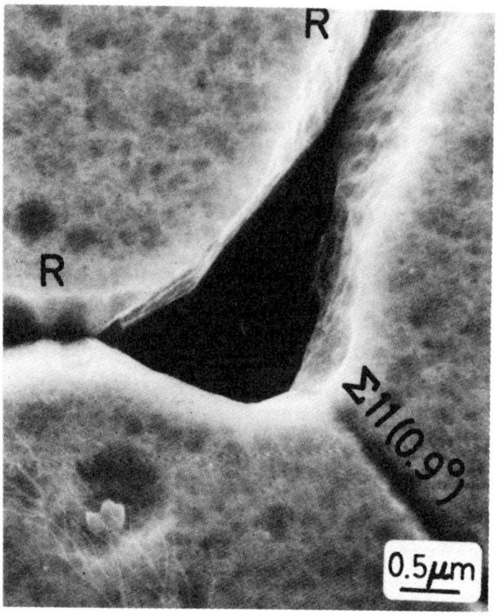

a

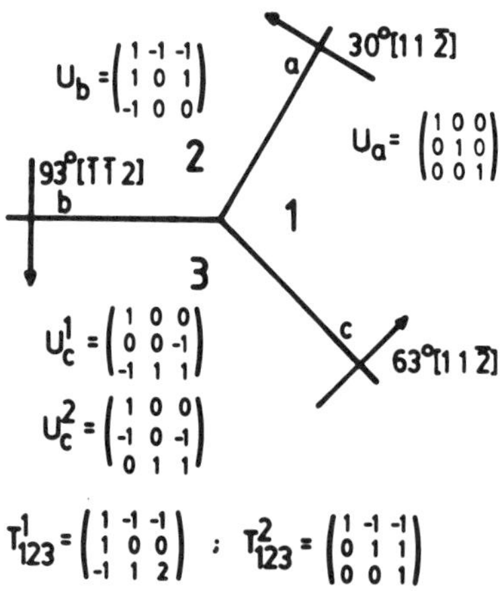

b

Fig. 8(a) Scanning electron micrograph of the corrosion morphology (following potentiostatic etching of 99.999% Ni at 1200 mV_{SCE}) at the intersection of a Σ11 interface with two non-CSL boundaries. (b) Corresponding orientation relationships and determined U-matrices resulting in a U-line arrangement (i.e., $T_{123} \neq I$) (Ref 35)

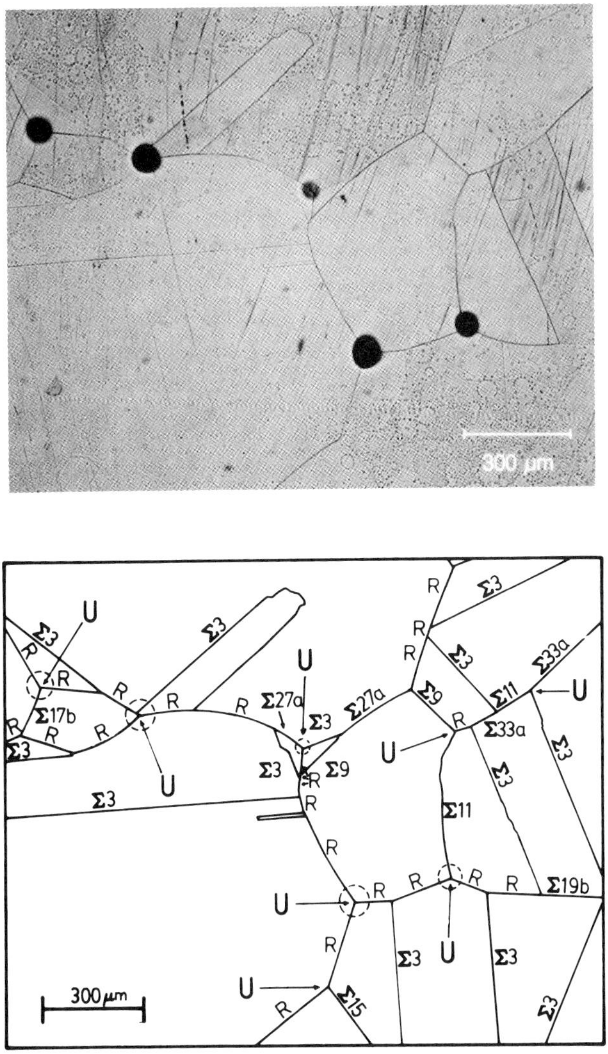

a

b

Fig. 9(a) Optical micrograph of a pitted surface of 99.999% Ni following potentiostatic corrosion in chloride-sulphate media and subsequent light etching to reveal grain-boundary positions. (b) Schematic representation of the determined grain-boundary and triple-junction character distribution associated with (a) (Ref 35)

Table 3 Experimentally determined triple-line character distributions

Material	Method	No. of triple junctions	I, %	U, %	Ref
99.999% Ni, Strain annealed	SEM/ECP	68	76.47	23.53	35
Cu-8.5% Al, recrystallized	TEM	14	57.14	42.86	39

this system, the junction is formed by the intersection of a Σ3 boundary with a Σ27 boundary and a non-CSL-related interface (R). Values in parentheses following the Σ values represent the angular deviation from the exact CSL. In Fig. 8, extensive corrosion is observed at the triple junction formed by the intersection of a Σ11 boundary with two non-CSL-related interfaces. Using Bollmann's theory of triple lines, a nodal balance of intersecting grain-boundary dislocation arrays is calculated in the first case (termed I-line). No nodal balance is found in the second case (termed U-line). Defects (i.e., surplus dislocations) are considered to be present in the triple line that corroded preferentially. These triple-line defects (i.e., Bollmann's U-lines) are higher order dislocations or disclinations, which are not characterized by a displacement vector (i.e., Burgers vector) but only by a rotation. As seen in Fig. 9 (Ref 35), U-lines in 99.999% Ni have been found to have distinct electrochemical characteristics, as confirmed by preferential pit nucleation at these sites in chloride-sulphate media. On the basis of pitting susceptibility at various electrochemical potentials, it was determined that U-type intersections have a distinct pitting potential (E) such that $E_{\text{U-line}} < E_{\text{grain boundary}} \approx E_{\text{I-line}} < E_{\text{matrix}}$.

Support for Bollmann's disclination treatment of triple junctions has also been obtained through experimental identification of triple junctions having I and U crystallographic characteristics in both high-purity Ni (Ref 33-35) and Cu-8.5% Al (Ref 39); transmission electron microscopy (TEM) confirmation of a grain-boundary dislocation (GBD) balance at an I-type intersection in Cu-3% Si (Ref 40); and computer simulation of triple-junction stress fields, demonstrating uncompensated stresses at U-type junctions (Ref 41). In addition, preliminary experimental results obtained by scanning tunneling microscopy indicate that certain triple junctions can display a significant line energy (Ref 42).

Triple junctions have been shown to influence several other material properties, including ductility (Ref 43, 44), grain-boundary inelasticity (Ref 45), diffusion-induced grain boundary migration (DIGM) and diffusion-induced recrystallization (DIR) (Ref 46), internal friction (Ref 47), and grain-boundary sliding (Ref 48).

The triple-line character of interfaces, as given by Bollmann (Ref 36-38), is another form of interface control producing different structures at triple junctions. The overall contribution of triple junctions to properties of materials is dependent upon the triple-line character distribution (i.e., disclination fraction), the total triple-line content in the material, and triple-line chemistry.

The triple-line character distribution is primarily a function of crystallographic texture, with the frequency of U-type junctions (i.e., disclinations) increasing with increased system randomness (Ref 36). It has been shown by computer simulation (see Table 2) (Ref 49) that 96% of the triple junctions in a random crystal distribution would be U-lines, but this can be reduced to about 75% with the application of a suitable fiber texture (for example, within 10% of ⟨100⟩). Furthermore, the defect character of these disclinations (i.e., U-lines) is expected to be enhanced with increased crystal randomness, due to the reduced probability of, first, long-range one-dimensional periodicity (i.e., Π), which can arise through low index planar continuity across the adjoining

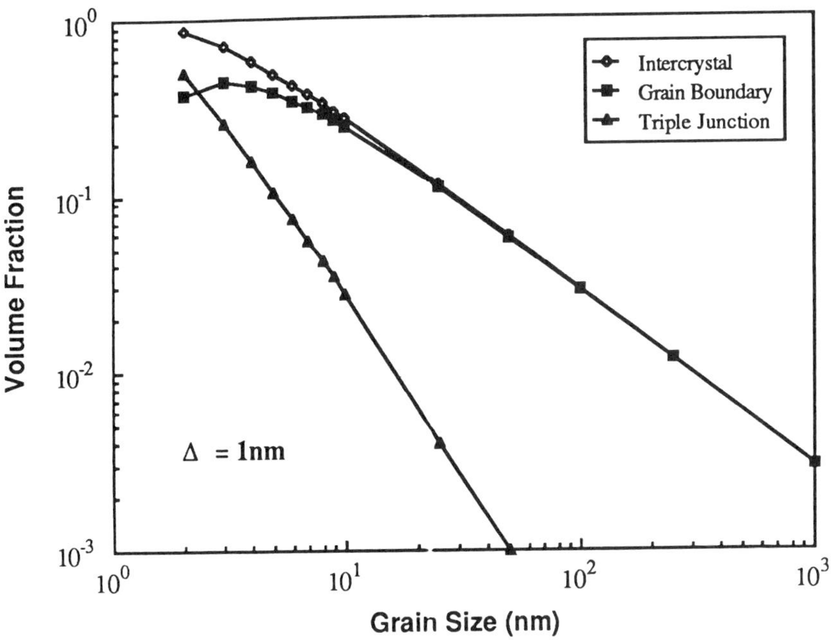

Fig. 10 The effect of grain size on calculated volume fractions for intercrystalline regions, grain boundaries, and triple junctions, assuming a grain-boundary thickness (Δ) of 1 nm (Ref 52)

lattices at the junction (Ref 50), and, second, CSL/DSC relaxations (i.e., Σ) at the adjoining interfaces (Ref 51).

Table 3 summarizes experimentally determined triple-line character distributions in conventional polycrystalline materials. As shown in this table, the percentage of experimentally observed U-lines (i.e., triple-line disclinations) is substantially less than that predicted in Table 2 even for strongly oriented materials. As discussed by Bollmann (Ref 36), energetic considerations, not accounted for in Table 2, may result in the preferential elimination of such high-energy disclinations during recrystallization and grain growth.

The triple-line content in a polycrystalline material is dependent on grain size and crystal geometry. Using a three-dimensional treatment involving tetrakaidecahedral grains, the present authors have shown (Ref 52) that the volume fraction of a material associated with triple junctions can be enhanced by decreasing the grain size (Fig. 10). This effect is most pronounced in the nanocrystal grain size range (≤20 nm) where the triple-junction volume frac-

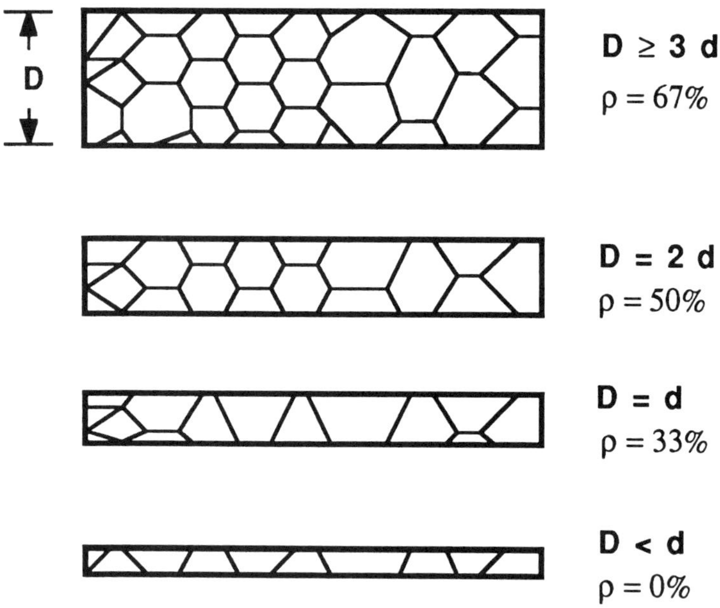

Fig. 11 Schematic representation of the effect of reducing a wire (or rod) diameter (D) relative to the grain size (d) on the ratio of triple junctions to grain boundaries in the microstructure (ρ)

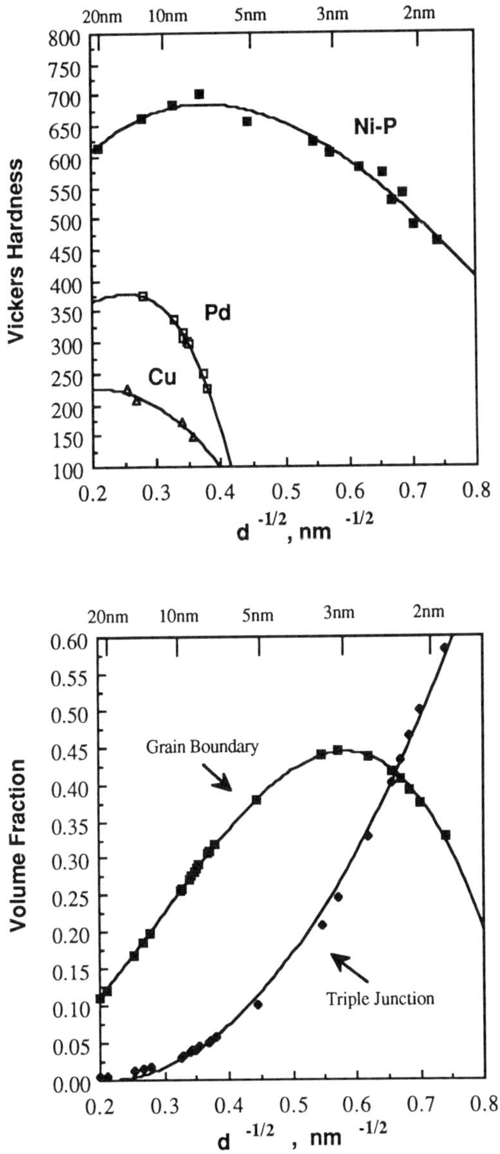

a

b

Fig. 12(a) Room-temperature Vickers hardness measurements as a function of grain size (d) for nanocrystalline Pd (Ref 60), Cu (Ref 60), and Ni-P (Ref 61). (b) Calculated grain-boundary and triple-junction volume fractions as a function of grain size using a boundary thickness of 1 nm (Ref 44) tion becomes comparable to that associated with the grain boundaries and matrix. Triple junctions can be eliminated entirely from a polycrystalline material by producing an "equi-axed to bamboo" transition in grain shape (Ref 43). This is seen in Fig. 11, where the transition is achieved, without recrystallization, by thinning (for example, electrolytic) of a polycrystalline wire (or rod) to a diameter (D) less than the average grain size (d) of the material. Finally, triple-line chemistry is expected to be dependent on both the type and quantity of solute in the material because triple junctions may, under certain conditions (for example, as disclinations), act as sinks for both matrix and grain-boundary solutes (Ref 35).

Nanocrystalline Materials

Nanocrystalline materials are single-phase or multiphase polycrystals, with a crystal size on the order of a few nanometers, typically 5 to 20 nm. As a result, as seen previously in Fig. 10, about 50 vol% of the material consists of intercrystalline regions, i.e., grain boundaries and triple junctions. The nanocrystalline materials most often studied are those prepared by a vapor condensation and compaction method (Ref 53); other methods include mechanical alloying (Ref 54) and electroplating (Ref 55).

Nanocrystalline materials, having a crystal size of less than approximately 10 nm, have been shown to possess unusual properties such as enhanced diffusivity (Ref 56) and solid solubility (Ref 57) and low-temperature ductility in conventionally brittle ceramics (for example, TiO_2) (Ref 58, 59).

Figure 12(a) shows the results of room-temperature Vickers hardness measurements on nanocrystalline Pd (Ref 60), Cu (Ref 60), and Ni-P (Ref 61). Hardness values are plotted as a function of $d^{-1/2}$ where d is the average grain diameter. As shown here, initial increases followed by significant decreases in hardness are noted with decreasing grain size in the nanocrystal range ($d \leq 20$ nm). The observed decreases in hardness are contrary to Hall-Petch behavior (Ref 62, 63). Figure 12(b) shows the corresponding grain-boundary and triple-junction volume fractions in this grain size range, calculated using a grain-boundary thickness of 1 nm (Ref 52). The onset of decreasing hardness (i.e., deviation from Hall-Petch behavior) in these systems occurs at grain sizes where triple lines begin to comprise a significant fraction of the bulk specimen volume (i.e., $\leq 1\%$). As shown here, the volume fraction of grain boundaries continues to increase throughout most of the grain size range where softening is observed, indicating that deviations from Hall-Petch behavior are not a result of decreasing grain-boundary volume fraction. The observed phenomena are in general agreement with the triple-line softening effects reported elsewhere (Ref 43, 44), in which increasing triple-junction content enhances bulk ductility in polycrystalline materials. Large triple-junction volume fractions in nanocrystals may also account for the substantial low-temperature ductility of nanocrystalline ceramics (Ref 57, 58).

In considering the role of triple junctions in bulk material softening, it is important to consider the specific triple-line disclination content (i.e., U-lines), since only these junctions are expected to have defect character. Structural considerations may account for the variations in behavior shown in Fig. 12(a), in which the decrease in hardness for Cu and Pd commences at a larger grain size and falls much more rapidly with decreasing grain size than it does for Ni-P. An electroplating technique was used to produce nanocrystalline Ni-P, yielding a substantial nanocrystal fiber texture (Ref 55) whereas the vapor condensation and compaction method used for the Cu and Pd yields a highly random crystal orientation distribution (Ref 64). At a given grain size (i.e., constant total triple-line volume fraction), the triple-line disclination content is expected to be higher with the randomly oriented Cu and Pd specimens, resulting in enhanced ductilization effects.

Thus, the mechanical behavior of nanocrystalline materials can be considered to be dependent on both the quantity and specific structure of triple junctions (Ref 44, 65). Consideration of triple junctions as distinct crystalline defects, comprising a significant fraction of the total volume of nanocrystalline systems, may allow a better understanding of the structure and unusual properties of these novel materials.

Summary

Results presented here demonstrate that the bulk properties of polycrystalline materials are strongly influenced by the structure, chemistry, and quantity of the intercrystalline component, which is considered here to consist of distinct grain-boundary and triple-junction defects. The structure and chemistry of grain boundaries can be controlled through various processing considerations, for example, solute control, controlled precipitation, texture, prestrain, etc., with the primary goal being the increase in the frequency of low-Σ CSL interfaces. The defect character of triple junctions (i.e., triple-line disclination) has been shown to be primarily affected by crystallographic texture; however, energetic and solute considerations are also expected to have an influence. The intercrystalline quantity is controlled through grain size considerations, leading to the novel class of nanocrystalline materials in which substantial volume fraction contributions from the intercrystalline component (i.e., $\leq$ 50%) can result in unique material properties. Finally, it has been demonstrated that the relative quantity of triple-junction defects in the intercrystalline component can be controlled through grain shape considerations (triple junctions can be eliminated in materials having a bamboo microstructure) and can possibly account for several of the unique properties of nanocrystalline materials (for example, enhanced ductility, solubility, and diffusivity).

Acknowledgments

This paper is dedicated to the memory of Professor Bruce Chalmers, who provided many of the initial concepts utilized here. Financial support from the Natural Sciences and Engineering Research Council of Canada is gratefully acknowledged.

References

1. T. Watanabe, *Res. Mech.*, Vol 11, 1984, p 47
2. M.L. Kronberg and F.H. Wilson, *Trans. AIME*, Vol 185, 1949, p 501
3. D.G. Brandon, *Acta Metall.*, Vol 14, 1966, p 1479
4. G. Palumbo and K.T. Aust, in *Atomic-Level Properties of Interface Materials*, D. Wolf and S. Yip, Ed., Chapman and Hall, London, in press
5. D.H. Warrington and M. Boon, *Acta Metall.*, Vol 23, 1975, p 599
6. K.T. Aust and J.W. Rutter, *Trans. TMS-AIME*, Vol 215, 1959, p 119
7. R.V. Belluz and K.T. Aust, *Metall. Trans. A*, Vol 6, 1975, p 219
8. G. Palumbo and K.T. Aust, in *Recrystallization '90*, T. Chandra, Ed., American Institute of Mining, Metallurgical, and Petroleum Engineers, 1990, p 101
9. K.T. Aust and J.W. Rutter, *Trans. TMS-AIME*, Vol 218, 1960, p 1023
10. K.T. Aust, *Trans. TMS-AIME*, Vol 221, 1961, p 758
11. G. Ferran, G. Cizeron, and K.T. Aust, *Acad. Sci. Paris*, Vol 257, 1963, p 3595
12. G. Ferran, G. Cizeron, and K.T. Aust, *Mem. Sci. Rev. Metall.*, Vol 54, 1967, p 1067
13. V. Randle and A. Brown, *Philos. Mag. A*, Vol 59, 1989, p 1075
14. V. Randle and B. Ralph, *J. Mater. Sci.*, Vol 22, 1987, p 2535
15. V. Randle, B. Ralph, and D. Dingley, *Acta Metall.*, Vol 36, 1988, p 267
16. T. Watanabe, *J. Phys.*, Vol 46, 1985, p C4-555
17. T. Watanabe, *Trans. Jpn. Inst. Met.* (Suppl.), Vol 27, 1986, p 73
18. M.W. Grabski, *J. Phys.*, Vol 46, 1985, p C4-567
19. K.T. Aust and J.W. Rutter, *Trans. TMS-AIME*, Vol 224, 1962, p 111
20. P.J. Wilbrandt and P. Haasen, *Z. Metallkd.*, Vol 71, 1980, p 385
21. A. Berger, P.J. Wilbrandt, and P. Haasen, *Acta Metall.*, Vol 31, 1983, p 1433
22. P. Haasen, in *Recrystallization '90*, T. Chandra, Ed., American Institute of Mining, Metallurgical, and Petroleum Engineers, 1990, p 17
23. J. Harase, R. Shimuzu, K. Kuroki, T. Nakayama, T. Wada, and T. Watanabe, *Trans. Jpn. Inst. Met.* (Suppl.), Vol 27, 1986, p 563
24. H. Makita, S. Hanada, and O. Izumi, *Trans. Jpn. Inst. Met.*, Vol 28, 1987, p 447
25. M. Sharp and C.G. Dunn, *Trans. AIME*, Vol 4, 1952, p 42

26. ME.D. Hondros, *Trans. Jpn. Inst. Met.* (Suppl.), Vol 27, 1986, p 393
27. K.T. Aust and G. Palumbo, *Trans. Jpn. Inst. Met.* (Suppl.), Vol 27, 1986, p 995
28. K.T. Aust, J.S. Armijo, and J.H. Westbrook, *Trans. ASM*, Vol 59, 1966, p 544
29. S.J. Thorpe, B. Ramaswami, and K.T. Aust, *Acta Metall.*, Vol 32, 1984, p 1297
30. G. Palumbo and K.T. Aust, *J. Phys.*, Vol C5-45, 1988, p 569
31. G. Palumbo and K.T. Aust, *Acta Metall.*, Vol 38, 1990, p 2343
32. K.T. Aust and O. Iwao, Localized Corrosion, *NACE-3*, 1974, p 62
33. G. Palumbo and K.T. Aust, *Scripta Metall.*, Vol 22, 1988, p 847
34. G. Palumbo and K.T. Aust, in *Proc. Int. Symp. Adv. Struct. Mater.*, D. Wilkinson, Ed., Vol 9, 1989, p 227
35. G. Palumbo and K.T. Aust, *Mater. Sci. Eng. A*, Vol 113, 1989, p 139
36. W. Bollmann, *Philos. Mag. A*, Vol 49, 1984, p 73
37. W. Bollmann, *Philos. Mag. A*, Vol 57, 1988, p 637
38. W. Bollmann, *Mater. Sci. Eng. A*, Vol 113, 1989, p 129
39. W. Bollmann and H. Guo, *Scripta Metall.*, Vol 24, 1990, p 709
40. L.M. Clarebrough and C.T. Forwood, *Philos. Mag. A*, Vol 55, 1987, p 217
41. W. Bollmann, *Mater. Sci. Eng. A*, submitted
42. P. Fortier, G. Palumbo, G.D. Bruce, W.A. Miller, and K.T. Aust, *Scripta Metall.*, submitted
43. V.B. Rabukhin, *Fiz. Met. Metalloved.*, Vol 61, 1986, p 149
44. G. Palumbo, U. Erb, and K.T. Aust, *Scripta Metall.*, in press
45. V.B. Rabukhin, *Fiz. Met. Metalloved.*, Vol 55, 1983, p 178
46. S.A. Hackney, *Scripta Metall.*, Vol 22, 1988, p 1255
47. G.M. Ashmarin, M.Yu. Golubev, N.I. Naumova, and A.V. Shalimova, *Fiz. Met. Metalloved.*, Vol 67, 1989, p 110
48. S. Miura, S. Hashimoto, and T.K. Fujii, *J. Phys.*, Vol C5-49, 1988, p 599
49. E.G. Doni, G. Palumbo, and K.T. Aust, *Scripta Metall.*, submitted
50. G. Palumbo and K.T. Aust, *Scripta Metall.*, Vol 24, 1990, p 1771
51. E.G. Doni and G.L. Bleris, *Phys. Status Solidi (a)*, Vol 110, 1988, p 393
52. G. Palumbo, S.J. Thorpe, and K.T. Aust, *Scripta Metall.*, Vol 24, 1990, p 1347
53. R. Birringer, U. Herr, and H. Gleiter, *Trans. Jpn. Inst. Met.* (Suppl.), Vol 27, 1986, p 43
54. E. Hellstern, H.J. Fecht, Z Fu, and W.L. Johnson, *J. Appl. Phys.*, Vol 65, 1989, p 305
55. G. McMahon and U. Erb, *Microstruct. Sci.*, Vol 17, 1989, p 447
56. J. Horvath, R. Birringer, and H. Gleiter, *Solid State Commun.*, Vol 62, 1987, p 319
57. R. Birringer, H. Hahn, H. Hofler, J. Karch, and H. Gleiter, Diffusion and Defect Data, in press
58. R.W. Siegel and H. Hahn, Current Trends in Physics of Materials, in *Nanophase Materials*, M. Youssef, Ed., World Scientific Publisher, Singapore, in press
59. J. Karch, R. Birringer, and H. Gleiter, *Nature*, in press
60. A.H. Chokshi, A. Rosen, J. Karch, and H. Gleiter, *Scripta Metall.*, Vol 23, 1989, p 1679
61. D. Ostrander and U. Erb, *Scripta Metall.*, submitted
62. E.O. Hall, *Proc. Phys. Soc. London*, Vol B64, 1951, p 747
63. N.J. Petch, *J. Iron Steel Inst.*, Vol 174, 1953, p 25

64. H.E. Schaefer, R. Wurschum, R. Birringer, and H. Gleiter, *Phys. Rev. B*, Vol 38, 1988, p 9545

65. D. Ostrander, U. Erb, G. Palumbo, K.T. Aust, D. Lahaie, J.D. Embury, and Z.S. Basinski, in *Proceedings of the Acta Metallurgica Conference on Materials with Ultrafine Microstructures*, in press

2

Grain-Boundary Structure and Behavior in Zinc

Kisoo Shin and A.H. King
Department of Materials Science and Engineering
State University of New York at Stony Brook
Stony Brook, New York 11794-2275

In this chapter, calculations of dislocation spacing for grain boundaries in zinc are presented, and it is shown that the spacings may be large enough to be resolved in transmission electron microscopy even at extreme deviations from ideal coincidence misorientations. This effect is the result of the need for the dislocation arrays to accommodate slight differences between the ideal and actual axial ratios, in addition to the difference between the ideal and actual misorientations. The importance of the axial ratio accommodation in determining the dislocation structure is tested by observing grain boundaries at different temperatures, where the actual axial ratios differ. The energy change of the dislocation arrays with varying temperature is shown to be sufficient to drive transformations of the interfacial structure. These transformations may explain the sudden changes in grain-boundary sliding behavior with changes of temperature, observed by Watanabe *et al.* (Ref 1). Subsequent observations of sliding behavior as a function of misorientation may also be rationalized within the framework of the structure calculations presented here.

Detailed transmission electron microscope (TEM) observations of dislocation structure in large-angle grain boundaries between cubic crystals have been reported since the early 1970s (for example, Bollmann, Michaut, and

Sainfort, Ref 2; Clark and Smith, Ref 3; Sun and Balluffi, Ref 4; and Forwood and Clarebrough, Ref 5). In almost all cases, these observations have been related to the existence of coincidence site lattices (CSLs) in which the volume of the CSL primitive cell is a small multiple of the volume of the primitive cell of the lattice. The ratio of these volumes is denoted by Σ, and small values correspond to short-period ordered structures of the grain boundary. However, these structures are formed only at certain precise values of misorientation. Conventional wisdom has it that certain periodic-structured boundaries have low energy and that these structures are preserved through small deviations of misorientation by the insertion of dislocation arrays, in spite of the additional energy associated with these arrays. According to Brandon (Ref 6), this model is held to provide a good description of the grain-boundary structure over some small range of misorientation around the ideal coincidence value, given by a function of the form:

$$\Delta\theta = 15^\circ/\sqrt{\Sigma} \qquad \text{(Eq 1)}$$

Outside this range the dislocations in the boundary are considered to be so close together that they lose their individual physical identity. The question of what constitutes individual physical identity is now less clear than in 1966 when Brandon reported on his criterion, since the ability to detect and resolve the structures has improved to the point of atomic resolution; but that is not the subject of this particular discussion.

Observations of grain-boundary dislocation structure have been reported in much smaller numbers for hexagonal materials than for the cubic crystals. Matters of interpretation are somewhat more complicated for hexagonal materials than for cubic ones because the formation of three-dimensional CSLs is generally forbidden where the square of the axial ratio is an irrational fraction. This problem has been overcome by the introduction of a constrained coincidence model in which short-period grain-boundary structures can be formed by changing the axial ratio of the natural crystals by a few percent to obtain a material with the necessary value of c^2/a^2. The structures created in this manner are termed constrained coincidence site lattices (CCSLs). Just as deviations from the ideal misorientation are accommodated by dislocation arrays in CSL models, deviations in both misorientation and axial ratio can be accommodated in the CCSL model. Since for the general case the natural c/a ratio does not correspond to that required to form a CCSL, there will always be a dislocation array in a coincidence-related grain boundary in a hexagonal material, even at the precise coincidence misorientation, except for certain especially symmetrical cases.

Tables of coincidence systems have been produced for axial ratios close to those of hexagonal close packed (HCP) metals (Chen and King, Ref 7). Detailed observations in certain cases have resulted in excellent agreement between the observed and theoretical structures based on the tabulated data for particular grain boundaries (Chen and King, Ref 8; Antonopoulos, Delavignette, Karakostas, Komninou, Laurent-Pinson, Lay, Nouet, and Vicens, Ref 9; Lartigue, Carry, and Priester, Ref 10). This provides strong evidence that short-period boundaries embody low energy in hexagonal metals, as in cubic cases.

Two interesting problems remain, however. In certain investigations of HCP metals, dislocation structures are observed in grain boundaries that do not appear to correspond to any known low-Σ CCSL (Loberg and Smith, Ref 11; Chen and King, Ref 8). It is also found that observed boundaries that do correspond to known CCSLs are very frequently close in misorientation and axial ratio to CCSLs of more than one Σ-value; thus the choice of reference structure may be highly equivocal. These two problems severely limit our experimental understanding of what constitutes a low-energy boundary in HCP metals. This discussion attempts to resolve some of these issues.

Calculations of Interfacial Dislocation Structure

Provided that we can identify the reference structure for a particular grain boundary and know the Burgers vectors of the dislocations accommodating any deviation from that structure, we can calculate the required orientation and spacing of the dislocations through the application of Bollmann's O-lattice formalism, or, more strictly, for large-angle grain boundaries, the O2-lattice formalism (Bollmann, Ref 12). For the case in which not only a deviation in misorientation but also a change of axial ratio is accommodated, the O2-lattice equation is modified to the form given by Shin and King (Ref 13):

$$(E^{-1} - R_c \cdot E^{-1} \cdot R_{exp})\mathrm{X(O2)} = \mathrm{b}^{dsc} \qquad \text{(Eq 2)}$$

where R_c is the rotation matrix describing the exact coincidence rotation for the material with an ideal c/a ratio, R_{exp} is the actual (or experimentally determined) rotation matrix, X(O2) is the O2-lattice vector corresponding to the Burgers vector b^{dsc}, where dsc is displacement shift complete, and E is the matrix relating the ideal (c) and natural (n) materials:

$$E = \begin{bmatrix} 1 & 0 & 0 \\ 0 & 1 & 0 \\ 0 & 0 & \dfrac{(c/a)_n}{(c/a)_c} \end{bmatrix}$$

This equation differs from that used by Chen and King (Ref 8) in that it is formulated in an orthonormal coordinate system related to the *natural* crystal 1 coordinate system. Chen and King used a method that employed the *constrained* crystal 1 lattice as the reference frame: the constrained crystal and the natural crystal are fully equivalent when the appropriate transformation is

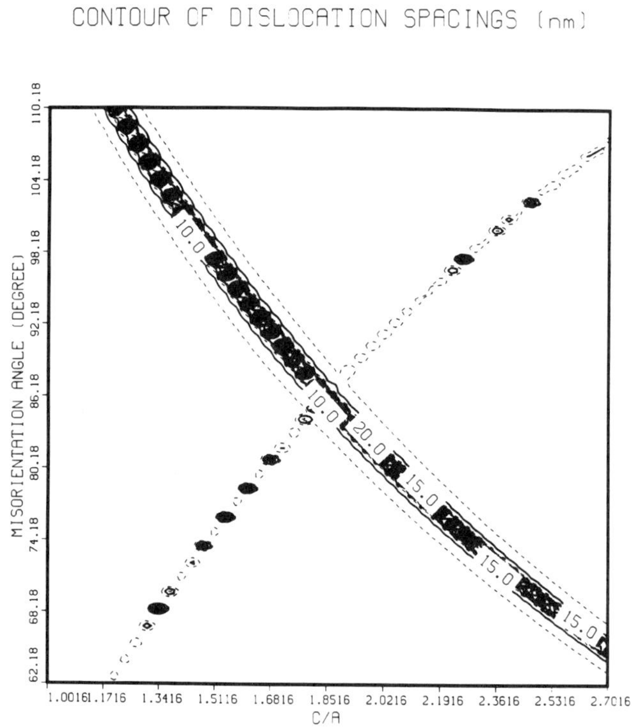

Fig. 1 Contour plot showing calculated dislocation spacings for a $(21\bar{1}0)$ boundary as a function of misorientation and axial ratio. The calculations are based on the Σ15 structure, with $b = a/15[\bar{5}\,\bar{2}\,7\,\bar{3}]$. The contours are drawn at 5 nm intervals, so the regions enclosed within the broadest (dashed) contours should correspond to resolvably spaced dislocations.

applied. As with all O-lattice or Bilby-Frank formulations, the dislocation networks that are predicted are those necessary to cancel the long-range strain fields if the interfaces are of infinite extent.

Using this equation, one can determine the spacing of any array of dislocations in a coincidence-related grain boundary at any misorientation or c/a ratio (noting that the axial ratios of zinc and cadmium are both strongly temperature dependent). Figure 1 is a plot of dislocation spacing as a function of both misorientation and axial ratio for a set of $\frac{a}{15}[\bar{5}\,\bar{2}\,7\,\bar{3}]$ dislocations in a Σ15 grain boundary. The initial data were taken from published CCSL tables for hexagonal crystals. This figure shows that the dislocation spacing can be very large even at significant deviations from the exact coincidence misorientation and axial ratio. The dislocations in question could not be described as "plane matching" dislocations (Pumphrey, Ref 14) because they do not have Burgers vectors perpendicular to any low-index plane. These dislocations would be readily resolvable using conventional TEM techniques at misorientations deviating 20° and axial ratios different by as much as 50% from that required to create an exact coincidence boundary. This misorientation difference greatly exceeds that which would be deduced from Brandon's criterion. In many, though not all, cases, multiple dislocation arrays may be observable at large deviations from the exact coincidence parameters. The large dislocation spacings at great deviations from coincidence are possible because of a canceling between the dislocation content needed to accommodate the misorientation error and that required to accommodate the axial ratio constraint.

This rather surprising result indicates that observations of dislocation structures in boundaries apparently far from coincidence may still be explainable in terms of a CCSL model, thereby rationalizing the observations by Loberg and Smith (Ref 11) and Chen and King (Ref 8). Now however, a much wider range of coincidence systems, covering a much wider range of c/a values, must be considered. This clearly presents a dilemma because one can no longer identify the correct coincidence system simply on the basis of proximity in misorientation space, and the number of coincidence systems to be considered is very large even if restricted to low-Σ systems.

When considering grain boundaries close to a constrained coincidence system, within the Brandon limit of misorientation error, for example, the choice of coincidence system (or reference structure) is still problematic because of the clustering of CCSLs in misorientation space. It has been shown that the choice of different reference structures has essentially no effect on the

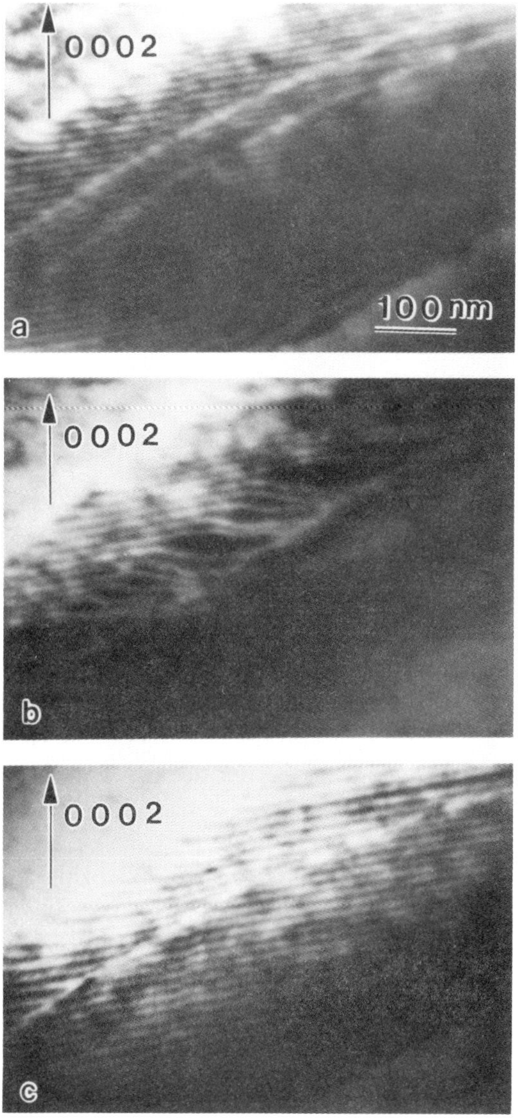

Fig. 2 Transmission electron micrographs of a grain boundary in zinc. (a) At room temperature. (b) At liquid nitrogen temperature. (c) At room temperature again. The misorientation is 86.7°/[2.0, –0.966, –1.035, 0.012].

Table 1 Deviations between measured and computed dislocation line directions at room temperature for b = (1/3)[1 $\bar{2}$ 1 0] in the boundary shown in Fig. 2 Σ32 represents a much better match than does any other CCSL system

Σ13	Σ15	Σ17	Σ24	Σ28	Σ32
73.3°	81.6°	84.0°	72.1°	75.6°	9.8°

available Burgers vectors for interfacial-line defects, but their calculated orientation and spacing are both strongly dependent upon that choice, and this effect has been used to identify the actual constrained coincidence system appropriate to an observed grain boundary (Ref 8). Furthermore, because of the temperature dependence of the c/a ratio (which is particularly pronounced in zinc and cadmium), the dislocation content of a grain boundary, particularly the component that accommodates the axial ratio deviation, changes with temperature. The authors have previously shown that this variation of dislocation content causes a temperature-dependent change in the relative elastic strain energies associated with the boundaries of the different CCSL systems in a cluster (Shin and King, Ref 13). Now as the relative energies of the boundaries vary with temperature and misorientation, it is reasonable to expect that phase transformations between the different CCSL structures may be observable.

Experimentation

Polycrystalline zinc sheet was cold rolled to a thickness of 0.23 mm, and 3 mm disks were punched from it. These disks were annealed at 85 °C *in vacuo* for 30 min to produce an experimentally convenient grain size and well-equilibrated grain-boundary structures. The specimens were then double-jet electropolished in a 10% nitric acid solution in methanol at –10 °C. Electron microscopy was performed at 120 kV using a Philips CM12 equipped with a Gatan liquid-nitrogen-cooled double-tilt stage. Grain boundaries were observed at temperatures ranging from approximately 140 to 300K.

Zinc twins profusely, providing large numbers of boundaries that can be described variously as reflections in the {01$\bar{1}$2} plane, 180° rotations about the {01$\bar{1}$2} plane normals, or rotations of approximately 85.5° about ⟨2 $\bar{1}$ $\bar{1}$ 0⟩. Precise measurements reveal that the exact twin misorientation is never observed and that the boundaries always contain accommodating dislocation arrays. There are several CCSL systems close to the observed boundary parameters, including Σ13, Σ15, Σ17, Σ24, Σ28, and Σ32. Chen and King (Ref 8) were able to show that observed dislocation line directions and spacings in one such

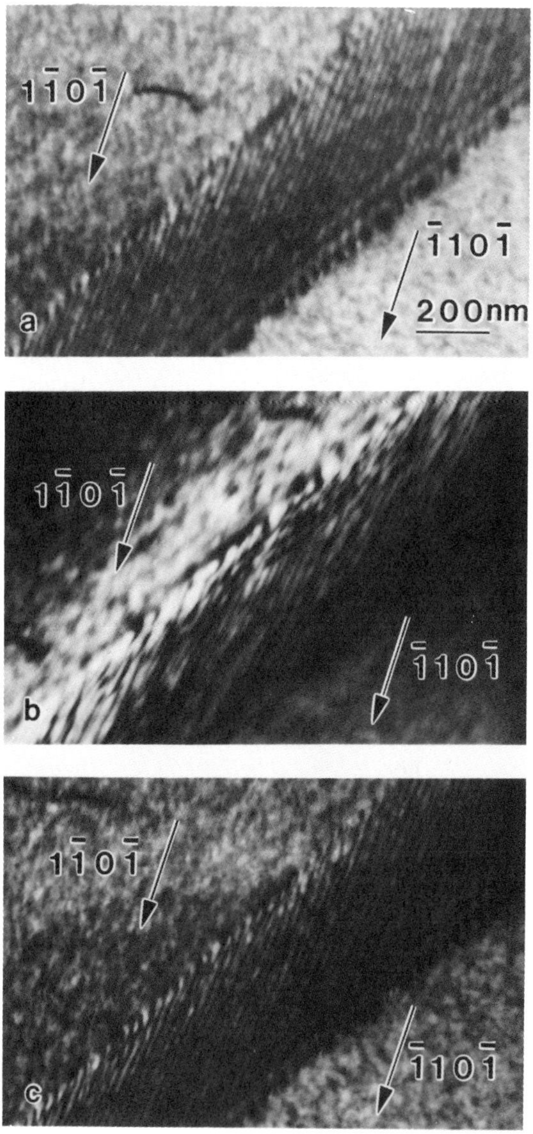

Fig. 3 Transmission electron micrographs of a grain boundary in zinc. (a) At room temperature. (b) At liquid nitrogen temperature. (c) At room temperature again. The misorientation is 86.2°/[2.0, –0.983, –1.017, 0.0].

boundary matched the calculated directions and spacings for only one of the possible CCSL systems. Because the exact coincidence misorientations and axial ratios for the six CCSLs are all slightly different, the possibility arises that boundaries of slightly different misorientations will adopt different periodic structures. The authors have used the same technique to detect structural transformations that occur as a result of small changes of axial ratio induced by changes of temperature.

Figure 2 shows a boundary characterized as 86.7°/[2.0, –0.966, –1.035, 0.012], first at room temperature, then at liquid nitrogen temperature, and then, again, at room temperature. It can be seen that the interfacial dislocations are straight and parallel in the first and last images, indicating a well-equilibrated structure. The orientation and spacing of the dislocation lines are a good match with the calculated values corresponding to Σ32, as indicated in Table 1, which gives the angles between the calculated and observed line

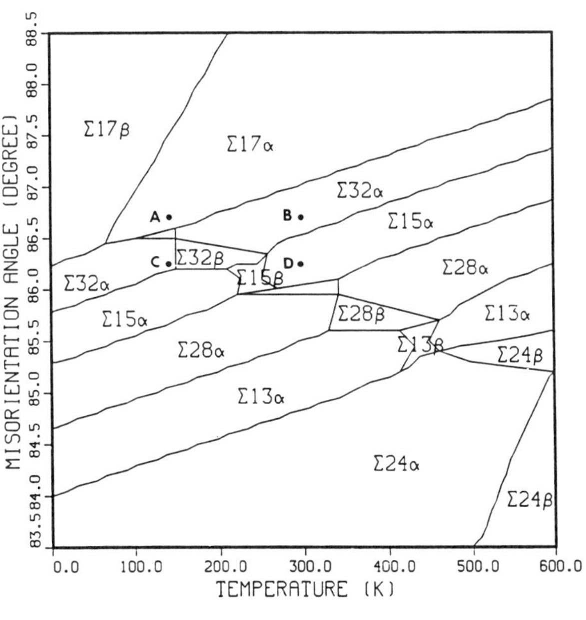

Fig. 4 Diagram indicating the lowest energy dislocation arrays for misorientations close to the twin. The α structures contain lattice dislocations, while the β structures are made up of primitive dsc dislocations. Points A and B refer to the grain boundary shown in Fig. 2, and points C and D refer to the boundary shown in Fig. 3.

Table 2 Deviations between measured and computed dislocation line directions at room temperature for b = (1/3)[1 $\bar{2}$ 1 0] in the boundary shown in Fig. 3. Σ15 represents a better match than does any other CCSL system.

Σ13	Σ15	Σ17	Σ24	Σ28	Σ32
23.3°	5.3°	49.8°	30.4°	11.4°	29.2°

directions. The micrograph taken at liquid nitrogen temperature indicates severe disruption to the dislocation array, which would normally be associated with a large increase in the interfacial energy (as shown, for example, by the calculations of King and Smith, Ref 15). It is inferred that the low-temperature image represents a grain boundary in the process of transforming to a new equilibrium structure, corresponding to a different CCSL.

A similar transformation is shown in Fig. 3. In this case the misorientation is 86.2°/[2.0, –0.983, –1.017, 0.0], and the observed structure matches well the Σ15 computed line direction, as shown in Table 2. Again a reversible disruption of the dislocation array in the boundary is found.

Figure 4 summarizes strain energy estimates for grain boundaries close in misorientation and axial ratio to the observed twins. The estimates are based on the Read-Shockley formula, and calculations were made for all possible dislocation arrays in all of the CCSL systems in the cluster. The lowest energy dislocation array is indicated at each point on this phase diagram. Points A and B in the diagram correspond to the boundary in Fig. 1, at liquid nitrogen and room temperatures, respectively. It can be seen that the predicted structure at room temperature (considering strain energy only) is Σ32, while a transformation to Σ17 is expected at about 190K. Similarly, points C and D represent the second boundary (shown in Fig. 3) at liquid nitrogen and room temperatures. Again, the room temperature structure is expected to be Σ15, agreeing with the observation, and this is expected to transform to Σ32 at low temperature.

In no case have the authors been successful in obtaining an equilibrated structure at liquid nitrogen temperature because the motion of the dislocations is a glide-plus-climb process, which depends upon the transport of point defects to and from the specimen surfaces. Approximations of the time necessary to equilibrate a boundary in a thin foil of zinc reveal that while it can be achieved in a few minutes at room temperature, several days are required at 140K. It is not possible to perform this experiment at elevated temperatures because of the complicated factors of grain-boundary migration and specimen oxidation. Thus it has not been possible to confirm a match with the transformed structure.

Implications of Findings

The finding (illustrated in Fig. 1) that dislocation structures may have resolvable separations at extreme deviations from the exact coincidence parameters presents significant challenges to the experimenter and new questions for the theoretician. For the experimenter, the problem is one of associating an observed grain-boundary structure with some kind of periodic reference structure, which is usually considered to be a CSL or CCSL. The conventional approach to this problem has always been to compare the measured grain-boundary parameters with tables of known "coincidence misorientations." When a coincidence system is found within a small deviation of the experimental parameters, it is presumed that the grain-boundary structure relates to that system. The accepted deviation used by most experimenters is either a fixed value such as 2 or 3°, or the Brandon limit. It is now clear that for grain boundaries in HCP materials, and for other cases involving constrained lattice coincidence, this search method is insufficient. In principle, the only fully reliable search method for determining the appropriate reference structure would be to compute all possible dislocation spacings (including dsc dislocations and combinations of dsc dislocation up to the magnitude of lattice dislocations) needed to accommodate the misorientation and axial ratio error for every CCSL at the measured misorientation and boundary plane. The magnitude of this task is daunting, to say the least. Thus far, CCSL misorientations have been tabulated for only a relatively small range of c/a ratios within a few percent of the actual values for HCP metals (Chen and King, Ref 7), and it is clear that information over a much wider range is required. The total number of CCSL systems that will have to be considered will be extremely large, even if any search is restricted to a small value of Σ. The authors have yet to assess the range of deviation of other parameters, notably the axis of misorientation, which must be considered. For these reasons, they have not yet been able to associate the structures observed by Loberg and Smith (Ref 11) or Chen and King (Ref 7) with any CCSL system.

The authors' finding, of course, also begs the question of whether high-deviation boundaries of this kind do, in fact, have what might be described as special structures. Although the authors' calculations reveal that certain arrays of dislocations may be sufficiently well spaced to allow imaging by conventional TEM, other arrays may be so close together that their cores are considered to be completely overlapping. In effect, the interfacial structure might then be considered random or amorphous in the direction parallel to the more widely spaced dislocation lines, while there is some periodicity perpendicular to them. The exact nature of these structures deserves the attention of computer simulation studies.

If it is true that large-deviation boundaries exhibit one or two dimensionally periodic structures in HCP materials, the fraction of all grain boundaries that may be considered special could be much larger than is the case with cubic materials. This has significant implications for the relevance of studies of grain boundaries to general grain-boundary problems, which has for many years been a problem relative to grain-boundary studies in cubic materials.

The problems are equally daunting for interfaces that are much closer to the parameters required to form coincidence site lattices. The appropriate choice of CCSL is usually unclear, and although the Burgers vectors of the potential dislocations are not usually affected by this choice, the line directions and spacings of the dislocations required to accommodate the deviation are very sensitive to it. The range of step heights available to the dislocations is also affected by the choice of coincidence system (King and Smith, Ref 16) although the smallest value of the step height generally does not change from one CCSL to another.

When an array of dislocations is formed in a CCSL-related boundary, its spacing and orientation are determined by the need to accommodate deviations from the ideal misorientation and axial ratio. As the axial ratio changes, the dislocation content of the interface changes and thus also its energy. Such changes may be engendered by changing the temperature of the specimen, as the authors have attempted in their experiments. They have found that the structure certainly changes with temperature, at least in terms of the dislocation orientation and spacing. It is likely that the observed effects are the result of transformations from one CCSL structure to another, and this possibility has been assessed by computing the total energy of the calculated dislocation arrays for each CCSL at each misorientation and axial ratio (Shin, Ref 17). While these computations are relatively unsophisticated, relying only on a standard Read-Shockley formulation, with temperature-independent, isotropic elastic constants, they do reveal an interesting potential for transformation. The lowest energy dislocation array at room temperature in the case of the boundary shown in Fig. 2 would be associated with the Σ32 CCSL, which also provides the best match with the line directions determined experimentally. At liquid nitrogen temperatures, however, the lowest energy dislocation array would be that associated with the Σ17 CCSL. In the case of Fig. 3, the lowest energy structure at room temperature is Σ15 (agreeing with the observed array orientation and spacing), whereas at 140K it is Σ32. As discussed above, it was not possible to observe the dislocation arrays approaching equilibrium at the depressed temperature of this experiment, but it is a strong possibility that the authors have observed the transformation of the boundary structure from one CCSL to another in both of these cases. Transformations of this type may have profound influences on temperature-dependent grain-boundary phenomena, including grain-boundary migration and deformation.

Finally, it should be noted that the change of the axial ratio as a function of temperature (a thermal strain effect) induces interfacial thermal stresses. The changes of the dislocation arrays that have been observed may be described as responses to these thermal stresses, but this is fully equivalent to the treatment, which solves for those arrays that give no long-range stresses at any point in the space defined in Fig. 4. An important point concerning these boundaries is that there are multiple solutions for the stress-relieving arrays, each corresponding to a different CCSL system or to a different choice of Burgers vectors within a single CCSL.

Influence of Grain-Boundary Structure on Sliding Behavior

The sliding of grain boundaries in zinc has been studied extensively and carefully using controlled bicrystals, and perhaps more is known about this phenomenon in this material than about any other grain-boundary-related behavior in any other material (Watanabe *et al.*, Ref 1, 18; Valiev and Khairullin, Ref 19). While many different effects occur simultaneously during most sliding experiments, certain phenomena appear to be related to the existence of coincidence-related structures in the boundaries. Watanabe *et al.* (Ref 1), for example, identified a clear minimum in the sliding rate associated with the Σ9 symmetrical-tilt misorientation. Without a detailed model of the sliding mechanism, however, it is difficult to determine why a minimum sliding rate would be associated with a CCSL boundary. One might postulate, for example, that the low density of intrinsic, or structural, dislocations in the boundary gives rise to the low sliding rate, but this would assume that these defects are involved in the sliding process itself. An alternative interpretation, relying upon extrinsic, or run-in, dislocations as the effectuating defects, might be concerned with the availability of Burgers vectors lying in or near the grain-boundary plane. When the run-in lattice dislocations can dissociate only into a small number of well-defined dsc dislocations, the likelihood of creating defects that can move by pure glide is diminished; thus sliding might be expected to be more dependent upon grain-boundary diffusion for CCSL

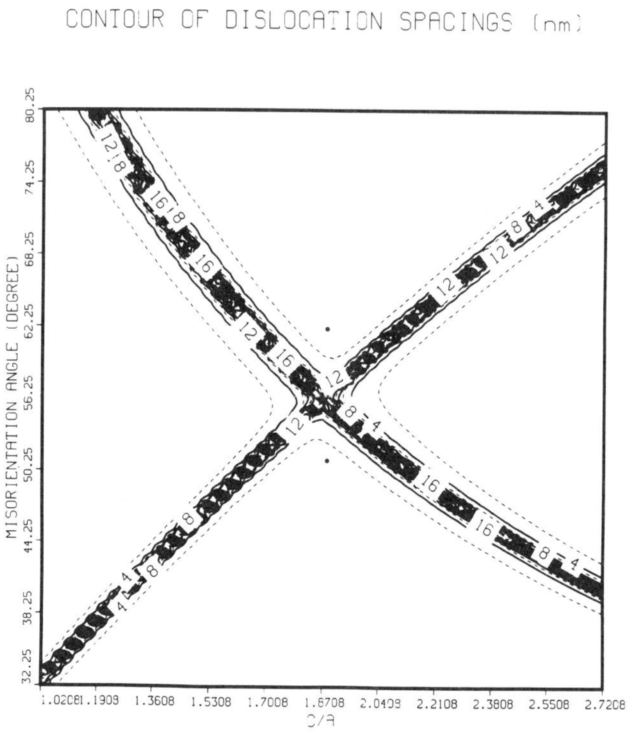

Fig. 5 Contour plot showing calculated dislocation spacings for $(10\bar{1}0)$ twist boundaries as a function of misorientation and axial ratio. The calculations are based on the Σ9 CCSL, with the $(1/3)[1\bar{2}10]$ Burgers vector. The upper point in the figure corresponds to the misorientation 62°, and the lower point to 51°, both with a c/a ratio of 1.8831, corresponding to zinc at the temperature 623K, as used by Watanabe *et al.* (Ref 18).

boundaries. Watanabe *et al.* (Ref 18) have shown that for boundaries not close to a CCSL system the sliding rate is well predicted on the basis of an extrinsic dislocation model, but there is a clear effect of grain-boundary structure that extends over a range of misorientation much greater than the Brandon range. The Σ9 structure seems to exert its influence up to 15° from the coincidence misorientation, confirming some of the authors' conclusions regarding the range over which the structure may be preserved by the insertion of dislocation arrays.

A second important feature of these sliding experiments (Watanabe *et al.*, Ref 1) is that the slope of the sliding rate-versus-temperature curve for each grain boundary shows a marked change at a well-defined temperature, which would seem to indicate the occurrence of a phase transformation. It is found that the temperature of this transformation depends upon the grain-boundary misorientation. An interpretation of this result can be found in Fig. 4, in which lines separating different stable phases cut across the misorientation-temperature plane at an angle; thus the temperature of the phase transformations observed by the authors seem indeed to be determined by the misorientation. It is thought, therefore, that the phase transformations detected by Watanabe *et al.* are of the same nature as the ones observed by means of electron microscopy by the present authors. The effect of these phase transformations on the sliding rates derive from the changing orientation and spacing of the intrinsic grain-boundary dislocations, both of which are discontinuous at the phase transformation. Such changes in the structure might affect either the impedance to motion of extrinsic dislocations or the effective density of intrinsic ones. The exact mechanism of the effect therefore remains obscure.

Most recently, Watanabe and coworkers have reported a study of grain-boundary sliding in $\langle 10\,\bar{1}0\rangle$ twist bicrystals of zinc (Watanabe *et al.*, Ref 18). In this case, the time taken to reach a particular creep strain was measured as a function of misorientation, and a generally increasing trend with increasing twist misorientation angle was found. The most striking feature of the curve, however, is a singularity at approximately the $\Sigma 9$ misorientation. As the misorientation increases through this value, the time required to reach any particular creep strain drops suddenly, by as much as two orders of magnitude. Otherwise stated, the strain rate suddenly increases at the $\Sigma 9$ misorientation but then resumes its tendency to decrease as misorientation increases. Again it is evident that the CCSL exerts some special effect on the boundary structure and its behavior, but this time the effect is very differently expressed. The authors calculated the intrinsic dislocation structure for the $\Sigma 9$ twist boundary and found that the boundaries that exhibited the extremes of behavior in the experiments (slowest straining at 51°, fastest at 62°) had well separated dislocations, as shown in Fig. 5. The principal difference between the two structures is that the dislocations in the 62° boundary were rotated by approximately 180° relative to those in the 51° boundary. Effectively, then, the Burgers vectors of the intrinsic dislocations are reversed upon passing through the CCSL misorientation, indicating that while they might be particularly well aligned to promote sliding in the one case, they would exert an opposing effect in boundaries on the other side of the CCSL misorientation. Now the mere reversal of the Burgers vectors would not affect the grain-boundary sliding, rate if the intrinsic dislocations were the effectuating defects for grain-boundary sliding, since they would merely move in the opposite direction, presumably with equal mobility. The observed effect is more likely to result from the reversal of the nature of the interaction with lattice dislocations that run into the grain boundary. Again, however, such a model is valid only if the lattice dislocations have predominantly only one sign, but this is a condition that is typically met in the vicinity of a grain boundary in a specimen under strain. The reason for this lies in the distribution of dislocation sources: dislocations run into the boundary from sources anywhere in the grain interiors or at free surfaces. Because dislocations of the opposite sign run out of the boundary under the same stress conditions, they must be created by sources within the boundary plane itself, which would be relatively sparse. It is therefore concluded that there will typically be a predominance of dislocations running into rather than out of the grain boundary, and thus of dislocations with a single sign. Under such conditions, the reversal of the Burgers vectors of the intrinsic structural dislocations in the grain boundary will have a distinctly reversed effect on the interaction with the predominating lattice dislocations, and thus a shift from an opposing effect to a synergistic one, as observed. The critical nature of the interaction is not yet clear, since the effect could arise in many ways. For example, long-range interactions between lattice dislocations and grain-boundary dislocations may either repel or attract the lattice dislocation, from or toward the boundary, respectively. Alternatively, and possibly more important, the effect could arise because in one case the run-in dislocation dissociates into dsc dislocations that add extra defects to the intrinsic array, but when the Burgers vector is reversed, the same run-in dislocation represents the subtraction of dislocations from the intrinsic array. In either case, the uniform spacing of the array is disturbed, but in one case there is a net increase in the number of defects available to promote sliding, whereas in the other case there is a decrease. For the case in which the number of dsc dislocations decreases, however, the array should become easier to move, according to the theory of King and Smith (Ref 15). Once again, therefore, it is presently impossible to explain the observed effect with the desired level of detail.

Conclusions

Grain-boundary structure in zinc is well depicted by the CCSL model, which predicts a broad range of misorientations over which special properties may be observed. Grain-boundary phase transformations are predicted by the model and observed in practice. Although many features of grain-boundary sliding experiments in zinc can be correlated with particular grain-boundary structures (or structural transformations), our understanding of the causes of these structure-property relationships remains unsatisfactory and the subject of further research.

Acknowledgment

This work was supported by the National Science Foundation, under grant number DMR 8901994.

References

1. T. Watanabe, Y. Masayuki, S. Shima, and S. Karashima, *Philos. Mag. A*, Vol 40, 1979, p 667
2. W. Bollmann, B. Michaut, and G. Sainfort, *Phys. Status Solidi (a)*, Vol 13, 1972, p 637
3. W.A.T. Clark and D.A. Smith, *Philos. Mag. A*, Vol 38, 1978, p 367
4. C.P. Sun and R.W. Balluffi, *Philos. Mag. A*, Vol 46, 1982, p 49
5. C.T. Forwood and L.M. Clarebrough, *Acta Metall.*, Vol 30, 1982, p 1443
6. D.G. Brandon, *Acta Metall.*, Vol 14, 1966, p 1479
7. F.-R. Chen and A.H. King, *Acta Crystallogr. B*, Vol 43, 1987, p 416
8. F.-R. Chen and A.H. King, *Philos. Mag. A*, Vol 57, 1988, p 431
9. J.G. Antonopoulos, P. Delavignette, T. Karakostas, P. Komninou, E. Laurent-Pinson, S. Lay, G. Nouet, and J. Vicens, *Colloque Phys.*, Vol 51, 1990, p C1-61
10. S. Lartigue, C. Carry, and L. Priester, *Colloque Phys.*, Vol 51, 1990, p C1-985
11. B. Loberg and D.A. Smith, *J. Microsc.*, Vol 102, 1974, p 317
12. W. Bollmann, *Crystal Defects and Crystalline Interfaces*, Springer, 1970
13. K. Shin and A.H. King, *Mater. Sci. Eng. A*, Vol 113, 1989, p 121
14. P.H. Pumphrey, *Scripta Metall.*, Vol 6, 1972, p 107
15. A.H. King and D.A. Smith, *Met. Sci.*, Vol 14 (No. 57), 1980
16. *Acta Crystallogr. A*, Vol 36, 1980, p 335
17. K. Shin, Ph.D. thesis, State University of New York at Stony Brook, 1990
18. T. Watanabe, M. Yamada, and S. Karashima, *Philos. Mag. A*, 1991 (to be published)
19. R.Z. Valiev and V.G. Khairullin, *J. Phys.*, Vol 51, 1990, p C1-685

3

Grain-Boundary Segregation in Ordered and Disordered Alloys

C.L. Briant
General Electric Company
Research and Development Center
P.O. Box 8
Schenectady, New York 12301

In the study of grain-boundary segregation in ordered and disordered alloys, the primary factors controlling segregation in binary systems are the bulk composition of the segregant and the temperature at which segregation occurs. In ternary and multicomponent systems the interactions between the segregant and the additional elements can make segregation much more complex. However, segregation data can be explained by considering the activity of the element in solution and the effect the other elements have on this activity. This chapter examines these topics and proposes that segregation can be thought of in terms of a simple chemical reaction and further, that many of the factors that limit the solid solubility of a solute in a solvent increase the likelihood that it will segregate.

The compositions of grain boundaries in a material are often quite different from the composition of the matrix. This difference occurs as various solute elements diffuse to the boundary and become trapped there. This trapping occurs because it is energetically favorable for the solute to remain in the grain boundary rather than diffuse back into the matrix. In particular, impurity elements often become very enriched at the grain boundary. The process by which this enrichment occurs is known as grain-boundary segregation.

Grain-boundary segregation has been an active field of research since the early 1970s. This activity has resulted because of various engineering problems in which grain-boundary segregation has been of overriding importance (Ref 1-6) and because of the development of Auger electron spectroscopy (Ref 7), a technique that has given researchers the ability to examine segregation in much more detail than was possible in the past. The initial studies of grain-boundary segregation that used Auger electron spectroscopy examined multicomponent engineering alloys (Ref 8-11). These studies were significant because they provided solutions to important problems, but the chemical complexity of the materials made the segregation data difficult to interpret. Consequently, efforts were directed at studying grain-boundary segregation in binary and ternary alloys. These studies allowed researchers to elucidate the fundamental principles of grain-boundary segregation, principles which could then be used to interpret the data obtained on the multicomponent alloys.

This chapter discusses segregation in both disordered and ordered alloys. First, the thermodynamics of segregation in disordered alloys are considered. Then the results for the kinetics of segregation in these materials are presented and, in particular, the effect that the thermal history that is developed as an alloy is worked and annealed can have on this segregation. The same processes are then considered for ordered alloys and the results compared with those obtained for disordered alloys. Finally, a simple way of thinking about grain-boundary segregation is presented, as well as a method for using this framework to interpret many results.

Segregation in Disordered Alloys

In experiments performed to study segregation in disordered alloys, the sample is usually heat treated at a high temperature to homogenize the composition and eliminate any segregation that has occurred during processing. Intermediate heat treatments specific to the purpose of the study may be desirable. Finally the sample is aged at the segregating temperature, and after the time at temperature is complete, the sample is quenched. In all of the studies described below, the amount of segregation is examined by Auger electron spectroscopy. The use of this technique requires that the sample be fractured along the grain boundaries in an ultrahigh vacuum spectrometer to ensure that the surface to be analyzed is composed of grain boundaries.

Reference 12 contains a review of this technique and its applications to grain-boundary segregation. It should be noted that various researchers have used different methods to quantify these results. Some simply normalize the signal of interest by a signal from a primary matrix element. An increase in this ratio should represent an increase in segregation. Others have attempted to convert the signal to actual concentrations. No effort has been made to change the various data used here to a single scale. The reader is referred to the individual references to obtain further information on this point.

There are many factors that can affect grain-boundary segregation in disordered alloys. To understand these effects, it is best to start with binary systems in which the segregant is completely soluble in the solvent. In these alloys the

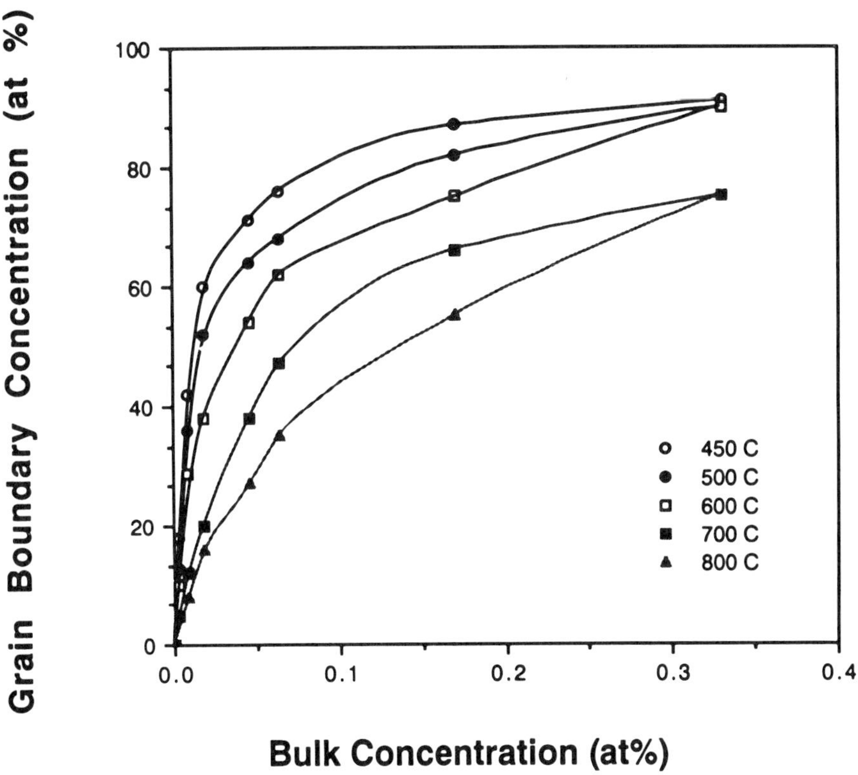

Fig. 1 Phosphorus segregation to grain boundaries in iron. The concentration of phosphorus on the grain boundaries is plotted as a function of the bulk phosphorus concentration. The different symbols represent different segregation temperatures. Source: Ref 13

equilibrium amount of segregation increases as the bulk concentration of the segregant increases. The equilibrium concentration of the segregant will also increase as the temperature at which segregation is allowed to occur decreases. Examples of these effects are shown in Fig. 1 (Ref 13).

If the segregant forms a precipitate with the host metal, the segregation behavior may be different. Once the solubility limit is exceeded, any additional segregant added to the system will go into a precipitate and will not be available to segregate to the grain boundaries. Figure 2 shows data that demonstrate this effect for antimony segregation in an Fe-3.5Ni alloy (Ref 14). As antimony is added to the Fe-Ni alloy up to a concentration of approxi-

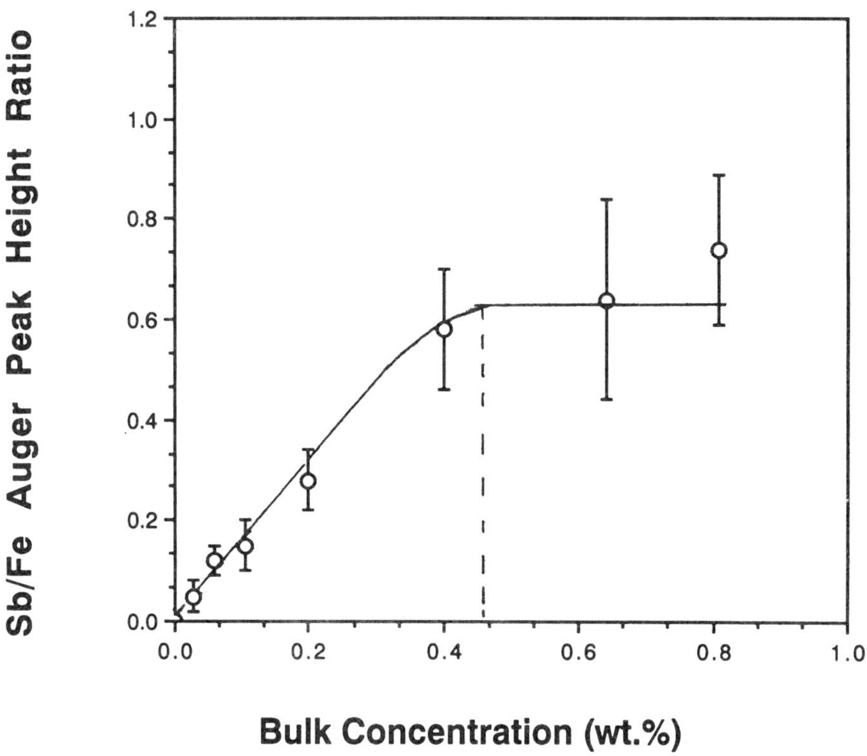

Fig. 2 Grain-boundary segregation of antimony in Fe-3.5Ni alloys. The concentration of antimony on the grain boundaries, represented as an Auger peak height ratio, is plotted as a function of the bulk antimony concentration. The dashed vertical line represents the solubility limit of antimony in this alloy. Source: Ref 14

mately 0.45 wt%, the amount of antimony on the grain boundaries increases. However, additions of antimony beyond approximately 0.45 wt%, the solubility limit at the segregation temperature used in this study (Ref 15), cause no additional segregation. The additional antimony goes into Fe-Ni-Sb precipitates, and the amount of antimony that is available to segregate in the Fe-Ni solid solution is unchanged.

The temperature dependence of segregation in systems in which the segregant forms a precipitate can also be different from that described above. At temperatures at which all of the segregant is in solution, the same principle will hold. However, in the temperature range in which precipitates form, two competing effects will occur. An increase in temperature will tend to cause desegregation but will also partially dissolve the precipitates. This dissolution

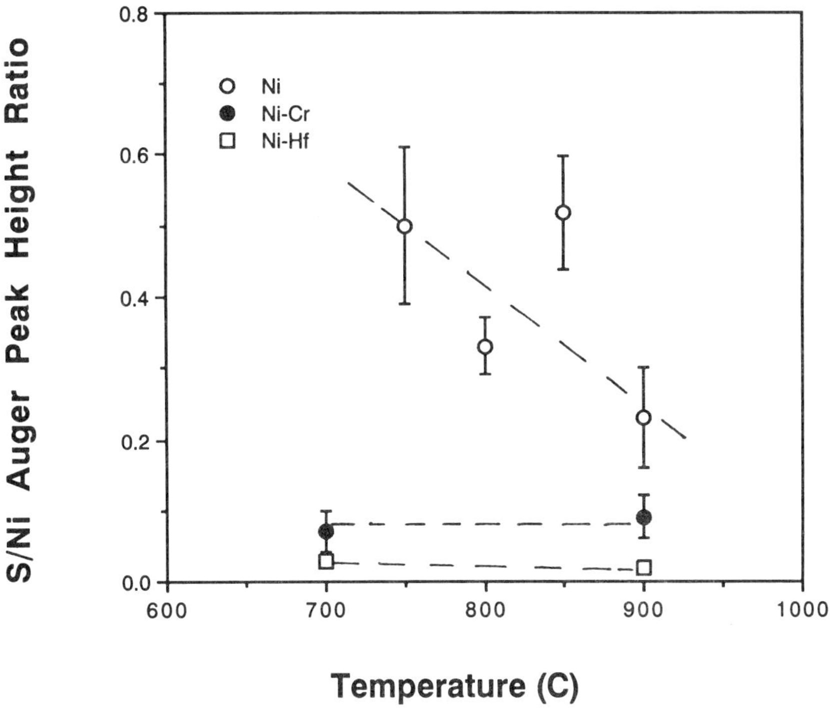

Fig. 3 The grain-boundary concentration of sulfur in Ni-S, Ni-Cr-S, and Ni-Hf-S alloys plotted as a function of the segregation temperature. The amount of sulfur on the grain boundaries is represented as a peak height ratio. Source: Ref 17

will increase the concentration of the segregant in solution and could thus increase segregation. If the temperature effect is dominant, a decrease in segregation with increasing temperature will still result. However, if the concentration effect is dominant, an increase in segregation with increasing temperature will be observed. Results show that the dominant effect depends on the system under study. In the Fe-S system, sulfur segregation increases with increasing temperature in the regime in which sulfides are present in the alloy (Ref 16). An increase in temperature partially dissolves the sulfides and increases the concentration of sulfur in solution. This increase dominates the normal desegregation tendency usually observed with an increase in temperature. In the Ni-S system, the opposite effect is observed (Ref 17). Even in the temperature-concentration range in which an increase in temperature partially dissolves sulfides, the temperature effect is overriding, and a decrease in segregation with increasing temperature takes place.

When a third element is added to a binary alloy, a number of possible effects can occur. One is that the third element will have no effect on segregation, and the amount of segregation will be identical to that observed in the binary alloy. This behavior has been reported for Fe-P-x alloys where x = Mo, V, W, and Cr (Ref 18-20). Another effect is that the third element can precipitate the segregant and thus lower or prevent its segregation. This effect has been observed in Fe-P-Ti alloys in which titanium precipitates the phosphorus as TiP (Ref 18) and in Fe-Cr-S and Fe-Mn-S alloys in which chromium and manganese precipitate the sulfur (Ref 21). Also, in nickel, chromium, and hafnium additions have been found to precipitate sulfur and reduce its segregation (Ref 17). An example of these effects is shown in Fig. 3.

Third elements can increase the amount of segregation over that observed in a binary alloy. The only well-documented example of this behavior is the increase in antimony segregation produced by nickel additions to iron (Ref 10, 14, 15, 22). An example is shown in Fig. 4. The plot in Fig. 4(a) presents data for two sets of samples containing similar amounts of antimony. One set of samples was alloyed with 1 wt% Ni, and the other with 3.5 wt% Ni. The results show antimony segregation to be greater in the alloy containing 3.5 wt% Ni. In contrast, the data in Fig. 4(b) show that antimony segregation has no effect on nickel segregation. The three alloys with 1 wt% Ni have similar amounts of nickel segregation, and the alloys with 3.5 wt% Ni have similar amounts of nickel segregation. Both sets are independent of the amount of antimony on the grain boundaries. As will be discussed below, the increase in antimony segregation caused by nickel additions results from the increase in the activity

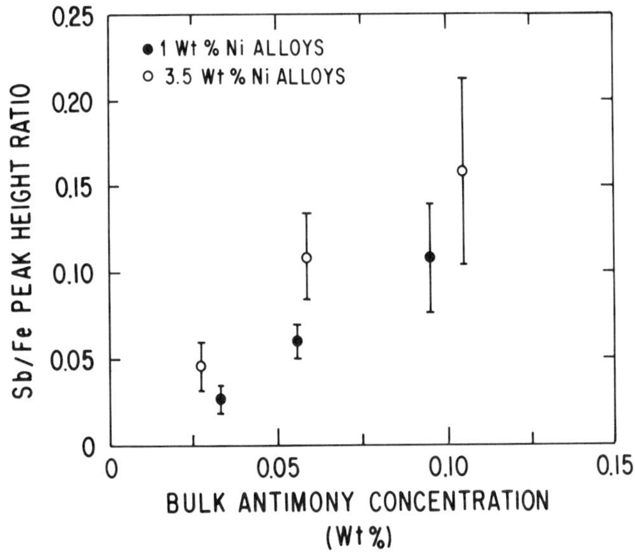

a

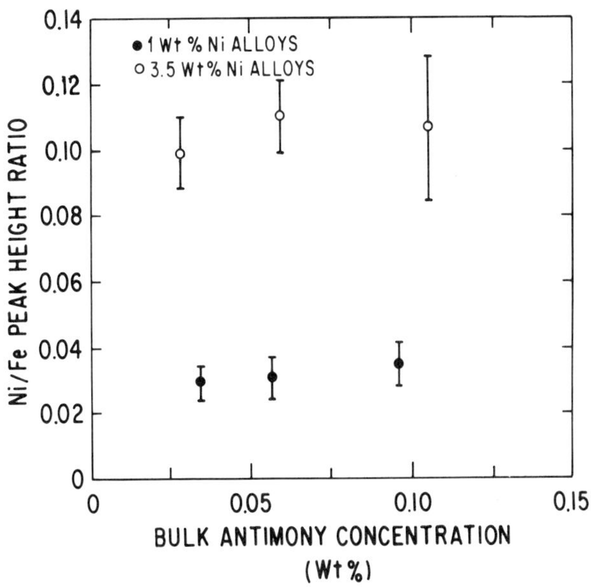

b

Fig. 4 (a) The antimony grain-boundary concentration, represented as an Auger peak height ratio, plotted as a function of the bulk antimony concentration in Fe-Ni alloys. The two different symbols represent two different nickel concentrations. (b) The nickel grain-boundary concentration, represented as an Auger peak height ratio, plotted as a function of the bulk antimony concentration in Fe-Ni alloys. The two different symbols represent two different nickel concentrations. Source: Ref 22 of antimony in solution, whereas antimony additions at these low levels have little effect on the activity of nickel in solution.

Finally, third elements can decrease segregation through site competition at the grain boundaries. A number of examples of this effect have now been found and include phosphorus and sulfur, phosphorus and carbon, sulfur and nitrogen, and antimony and carbon (Ref 13, 20-29). Figure 5 presents data showing the competition between phosphorus and carbon for sites at grain boundaries in iron.

These studies on binary and ternary alloys give the fundamental information needed to interpret results on multicomponent alloys. However, it is

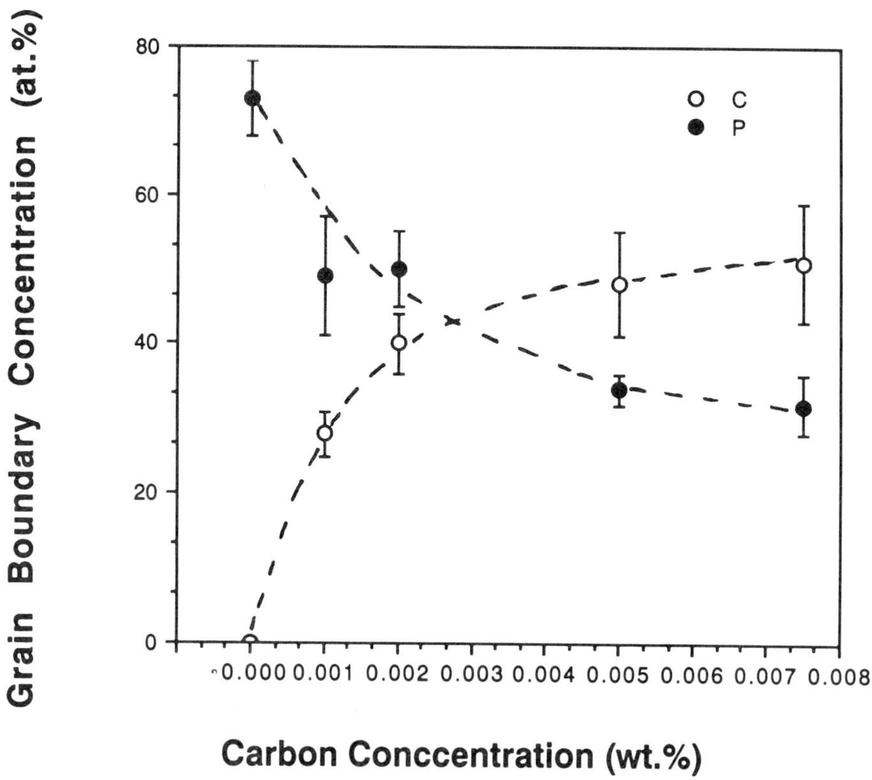

Fig. 5 Site competition between phosphorus and carbon at grain boundaries in iron. The grain-boundary concentrations of phosphorus and carbon are plotted as a function of the bulk carbon concentration in the Fe-C-P alloys. The alloy contained 0.17 at.% P. Source: Ref 13

important to consider all of the possible interactions described above. The following two examples will demonstrate this point.

It has been known for many years that phosphorus segregates readily in Fe-Cr-C alloys, and it had been proposed that additions of chromium actually enhance phosphorus segregation (Ref 11, 30). To determine whether this idea is correct, all of the binary interactions in the system must first be examined. Figure 6 shows the segregation data for phosphorus in high-purity iron and Fe-Cr alloys. The segregation is identical in both, and it must be concluded that chromium has no effect on phosphorus segregation. However, it has been established by many researchers that phosphorus and carbon compete for sites at the grain boundaries (Ref 13, 20, 24, 25). This fact is shown by the

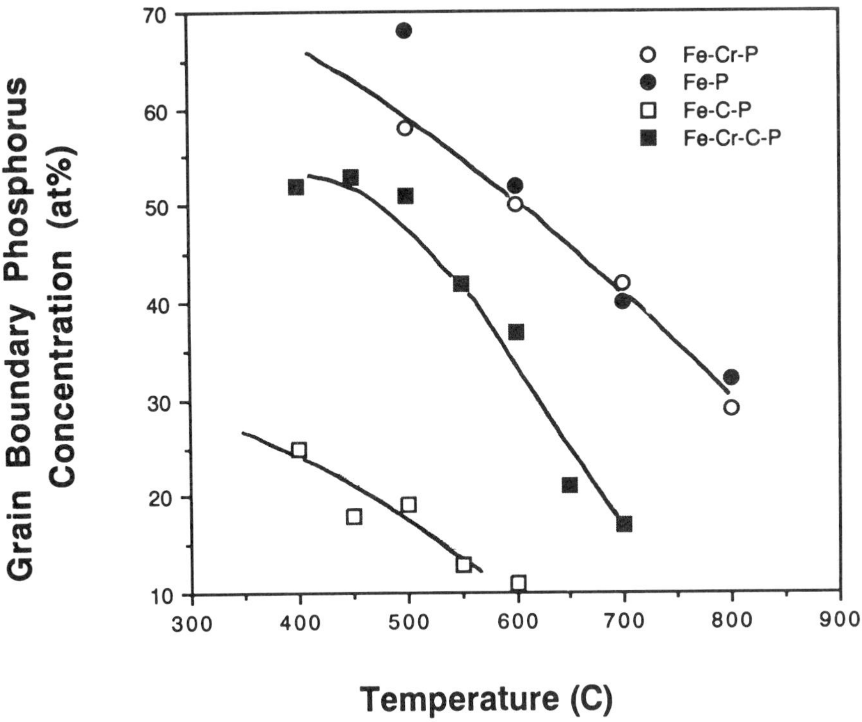

Fig. 6 The grain-boundary concentration of phosphorus in four different Fe-based alloys. The upper curve represents data for Fe-P and Fe-Cr-P alloys; the lowest curve, for Fe-C-P alloys; the intermediate curve, for Fe-Cr-C-P alloys. Source: Ref 13

decrease in phosphorus segregation in Fe-C-P alloys, also shown in Fig. 6. This same point is demonstrated in Fig. 5. The phosphorus segregation observed in the Fe-Cr-C-P alloy and shown in Fig. 6 is intermediate between the Fe-P alloy and the Fe-C-P alloy. The reason is that chromium additions precipitate some of the carbon as Cr-rich carbides (Ref 13). This precipitation reduces the concentration of unprecipitated carbon and diminishes site competition between phosphorus and carbon. However, since some carbon is left in solution, there will still be some site competition, and the amount of phosphorus segregation is not as great as in carbon-free alloys.

The other example to consider is antimony segregation in Fe-Ni-Cr-C-Sb, Fe-Ni-C-Sb, and Fe-Cr-C-Sb steels. The results for antimony segregation are

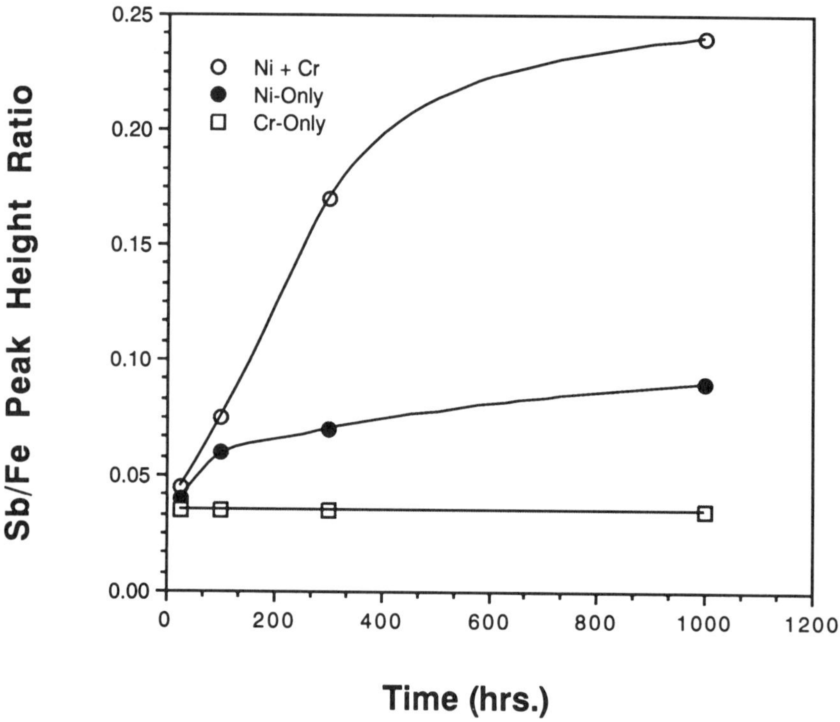

Fig. 7 Antimony segregation in three Fe-based alloys. The upper curve represents data for Fe-Ni-Cr-C-Sb steels; the intermediate curve, for Fe-Ni-C-Sb steels; the lowest curve, for an Fe-Cr-C-Sb steel. The antimony segregation is represented by the antimony-to-iron Auger peak height ratio. Source: Ref 31

plotted in Fig. 7. They show that antimony segregation is greatest in the Ni-Cr steel, next greatest in the Ni-only steel, and least in the Cr-only steel (Ref 31). The results presented in Fig. 4 show that a decrease in nickel concentration will decrease the segregation of antimony. Thus, the complete removal of nickel would probably cause the decrease in nickel segregation observed here. However, because chromium has been shown to have no effect on antimony segregation (Ref 32), it is necessary to determine why its removal causes the observed decrease. Previous work has shown that carbon competes with antimony for sites at the grain boundary just as it does with phosphorus (Ref 22). Thus the removal of chromium from the steel will allow much more carbon segregation and will decrease the antimony segregation. In addition,

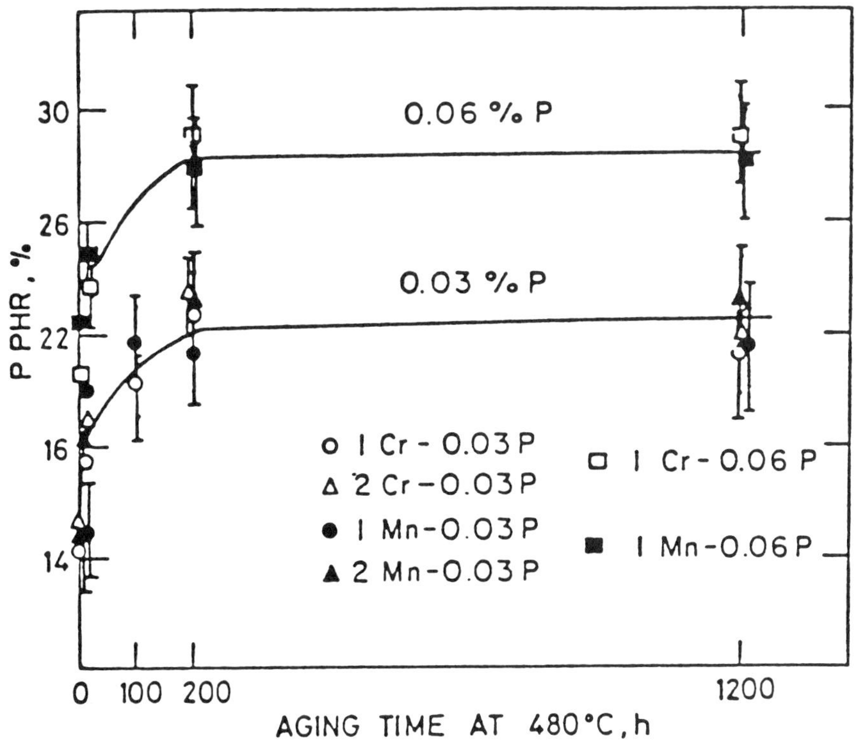

Fig. 8 The kinetics of phosphorus segregation in Fe-P-Cr and Fe-P-Mn alloys. The phosphorus-to-iron Auger peak height ratio is plotted as a function of aging time. The aging temperature was 480 °C. Source: Ref 20

the removal of chromium allows sulfur to segregate because Cr-S will no longer be formed (Ref 31).

The next subject to consider is the kinetics of grain-boundary segregation in these alloys. When an experiment on simple binary systems is carried out in the laboratory and the sample has been initially annealed so that at the start of the segregation process there is little or no segregant on the grain boundaries, the kinetics of segregation are very straightforward. There is an initial increase until the equilibrium concentration is reached, and then the amount of segregation remains constant. Figure 8 shows phosphorus segregation in Fe-Cr and Fe-Mn alloys. The length of time required to research the plateau depends on the diffusion coefficient of the segregant and the segregation temperature (Ref 10, 11, 20).

In multicomponent systems, the kinetics of segregation can be more complicated. For example, Fig. 9 shows the segregation of phosphorus in a

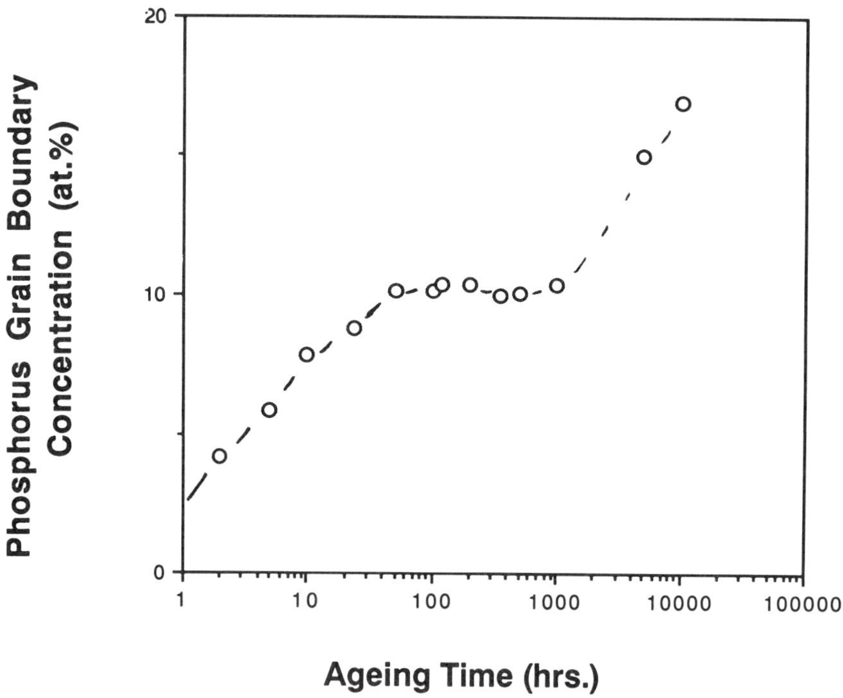

Fig. 9 Phosphorus segregation in a Ni-Cr-Mo steel. The phosphorus concentration on the grain boundaries is plotted as a function of the aging time. The aging temperature was 500 °C. Source: Ref 33

Ni-Cr-Mo steel (Ref 33). Initially, this steel had been tempered at 650 °C for a short time so that the only carbides precipitated were Fe-rich M_3C. When the sample was aged at 500 °C, phosphorus segregated to the grain boundaries and reached a plateau. However, after approximately 1500 h of aging, the segregation of phosphorus increased again. This second increase occurred because alloy-rich M_7C_3 carbides began to precipitate in the matrix, lowering the carbon activity and allowing more phosphorus to segregate. There are many other examples in the literature, particularly for steels, that show this type of segregation pattern as equilibrium is approached over hundreds or thousands of hours. The point to note is that because of the many interactions

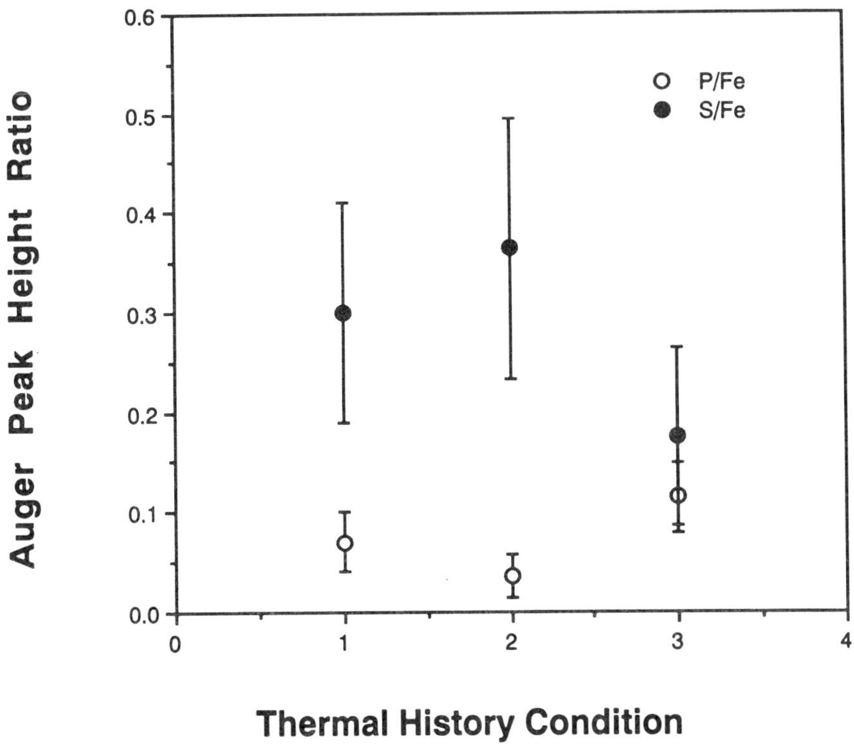

Fig. 10 Phosphorus and sulfur grain-boundary segregation for three different thermal history conditions. Condition 1 is the as-cast state. Condition 2 is the state after the cast material had been rolled, recrystallized at 657 °C, and aged at 480 °C. Condition 3 is the state after the cast material had been rolled, annealed at 1200 °C, annealed at 675 °C, and aged at 480 °C.

that occur over time in the multicomponent alloy, segregation may go through a number of plateaus before the final equilibrium concentration is reached.

Given the many interactions that can occur in an alloy, one usually finds that the amount of segregation depends on the complete thermal history of the sample. Segregation first occurs during solidification as various elements are ejected from the matrix and diffuse into the newly formed grain boundaries. If care is not taken to anneal these samples, this segregation may remain on the grain boundaries. An example of this can be seen in Fig. 10. The alloy used to make this plot contained iron, phosphorus, and sulfur. When the samples were cooled from the melt, which is designated condition 1 in Fig. 10, the amount

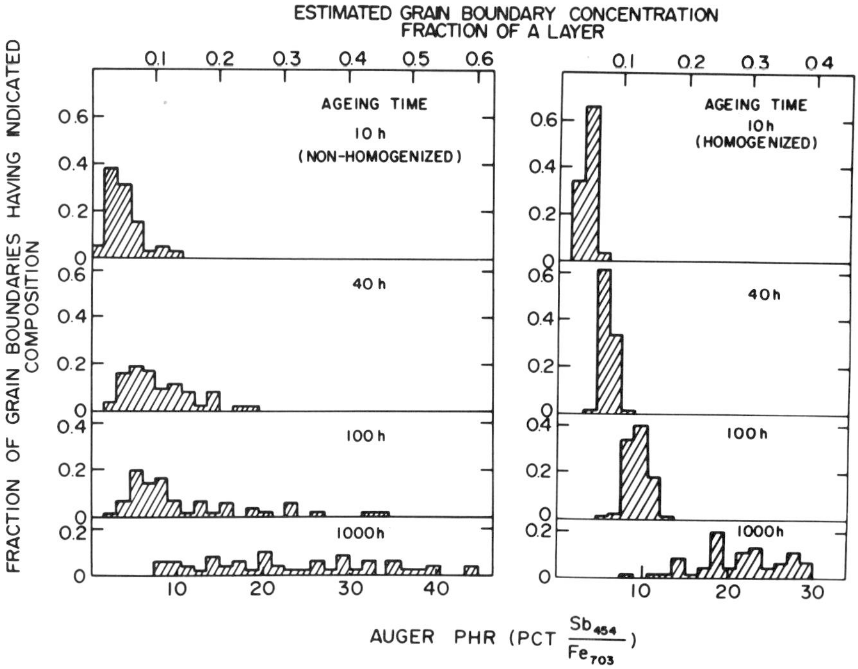

Fig. 11 The effect of compositional banding on the grain-boundary concentration of antimony in a nickel-chromium steel. The histograms give the fraction of grain boundaries with a given antimony concentration, which is represented by the antimony-to-iron Auger peak height ratio. The left panel represents samples that were only homogenized at 1025 °C for 1 h. This treatment did not rid the sample of compositional banding. The right panel represents samples that were homogenized at 1250 °C for 24 h, thereby eliminating banding. The final aging temperature was 480 °C for the times indicated. Source: Ref 35

of sulfur segregation to the grain boundaries was very high. Much of this sulfur was probably rejected into the boundary during the cool-down, since the kinetics of sulfide formation were probably more sluggish than segregation. If the samples were rolled, recrystallized at 675 °C, and aged at 480 °C, the amount of segregation was essentially unchanged. This treatment is labeled condition 2 in the plot. Only after the sample was rolled, recrystallized at 1200 °C for 48 h, annealed at 675 °C, and then aged at 480 °C did the amount of segregation change. The results in the figure show that after this treatment, which is labeled condition 3 in the plot, did sulfur segregation decrease and the amount of phosphorus segregation increase.

Another factor that affects segregation is compositional banding (Ref 34, 35). Banding occurs during solidification as temperature gradients and diffusion limitations produce different compositions of the solid that is forming throughout the ingot. If the segregant is banded, the local concentrations of the segregant vary throughout the ingot. Many studies have shown that these differences give rise to large variations in the amount of segregation and that these variations can be eliminated only through long-time, high-temperature anneals. An example of this effect is shown in Fig. 11 (Ref 35).

Segregation in Intermetallic Compounds

Much less work has been done on segregation in intermetallic compounds than on segregation in disordered materials. Much of the work that has been reported has simply addressed the question of whether or not an element segregated. Work has now shown that boron segregates in Ni_3Al, Ni_3Ga, Ni_3Si, Ni_3Ge (Ref 36-39), and NiAl (Ref 40); that sulfur segregates in Ni_3Al (Ref 38); and that carbon segregates in Ni_3Al and Ni_3Si (Ref 41, 42).

A great amount of attention has been directed at boron segregation in various $L1_2$ compounds because in some cases it improves the ductility of these compounds. Based on these studies, certain conclusions can be drawn. In Ni_3Al and Ni_3Ga, boron is soluble at concentrations greater than 1 at.%, and throughout that range boron segregation increases with increasing bulk concentration. An example is shown in Fig. 12. In Ni_3Si and Ni_3Ge, boron solubility is very low, and in most alloys that have been studied, borides were present on the grain boundaries. There was no measurable change in the amount of segregation with increasing boron content in these alloys because all additional boron goes into the precipitates. This effect is shown in Fig. 13. Finally, it has been observed that in Pt_3Ga, even at boron concentrations of 0.36 at.%, no borides were present in the alloy and no boron segregation was observed (Ref 43). A similar lack of segregation was observed for carbon in NiAl (Ref 40).

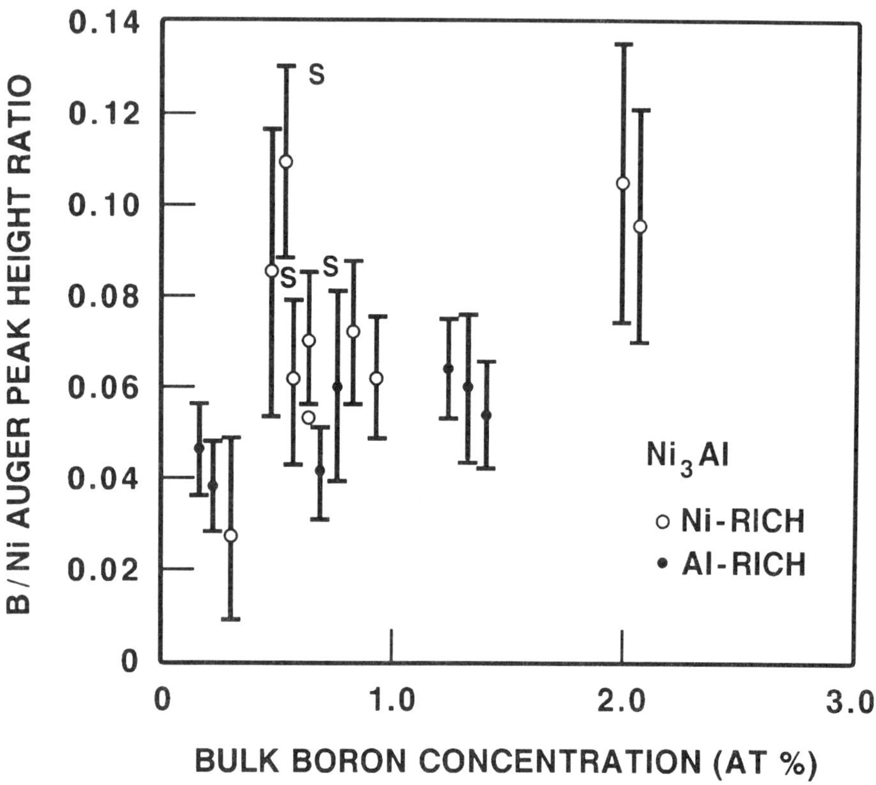

Fig. 12 Boron segregation in Ni_3Al. The grain-boundary boron concentration, represented by the boron-to-nickel Auger peak height ratio, is plotted as a function of the bulk boron concentration. The three data points marked with an S represent samples in which there was also sulfur segregation. Source: Ref 38

The temperature dependence of boron segregation has been studied only in Ni_3Al. Choudhary *et al.* (Ref 37) found that slowly cooling boron-doped samples caused an increase in segregation over that observed in rapidly quenched materials. However, Briant and Taub (Ref 44) did not find a measurable temperature dependence of segregation in rapidly solidified Ni_3Al. This point is yet to be resolved.

In general, the patterns observed in disordered alloys are also found in these materials. Many of the complex effects such as site competition and the

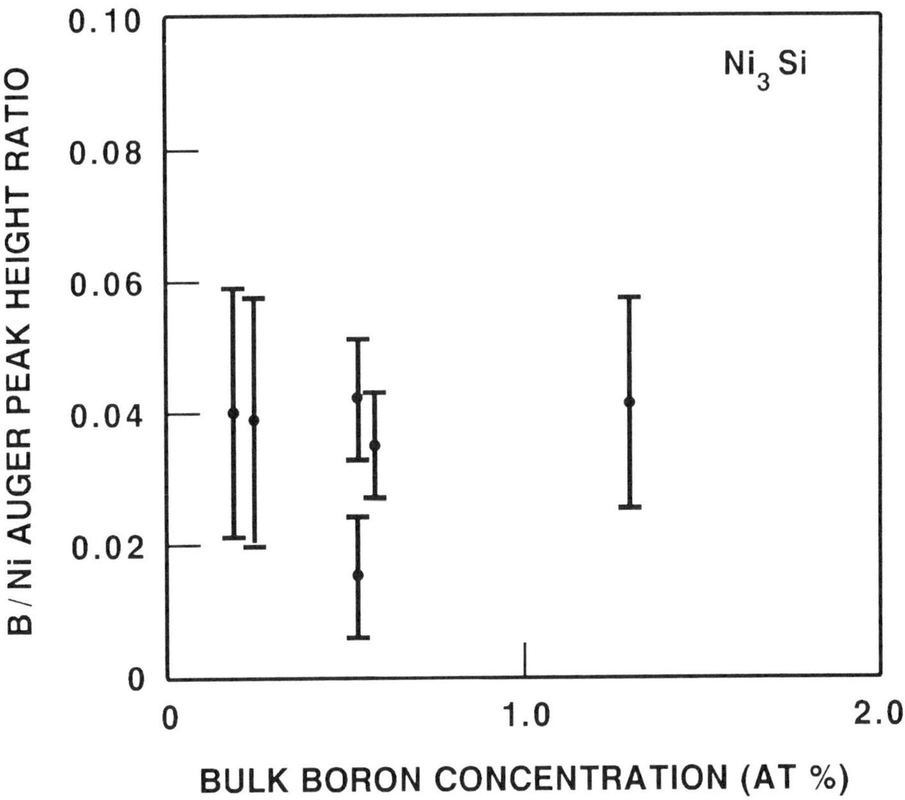

Fig. 13 Boron segregation in Ni_3Si. The grain-boundary boron concentration, represented by the boron-to-nickel Auger peak height ratio, is plotted as a function of the bulk boron concentration. Source: Ref 38

effects of fourth- and fifth-element additions have not yet been studied, but there is no indication that these will be different just because the compound is ordered.

Analysis of Grain-Boundary Segregation

The previous sections have presented an overview of grain-boundary segregation in both ordered and disordered alloys. This section considers several simple concepts that will help interpret these data.

First, a simple physical picture of the segregation process can be drawn. The material is heated to an elevated temperature so that diffusion can occur. The segregant then begins to move through the material. If it encounters a grain boundary, it has a high probability of becoming trapped there. With sufficient time, enough segregant reaches the grain boundary that its concentration becomes significantly enriched there. This process can be simply described by the chemical reaction:

$$I_b \rightleftarrows I_{gb} \qquad \text{(Eq 1)}$$

where I represents the segregant, b is bulk, and gb is the grain boundary. This chemical equation describes the distribution of a solute between two phases, and an equilibrium constant for this reaction can be written:

$$K = \frac{a^{I}_{gb}}{a^{I}_{b}} \cong \frac{[I]_{gb}}{[I]_{b}} \qquad \text{(Eq 2)}$$

where a denotes activity and the square brackets denote concentration. The rigorous definition of K would be in terms of activity, but since activities are often difficult to measure, concentrations can be used as an approximation. This idea then allows consideration of two important points. The first of these is what sort of segregation one should expect to observe in a given system and, second, what physical factors should determine the differences in segregation from one system to another.

For the first point, Eq 2 would suggest that segregation increases with increasing bulk concentration of the segregant until it reaches a solubility limit. In the simplest case, this limit would be when the solute begins to precipitate with the host metal or with another solute in the metal. Once precipitation begins, the activity or concentration in the matrix remains constant with increasing bulk concentration, and no change in the amount of segregation would be expected. This level of segregation will hold until a new phase begins to form, at which point the activity of the segregant in the matrix will change again. Thus by examining the phase diagram, some prediction about the grain-boundary segregation that occurs in a given system can be made.

Figures 1 and 2 show examples of these effects. The data reported in Fig. 1 are for phosphorus segregation in iron. In the alloys used for these measurements, the phosphorus concentration was well below the solubility limit in iron, and an increase in segregation with increasing bulk concentration was observed. However, the results in Fig. 2 are for a system in which antimony

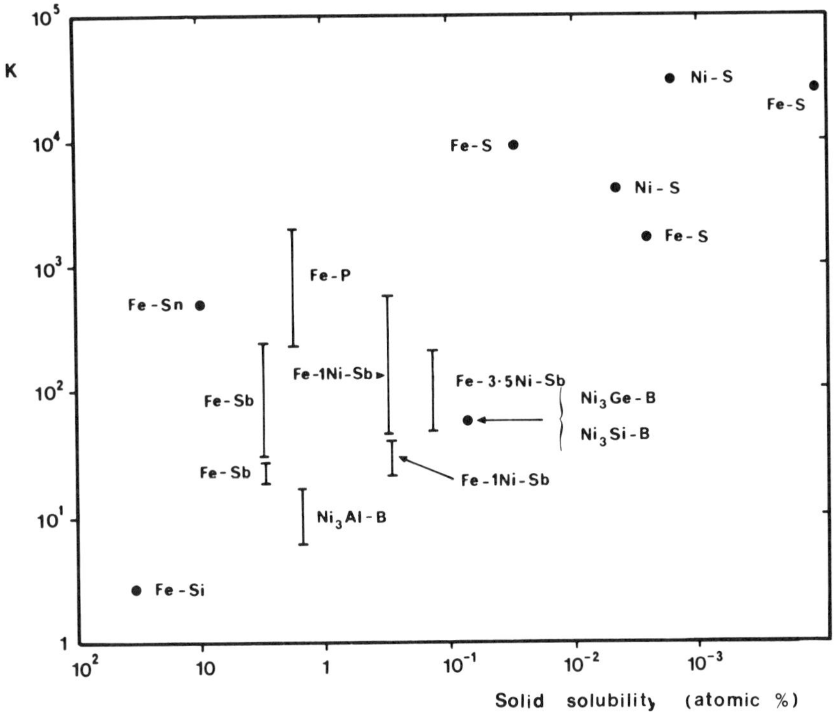

Fig. 14 The grain-boundary enrichment of various segregants (K) plotted as a function of their solubility in the alloy in which segregation was measured. The enrichment is calculated as the ratio of the grain-boundary concentration to the bulk concentration. Source: Ref 45, 46

was added to concentrations greater than the solubility limit. Once this limit was crossed, no additional segregation was observed. A similar result can be observed in Fig. 13 for boron segregation in Ni_3Si.

The results described for third-element additions can also be well explained by the ideas contained in Eq 2. First, third elements that cause precipitation of the segregant lower its activity in the matrix and cause a decrease in segregation, as is observed. Third elements that cause an increase in segregation most likely do so by increasing the activity of the segregant in solution. Those elements that have no effect on segregation have no effect on the activity of the segregant. Also, site competition occurs because of a change in activity in the grain boundary. Although complex equilibriums can be set up in multicomponent systems, this simple concept appears to be a way to think through the various segregation data that are reported.

The discussion above has indicated that the activity of the segregant in the matrix is of great importance in determining the causes of grain-boundary segregation. With this concept in mind, it is possible to ask what factors will cause segregation in the first place and why K varies from system to system. In the past, many researchers have suggested that atomic size affects the extent to which an atom will segregate. Atoms that do not fit well into the lattice may be able to find a site that is more accommodating to them in the grain boundary. However, chemical factors that affect segregation must also be present. In the lattice the segregant must form bonds with atoms arranged around it in certain fixed positions. Research has shown that this arrangement is probably not the one that would be preferred by the segregant and that more favorable arrangements for chemical bond formation can be found in the grain boundary (Ref 45). This chemical effect must also contribute to the extent to which a given element will segregate.

The various factors discussed above are very similar to the factors Hume-Rothery proposed as being important in determining the extent of the solid solubility of one element in another. Thus an inverse correlation between solid solubility and grain-boundary segregation might be expected, as first proposed by Hondros and Seah (Ref 46), since factors that limit solubility would tend to increase segregation. Figure 14 shows this correlation. Although there is scatter in the data, it can be observed that elements that have a low solubility have a greater segregation on the grain boundaries.

Conclusions

Grain-boundary segregation in disordered and ordered alloys have been reviewed. In both cases there are some principles that govern segregation. As long as the segregant is completely in solid solution, an increase in segregation with increasing bulk concentration and decreasing segregation temperature is expected. If the segregant begins to form a precipitate, the new equilibrium that is set up in the system must be considered in order to interpret the segregation data. Additional alloying elements can have a variety of effects on the segregation process depending on how they affect the activity of the element in solution. Indeed, all of these effects can be interpreted in terms of changes in the activity of the segregant in solution.

References

1. A.R. Ellery, Report 263, Research and Development Department, State Electricity Commission of Victoria, Australia, 1972

2. R.L. Bodnar and R.F. Cappellini, Research Department Publication, Bethlehem Steel Corporation, 1986
3. D.L. Newhouse, *Trans. AFS*, Vol 85, 1977, p 389
4. S.G. Druce and B.C. Edwards, Report Atomic Energy Research Establishment-R-9652, AERE Harwell, 1980
5. D. Newhouse, *Met. Prog.*, March 1967, p 98
6. J.M. Hodge and I.L. Mogford, *Proc. Inst. Mech. Eng.*, Vol 193, 1971, p 93
7. L.A. Harris, *J. Appl. Phys.*, Vol 39, 1968, p 1428
8. D.F. Stein, A. Joshi, and R.P. LaForce, *Trans. ASM*, Vol 62, 1969, p 776
9. J.R. Rellick, C.J. McMahon, Jr., H. Marcus, and P.W. Palmberg, *Metall. Trans.*, Vol 2, 1971, p 1492
10. H. Ohtani, H.C. Feng, C.J. McMahon, Jr., and R.A. Mulford, *Metall. Trans. A*, Vol 7A, 1976, p 87
11. R.A. Mulford, C.J. McMahon, Jr., D.P. Pope, and H.C. Feng, *Metall. Trans. A*, Vol 7A 1976, p 1183
12. C.L. Briant, in *Auger Electron Spectroscopy*, C.L. Briant and R.P. Messmer, Ed., Academic Press, 1988, p 111
13. H. Erhart and H.J. Grabke, *Met. Sci.*, Vol 15, 1981, p 401
14. C.L. Briant, *Mater. Sci. Technol.*, Vol 4, 1988, p 956
15. P. Gas, M. Guttmann, and J. Bernardini, *Acta Metall.*, Vol 30, 1982, p 1309
16. C.L. Briant, *Acta Metall.*, Vol 33, 1985, p 1241
17. R.A. Mulford, *Metall. Trans. A*, Vol 14A, 1983, p 865
18. H.J. Grabke, R. Moller, H. Erhart, and S.S. Brenner, *Surf. Interface Anal.*, Vol 10, 1987, p 202
19. D.Y. Lee, E.V. Barrerra, J.P. Stark, and H.L. Marcus, *Metall. Trans. A*, Vol 15A, 1984, p 1415
20. Y.-Q. Wang and C.J. McMahon, Jr., *Mater. Sci. Technol.*, Vol 3, 1986, p 207
21. C.L. Briant, *Acta Metall.*, Vol 36, 1988, p 1805
22. C.L. Briant, *Acta Metall.*, Vol 35, 1987, p 149
23. H. Sato and K. Sato, *J. Jpn. Inst. Met.*, Vol 41, 1977, p 458
24. S. Suzuki, M. Obata, K. Abiko, and H. Kimura, *Scr. Metall.*, Vol 17, 1985, p 1325
25. S. Suzuki, K. Abiko, and H. Kimura, *Scr. Metall.*, Vol 15, 1981, p 1139
26. S. Suzuki, S. Tanii, K. Abiko, and H. Kimura, *Metall. Trans. A*, Vol 18A, 1987, p 1109
27. G. Tauber and H.J. Grabke, *Ber. Bunsenges. Phys. Chem.*, Vol 82, 1978, p 298
28. H. Erhart and H.J. Grabke, *Scr. Metall.*, Vol 15, 1981, p 531
29. R.H. Jones, S.M. Bruemmer, M.T. Thomas, and D.R. Baer, in *Hydrogen Effects in Metals*, I.M. Bernstein and A.W. Thompson, Ed., The Metallurgical Society of American Institute of Mining, Metallurgical, and Petroleum Engineers, 1981, p 369
30. J. Yu and C.J. McMahon, Jr., *Metall. Trans. A*, Vol 11A, 1980, p 277
31. R.A. Mulford, C.J. McMahon, D.P. Pope, and H.C. Feng, *Metall. Trans. A*, Vol 7A, 1976, p 1269
32. C.L. Briant, *Metall. Trans. A*, Vol 20A, 1989, p 2170
33. B.C. Edwards, B.L. Eyre, and G. Gage, *Acta Metall.*, Vol 28, 1980, p 335

34. C.L. Briant, *Acta Metall.*, Vol 31, 1983, p 257
35. J. Kameda and C.J. McMahon, Jr., *Metall. Trans. A*, Vol 12A, 1981, p 31
36. C.T. Liu, C.L. White, and J.A. Horton, *Acta Metall.*, Vol 33, 1985, p 213
37. A. Choadhury, C.L. White, and C.R. Brooks, *Scr. Metall.*, Vol 20, 1986, p 1061
38. C.L. Briant and A.I. Taub, *Acta Metall.*, Vol 36, 1988, p 2761
39. I. Baker, R.A. Padgett, and E.M. Schulson, *Scr. Metall.*, Vol 23, 1989, p 1969
40. E.P. George and C.T. Liu, *J. Mater. Res.*, Vol 5, 1990, p 754
41. C.L. Briant and S.C. Huang, *Metall. Trans. A*, Vol 17A, 1986, p 2084
42. A.I. Taub and C.L. Briant, *Metall. Trans. A*, Vol 20A, 1989, p 2025
43. C.L. Briant, A.I. Taub, and E.L. Hall, *J. Mater. Res.*, Vol 5, 1990, p 2841
44. C.L. Briant and A.I. Taub, *Mater. Res. Soc. Symp.*, Vol 133, 1989, p 281
45. C.L. Briant, *Metall. Trans. A*, Vol 21A, 1990, p 2339
46. E.D. Hondros and M.P. Seah, *Scr. Metall.*, Vol 6, 1972, p 1007

Section II:

Interfaces in Ordered Alloys

4

Examination of Grain-Boundary Structure and Chemistry in Ordered Alloys

Ian Baker
Thayer School of Engineering, Dartmouth College
Hanover, New Hampshire

This chapter reviews recent experimental research on the structure and chemistry of grain boundaries in ordered alloys. The techniques used for grain-boundary examination are discussed, with emphasis on the limitations of each technique.

Most work on grain-boundary structure and chemistry in ordered alloys has been on $L1_2$ compounds, in particular Ni_3Al. Several researchers have noted that the grain boundaries in boron-doped nickel-rich Ni_3Al are enriched in nickel. Some researchers have found evidence of an fcc phase at the grain boundary in this material, while others have found no evidence of such a phase. In addition, there are some reports of nickel enrichment and the presence of a phase at the grain boundaries in boron-free nickel-rich Ni_3Al, and one report of grain-boundary nickel enrichment in boron-doped stoichiometric Ni_3Al. Possible reasons for the variety of results, in addition to differences in examination technique, are differences in the Ni:Al ratio, in the levels of boron and other interstitials, and in heat treatment.

There have been few studies of grain boundaries in B2 compounds. However, grain-boundary compositions in the B2 compounds NiAl and FeAl were usually found to be the same as those of the nearby matrix.

Many ordered intermetallic compounds suffer from brittle intergranular fracture at room temperature. The occurrence of this brittle fracture even in high-purity compounds in which no segregants can be found at the grain boundaries (Ref 1, 2) suggests that this is an intrinsic problem. The outstanding example of a remedy for brittle intergranular fracture in an intermetallic is the addition of boron to the $L1_2$-structured (ordered face-centered cubic) compound Ni_3Al (Ref 3-5). In this case, the improvement in ductility is dramatic, resulting in a change from fracture around the yield strength in unalloyed Ni_3Al to over 50% elongation at room temperature for the optimum composition Ni-24Al-0.48B (all compositions are in atomic percent) (Ref 5).

While this discovery represented a dramatic technological breakthrough, the approach to ductility improvement was intuitive empiricism. Unfortunately, boron does not simply act like a glue that sticks grain boundaries together. While boron doping also improves the ductility of the related $L1_2$ compounds Ni_3Si (Ref 6-8) and Ni_3Ga (Ref 8), it produces no ductility improvement in the isostructural compound Ni_3Ge (Ref 8) or in other ordered alloys such as the B2 (ordered body-centered cubic) compound NiAl (Ref 9), which also fractures intergranularly at the stoichiometric composition. Interestingly, other interstitials, such as Be and C, can improve the ductility of some $L1_2$ compounds (Ref 8, 10).

The above results have led to a growing interest in the structure and chemistry of grain boundaries in intermetallic alloys. The technological objective is to understand the mechanisms for brittle intergranular fracture from which some intermetallic compounds suffer and, hence, to prescribe a remedy to circumvent or delay these failure mechanisms.

The research on grain boundaries in ordered intermetallic compounds can be divided into two parts. First, there have been a number of studies (for example, Ref 11-22) that have attempted to model grain-boundary structures, energies, and chemistries (almost exclusively in $L1_2$ compounds). These studies have involved either geometric modeling or the use of interaction energies between atoms. The studies have primarily been on binary compounds, and the effect of boron on the boundary has been inferred in a rather *ad hoc* manner. Also, these studies have largely been concerned with the idealized stoichiometric composition and hence do not consider the situation, away from stoichiometry, in which one element is in excess. This is important since some $L1_2$ compounds do not even exist at the stoichiometric composition; for example, Ni_3Si exists only as a nickel-rich compound. It is also worth noting that not all of the predictions from these studies are in agreement.

The second body of research in this area has focused on measuring the chemistry and/or determining the structure (the two are often interlinked) of grain boundaries or the near-grain-boundary regions in intermetallic compounds. Much of this work has been on boron-doped alloys, alloys that have ternary additions or nonstoichiometric compositions. This has made it difficult to compare directly the results of the experimental studies with the results of the modeling studies. This second body of work is the subject of this review. The effect of grain-boundary chemistry/structure on the resulting mechanical properties is beyond the scope of this discussion.

$L1_2$ Compounds

Most experimental studies of the structure/chemistry of grain boundaries in intermetallic compounds have been on $L1_2$ compounds, the majority of these being on Ni_3Al, primarily in alloys containing boron. Thus, while this chapter reviews the experimentally determined chemistry/structure of grain boundaries in ordered alloys as a whole, it is largely devoted de facto to grain boundaries in Ni_3Al, which will be covered first. Data on other $L1_2$ compounds will then be considered, and, finally, some work on B2 compounds will be described. It is important to point out that the experimental studies of grain boundaries have in fact often been about the near-grain-boundary regions, not grain boundaries per se.

Rather than simply catalog the observations of various workers on the structure/chemistry of grain boundaries in Ni_3Al, the results of Baker and coworkers (Ref 23-27) will first be reviewed since these observations on the structure/chemistry of grain boundaries in Ni_3Al are possibly the most comprehensive in terms of alloy stoichiometry and boron content. Then the observations of other researchers will be compared to this work. In reviewing the data, the limitations of the various experimental techniques used to examine grain boundaries will be emphasized, as well as specific limitations of the results, for example, the examination of a relatively small number of samples. In addition to considering the observation technique used, care must be taken in comparing like with like; often slightly different compositions have been studied and different heat treatments have been used, factors that may be expected to give different results. Some studies have, in fact, used material that is clearly far from equilibrium, such as studies on as-cast material or rapidly-solidified material. Since these cannot help to elucidate the equilibrium structure/chemistry, they will not be considered here.

STEM (EDS) and CTEM. Baker and coworkers (Ref 23-27) examined thin foils of boron-free Ni_3Al, produced by the hot extrusion of rapidly solidified powders, using energy-dispersive x-ray microanalysis (EDS) on a field emission gun (FEG) scanning transmission electron microscope (STEM), a VG HB501. The incident electron probe was only about 1 nm in diameter; however, the size of the probe upon exiting the thin foil was up to 100 nm, thus limiting the resolution to considerably greater than 1 nm. All alloys were annealed, as disks, for 1 h at 1000 °C in flowing argon and furnace cooled prior to examination. Grain boundaries in both aluminum-rich (Ni-26%Al) and stoichiometric Ni_3Al were found to have compositions indistinguishable from that of the matrix. In contrast, in nickel-rich material (Ni-24%Al) some grain boundaries were found to be slightly nickel-enriched, while others were found to be strongly nickel-enriched (Fig. 1a). Conventional transmission electron microscope (CTEM) observations of the latter material showed that some grain boundaries contained a distinct second phase up to 20 nm in width in some places (Fig. 1b), while others contained either no phase (or one too narrow to resolve) or only small regions of second phase. Convergent-beam electron diffraction (CBED) was used to show that this phase had the same orientation as one of the grains but that it was disordered (fcc), as indicated by the lack of superlattice reflections in the CBED patterns (Fig. 2). Because of the size of the electron probe, greater than or equal to 7 nm, plus the broadening of the electron beam, at approximately (thickness)$^{3/2}$, no information was obtained about whether the change in degree of order at the interface was abrupt or more slowly varying. The disordered phase was epitaxial with one grain and thus was bounded by the grain boundary on one side and a coherent or semicoherent interface with the ordered matrix on the other side. The phase was most clearly observable under strong two-beam conditions and appeared to exhibit strain contrast at the interface (see, for example, Fig. 1b).

Similar examinations of boron-containing (~0.35 at.%) alloys, again, could find no differences between grain boundaries and the nearby matrix in the aluminum-rich composition, but in stoichiometric material, slight nickel enrichment of grain boundaries was observed (Fig. 3a), although there was no evidence of a second phase. That is not to say that no grain-boundary phase was present in this case. Clearly, no 10 to 20 nm thick grain-boundary phase was present, but, using CTEM, the presence of, for example, a 2 nm thick phase is not easily distinguishable from simple grain-boundary nickel enrichment. (It is worth noting that a reexamination of the exact same grain boundaries that showed nickel enrichment in stoichiometric Ni_3Al + B using a VG HB501 could not find this slight nickel enrichment using a Philips 400 FEG,

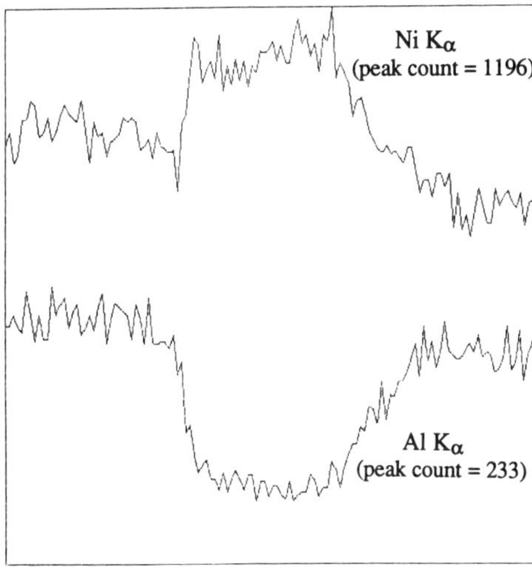

a

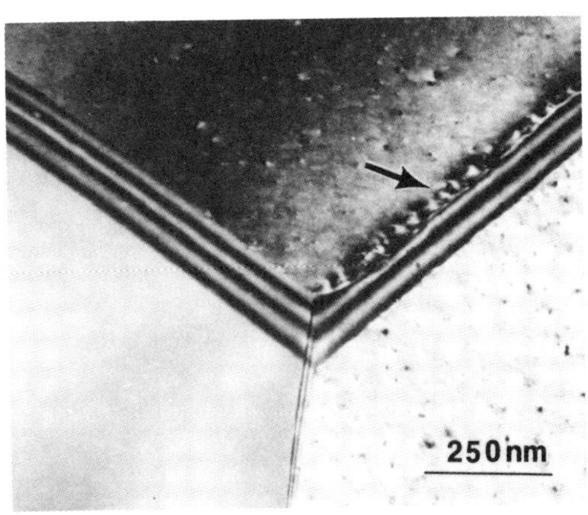

b

Fig. 1 (a) EDS line profile (128 channels, 1 nm wide, 5 s per channel) across a grain boundary. (b) Bright-field transmission electron micrograph of a triple point showing the presence of a phase (arrow) along one grain boundary in unalloyed Ni-rich Ni_3Al. Source: Ref 27 presumably because of the larger probe size and lower beam current in the latter instrument. It should be noted that boron itself is not easily detectable in a metal, even using a windowless EDS system.) In boron-doped nickel-rich Ni_3Al, a second phase (up to, but usually less than, 20 nm wide) was observed, using CTEM, on one side of every grain boundary (Fig. 4). Some grain boundaries had a wavy appearance. The phase was never observed on both sides of the grain boundary at the same point. Using EDS on a VG HB501 (Fig. 3b) and CBED on a CTEM (Fig. 4), the phase was found to be highly nickel enriched and fcc coherent with the adjacent matrix, respectively.

This second phase was also observed using the STEM-based Z-contrast imaging technique, in which a high-angle annular detector on a VG HB501 was used to image composition changes directly at grain boundaries with high resolution. The technique has the advantage that contrast reversals do not

a

b

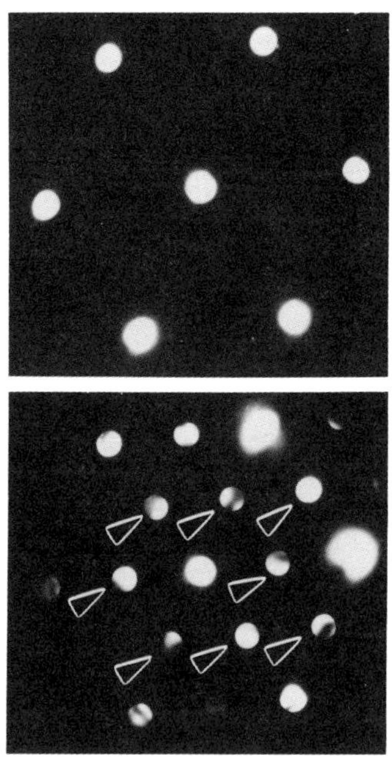

Fig. 2 CBED patterns (7 nm probe) from (a) grain-boundary phase and (b) adjacent matrix in unalloyed Ni-rich Ni_3Al. It should be noted that the superlattice reflections (see arrows) in (b) are absent in (a), indicating that the grain-boundary phase is disordered. Source: Ref 25

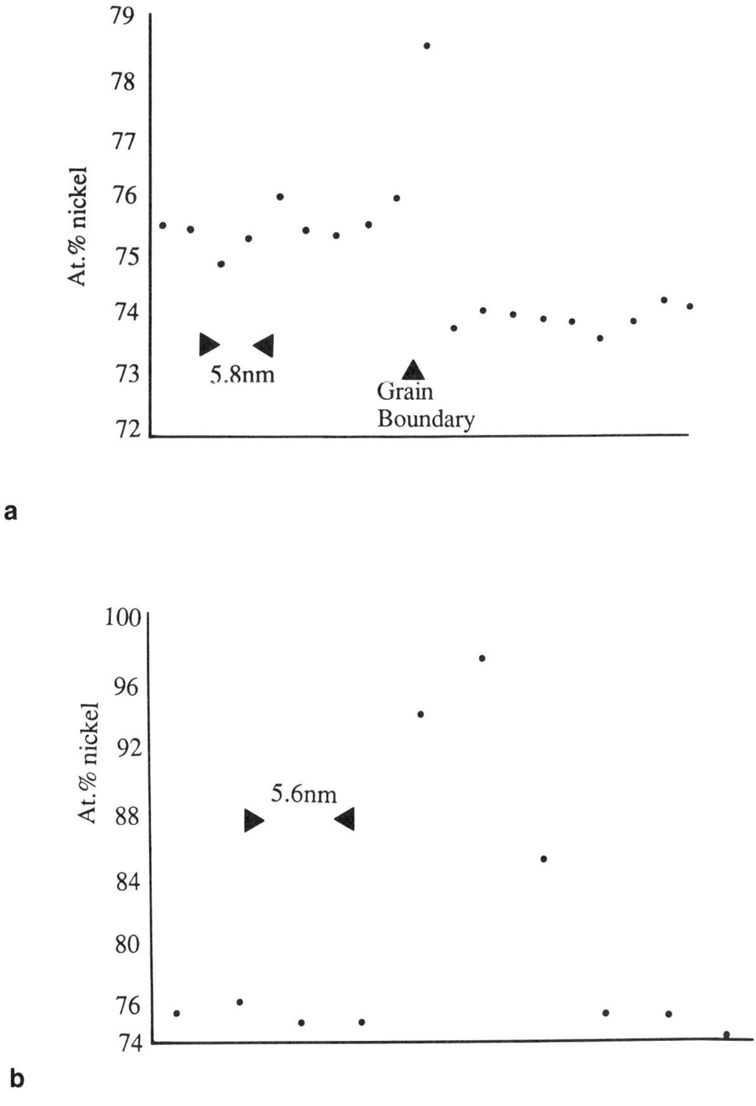

Fig. 3 Graph of at.% Ni obtained from an EDS multipoint profile (1 nm probe) across grain boundaries in B-doped (a) stoichiometric Ni_3Al and (b) Ni-rich Ni_3Al. Source: Ref 24, 27

occur with changes in sample thickness or with image defocus. Z-contrast STEM images of boron-doped nickel-rich Ni_3Al foils are shown in Fig. 5. These images are consistent with both the EDS and the CBED results. Figure 5(a) shows a boundary phase approximately 10 nm wide. Neither of the grains, nor the boundary phase, which is epitaxial with the upper grain, is channeling; thus the bright contrast of the boundary phase is consistent with the enrichment of nickel found by x-ray microanalysis. Figure 5(b) shows a high-resolution image of the interface between Ni_3Al and the epitaxial boundary phase imaged along ⟨001⟩. In this direction, the ordered phase projects as a square lattice of lattice parameter 0.356 nm, consisting of columns of pure aluminum,

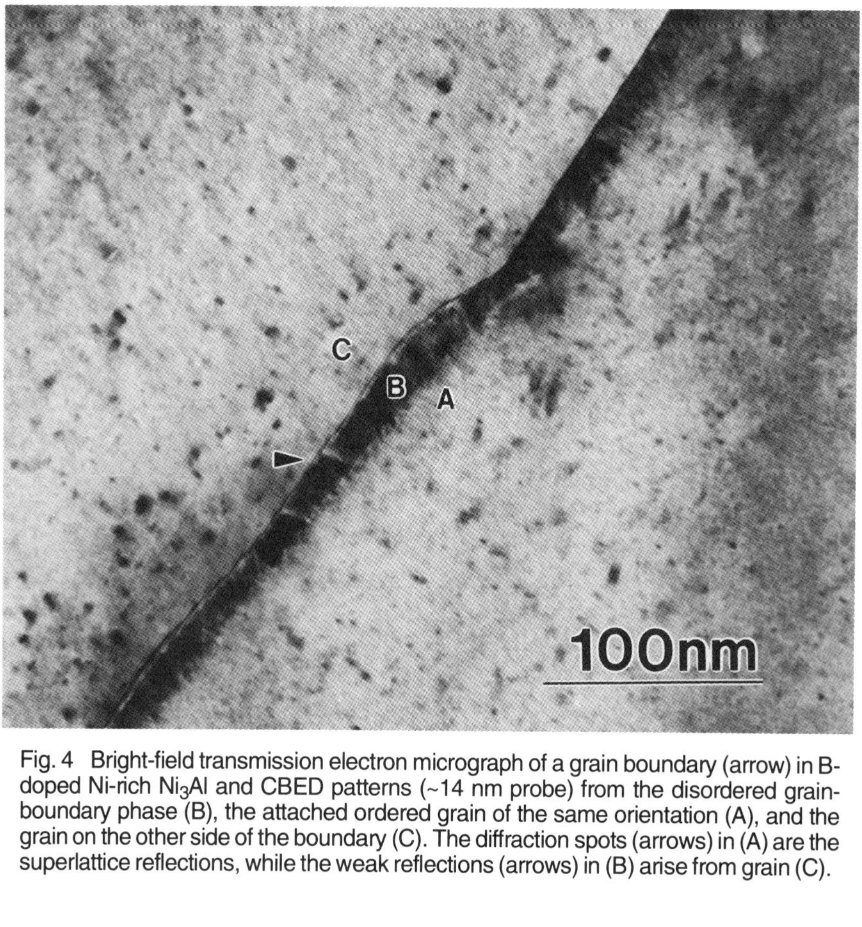

Fig. 4 Bright-field transmission electron micrograph of a grain boundary (arrow) in B-doped Ni-rich Ni_3Al and CBED patterns (~14 nm probe) from the disordered grain-boundary phase (B), the attached ordered grain of the same orientation (A), and the grain on the other side of the boundary (C). The diffraction spots (arrows) in (A) are the superlattice reflections, while the weak reflections (arrows) in (B) arise from grain (C).

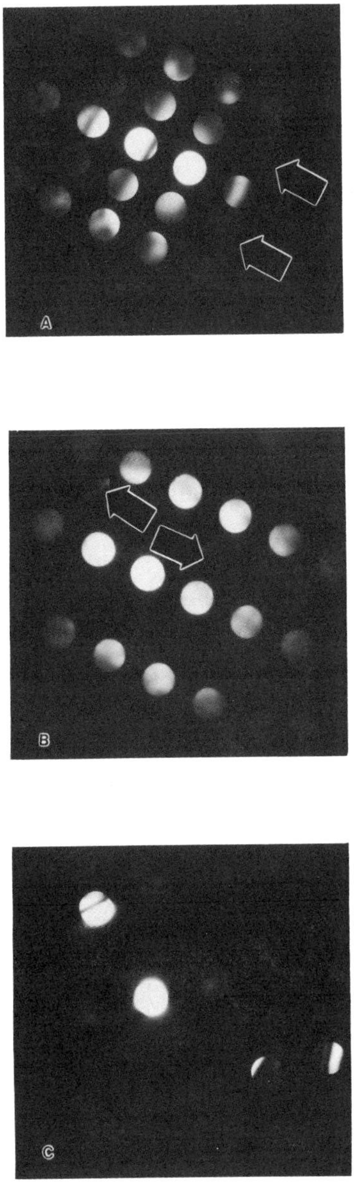

Fig. 4 (continued)

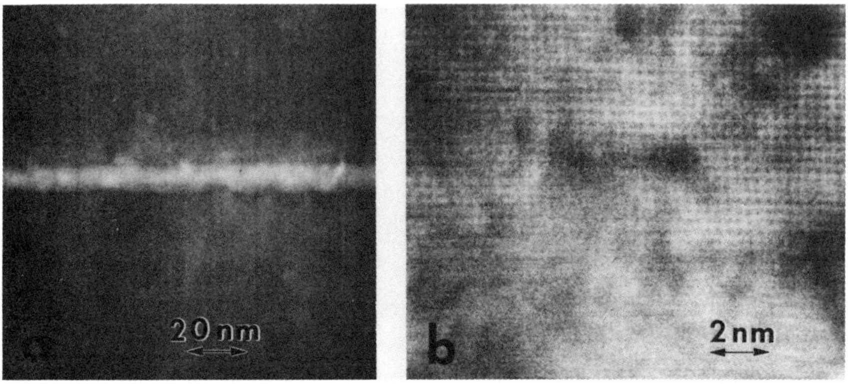

Fig. 5 Z-contrast images from a grain-boundary region in B-doped Ni-rich Ni_3Al. (a) Image showing that the (bright) grain-boundary phase is about 10 nm wide. (b) High-resolution image showing that the order in the lattice (top of the image) is not present in the grain-boundary phase (bottom of image). Source: Ref 27

with the three nickel columns at the positions (0,1/2), (1/2,0), and (1/2,1/2). Their separation of $a_0/2$ is below the resolution limit, therefore the Z-contrast image appears as a square mesh, the dark contrast corresponding to the aluminum columns. The absence of ordering in the boundary phase is consistent with the CBED results.

Figure 6 summarizes the results from grain boundaries in Ni_3Al, and Fig. 7 shows the structure observed on the grain boundaries in a boron-doped nickel-rich Ni_3Al containing the second phase. The latter structure is identical to that predicted in boron-doped nickel-rich Ni_3Al by the atomistic modeling of Chen *et al.* (Ref 20) by which they also predicted that nickel and boron cosegregation to the grain boundaries should occur.

Before leaving this work, two final comments are worth making on the EDS results. First, although many grain boundaries were surveyed using CTEM, the data on the chemistry of the grain boundaries were only for a maximum of five or six boundaries for each alloy. Nonetheless, it is unlikely, though possible, that the six grain boundaries in, for example, the boron-doped nickel-rich Ni_3Al were atypical since all grain boundaries exhibited a second phase and all measurements showed a large nickel enrichment. The second point concerns the quantification of the EDS data. While the data for boron-doped nickel-rich Ni_3Al clearly show nickel-enriched grain boundaries, the compositions are extraordinarily high. There are two possible reasons for this. First, the grain-boundary region was thicker than the adjacent matrix (the

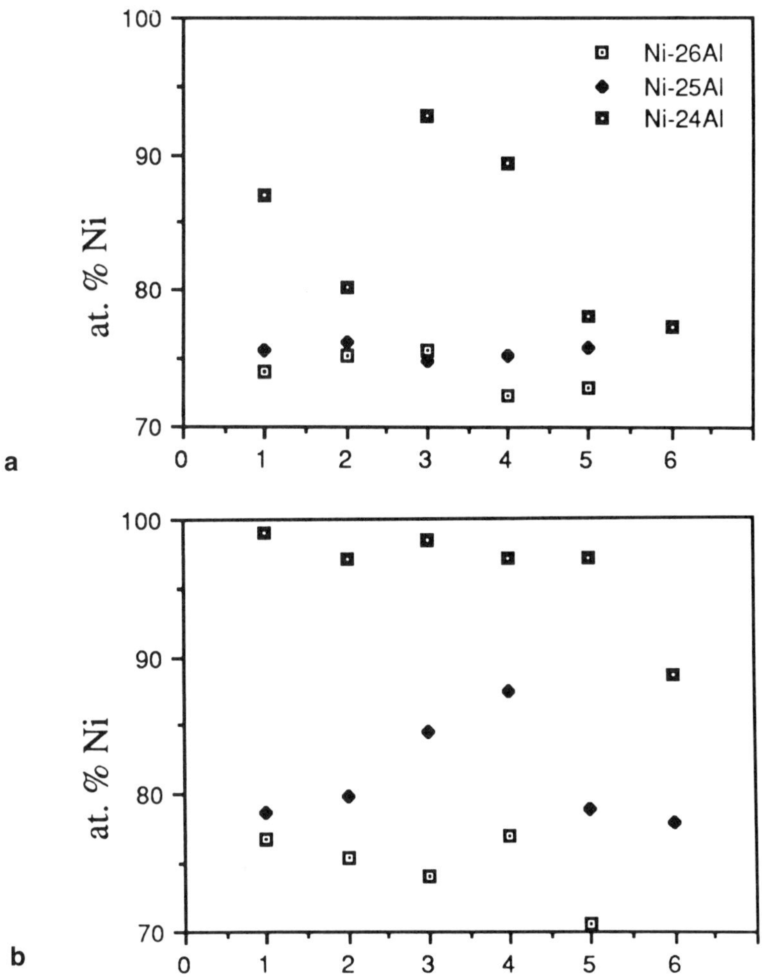

Fig. 6 Graphs of at.% Ni, as determined by EDS on a STEM (1 nm probe) at several different grain boundaries for (a) undoped and (b) B-doped Ni_3Al alloys. Source: Ref 25

thickness of the adjacent matrix was used for absorption correction) and, hence, a relatively higher yield of x-rays from the heavier element occurred, although the total number of x-rays from the grain-boundary regions was

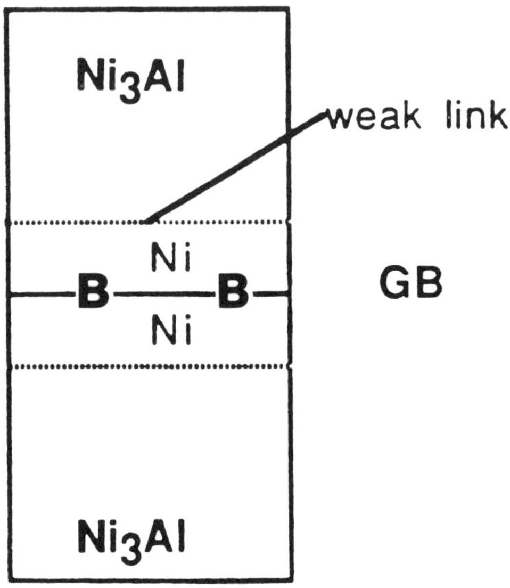

Fig. 7 A grain boundary in B-doped Ni-rich Ni_3Al. Source: Ref 20

similar to that from the adjacent matrices. Second, the Cliff-Lorimer factor, which is used for quantification, may be different for an ordered and a disordered material even if the composition remains the same.

High-Resolution TEM. Mackenzie and Sass (Ref 28) provided the direct evidence of disordering in the vicinity of a grain boundary, using high-resolution TEM (HRTEM) on a sample of boron-doped, nickel-rich Ni_3Al, Ni-24%Al-0.14%B. In a grain that was oriented such that the beam normal was [110] and the grain boundary of interest was tilted to be close to vertical, and in which only one grain was strongly diffracting, they showed that the ordered grain interior exhibited {100} and {110} lattice fringes, whereas in the vicinity of the grain boundary only {111} fringes were observed (Fig. 8). Simulated images showed that these observations were consistent with a disordered grain-boundary region being disordered and an ordered grain interior. No measurements of chemistry were made on the sample. It is worth noting that the disordered region was 1.5 to 2.5 nm wide, had a fairly abrupt interface (<0.4 nm) with the ordered matrix, and was observed on only one side of the grain boundary (although the other side of the boundary was not examined in detail). Subsequently, Kung and Sass (Ref 29, 30) showed, again using lattice imaging, that a disordered second phase was present on all the high-angle

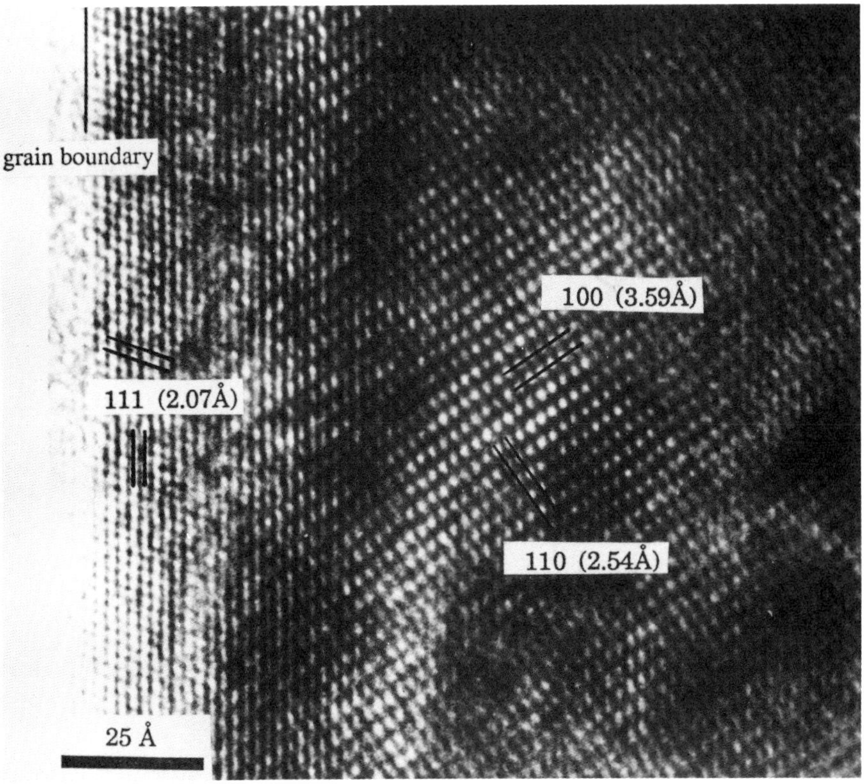

Fig. 8 HRTEM image showing the change of lattice fringe contrast in the vicinity of a near-vertical high-angle grain boundary in B-doped Ni-rich Ni_3Al. Source: Ref 28

boundaries examined in nickel-rich boron-doped Ni_3Al and, using EDS on a HB501, that these disordered regions were nickel enriched. Also, they found no evidence of disordering around low-angle boundaries.

The disordered region observed by Mackenzie and Sass (Ref 28) and Kung and Sass (Ref 29, 30) is clearly the same as the second phase observed by Baker and coworkers (Ref 23-27). One question that remains is why a smaller thickness (1.5 to 2.5 nm) of the second phase was observed, in comparison to the observations of Baker and coworkers (up to 20 nm). The two groups used material of the same Ni:Al ratio, 76:24, and similar heat treatments, annealing at 1000 °C in argon followed by a furnace cool (although obviously different furnaces cool at different rates). Two explanations can be proffered. First, the

Fig. 9 HRTEM image of a Σ5 grain boundary in Ni-23.2%Al-1%Ta-0.03%B. The superlattice crossfringes in the left-hand grain and the single set of superlattice fringes in the right-hand grain extend very close (≤0.5 nm) to the grain boundary. Courtesy of M.J. Mills, S.H. Good, and J.R. Whetstone

material examined by Baker and coworkers contained 0.35% B, whereas that examined by Mackenzie and Sass had only 0.14% B; thus the extent of the disordered region could depend on the boron concentration. This seems very likely since the disordering is due to excess nickel and some of the excess nickel at the grain boundary may arise from nickel and boron cosegregation to the grain boundary (see the section "Field Ion Microscopy" below). Second, other interstitials may also affect the nickel segregation and hence the width of the disorder region. While it is not known whether carbon and oxygen affect nickel enrichment at the grain boundaries (they may decrease or increase it), it is known that the levels of other interstitials in the work of Baker *et al.* (Ref 27) were of a similar magnitude to the amount of boron.

By contrast, Mills (Ref 31) could find no evidence of disorder in the vicinity of grain boundaries in directionally solidified boron-doped nickel-rich Ni_3Al. However, it has been pointed out (Ref 32) that all the grain boundaries were special boundaries. Mills (Ref 33) also performed a careful set of observations on nickel-rich Ni_3Al bicrystals containing 23.2%Al, 1.0%Ta, and 0.3%B that were annealed for 36 h at 1000 °C and then slowly cooled over 24 h. The bicrystals consisted of grains that contained an approximately common ⟨100⟩ that lay in the Σ5 (210) 36.9° boundary plane. For example, Fig. 9 shows a sample oriented so that neither grain was perfectly aligned along ⟨001⟩, but was tilted to minimize the Moiré fringe region at the boundary. Thus, the boundary is virtually "end on" in this image. In the left-hand crystal, superlattice crossfringes are present, while the right-hand grain exhibits one set of superlattice fringes. The superlattice fringes in both grains extend very close to the boundary plane ($\leq$ 0.5 nm) with no detectable change in contrast. These observations indicate that if disorder was occurring at the boundary, it was on a scale less than 0.5 nm in width. These observations, together with the results of Kung and Sass (Ref 29, 30), clearly show that low-angle and special boundaries do not contain any disordered phase, although the presence of disorder in the grain-boundary plane cannot be ruled out.

It is worth noting that while HRTEM can provide the highest resolution for determining the presence of a boundary phase (~0.5 nm), only a limited number of boundaries have been examined because of the exacting geometric requirements on the sample. Also, careful image simulation, as performed by Mackenzie and Sass (Ref 28) and Mills (Ref 31), is required in order to correctly interpret the images. To date, no HRTEM studies have been reported on boron-free or non-nickel-rich Ni_3Al.

Field Ion Microscopy. Sieloff, Brenner, and Burke (Ref 34) were the first to observe nickel and boron cosegregation to the grain boundaries of boron-doped nickel-rich Ni_3Al, Ni-23.5%Al-0.5%Hf-0.2%B. They performed atom probe field ion microscopy (APFIM) on as-cast material that was subjected to various thermomechanical treatments and given a final recrystallization anneal of either 20 min at 800 °C or 6 min at 900 °C (the cooling rate was not stated), after which the material had a grain size of about 5 μm. They showed that boron had segregated to the grain boundaries (or the near-boundary region) and that the aluminum and boron concentrations were inversely related to each other near the boron-enriched grain boundary. The nickel and boron-rich near-grain-boundary region showed nickel levels as high as 81.4 at.% and boron levels up to 1.7 at.% (Fig. 10). These values should be considered lower limits since the enriched zone may have been less than the projected aperture diameter (probe hole) of 3 nm. Although no substantial changes in composition were observed when the aperture was reduced to 1 nm diam, an error in the projected aperture area, somewhat analogous to the error in selected-area electron diffraction in TEM, indicates that the chemical information obtained from the aperture does not wholly coincide with the area imaged within the aperture. Some field ion microscope (FIM) images showed the

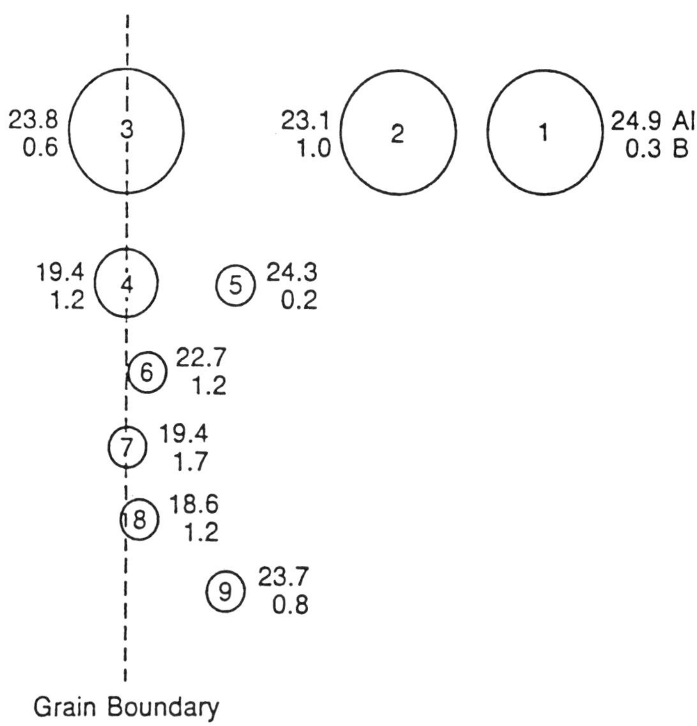

Fig. 10 APFIM results showing inverse correlation between Al and B and showing that the Al concentration decreases and the B concentration increases at the boundary in Ni-23.5%Al-0.5%Hf-0.2%B. Circles indicate relative size of areas of analysis; the smallest diameter was 1 nm. The positions of the circles are offset vertically for clarity. Source: Ref 34

boron that had segregated to the grain boundary as bright points, as, for example, in Fig. 11. In this case the distribution of boron along the boundary is clearly inhomogeneous.

In subsequent APFIM work (Ref 35, 36) it was concluded that nickel and boron cosegregation occurred on all high-angle grain boundaries in boron-doped nickel-rich Ni_3Al and that the segregation was found to occur up to 3 nm from some boundaries. In an analysis in which a grain boundary in boron-doped nickel-rich Ni_3Al touched the edges of {112} superlattice planes on both sides of the boundary, it was concluded (Ref 36) that the maximum width of a disordered layer was 1 to 3 nm, with the lower limit being more

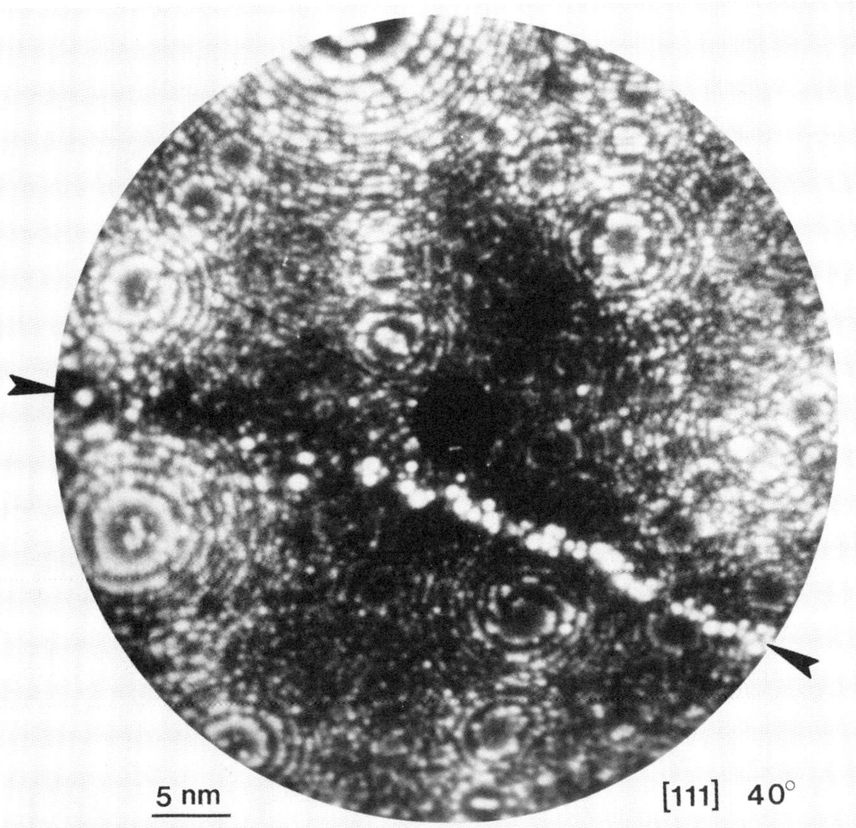

Fig. 11 FIM image showing brightly (boron) decorated grain boundary in Ni-24%Al-0.24%B. Courtesy of S.S. Brenner and H. Ming-Jian

likely. However, given the experimenters' observations of aluminum concentrations below 24 at.% up to 3 nm from the grain boundary and their observation that the single-phase field ends at approximately 24 at.% aluminum, one must draw the conclusion that a nickel-rich second phase containing boron is present up to 3 nm from the grain boundary. No interface was imaged between the phase and the nearby matrix. However, Brenner and Ming-Jian (Ref 36) showed that an interface, even if present, would not necessarily be observable since no interface was observed even at γ/γ' interfaces in a Ni-15.5%Al alloy in which the fcc matrix had a composition of 10 to 12 at.% Al and the $L1_2$ precipitates had a composition of about 24 at.% Al.

One further point is worth noting. In the published compositional profiles across grain boundaries in boron-doped nickel-rich Ni_3Al, the nickel enrichment was asymmetrical (Ref 34, 35); that is, little nickel enrichment was observed on one side of the boundary, but a stronger nickel enrichment was observed on the other side of the boundary and up to a greater distance. This observation is consistent with the results of Baker *et al.* (Ref 25-27) in which the nickel-rich phase was observed on one side of the boundary, and it suggests the possibility that the light nickel enrichment on the opposite side of the boundary is due to "error" in the information from the probe hole.

Boron-doped slightly aluminum-rich (Ni-25.2%Al-1%B) Ni_3Al (Ref 35) was also examined and shown to have grain boundaries enriched with boron, but no data on the Ni:Al ratios were given. No FIM work has yet been performed on boron-free nickel-rich Ni_3Al that is in an equilibrium state, but there have been limited observations on stoichiometric Ni_3Al. These observations show that the grain-boundary composition is the same as the matrix, in agreement with the EDS results quoted earlier (Ref 23, 24). This group (Ref 36) has also shown that carbon segregates to grain boundaries in nickel-rich Ni_3Al but that the effect is much weaker than boron segregation and that no nickel cosegregation occurs. Data on other compositions were from rapidly solidified material and thus were nonequilibrium.

The FIM results are in broad agreement with the observations on nickel-rich boron-doped Ni_3Al by Baker and coworkers (Ref 23-27) and Sass and coworkers (Ref 28-30); that is, all high-angle grain boundaries have a disordered nickel-rich phase on one side. The differences in the measured levels of nickel enrichment may be real or may be due to the use of different measuring techniques. The width of the grain phase was similar to those reported by Sass and coworkers (Ref 28-30). The differences in the width of the phase compared to the results of Baker and coworkers again may be due to different heat treatments or to slight differences in composition, for example, in interstitial contents.

CTEM/CBED/EDS. Conventional TEM examination of grain boundaries in Ni_3Al has been performed by a number of workers. The phase ($\lesssim$3 nm wide) observed in boron-doped nickel-rich Ni_3Al using HRTEM and APFIM could not be observed by Horton and Liu (Ref 32) even in similar materials, Ni-24%Al + 0.24 and 0.48%B, using diffraction contrast in a TEM. One possible reason is a difference in heat treatment. The material examined by HRTEM was furnace-cooled after annealing at 1000 °C, while the material examined using conventional TEM had a faster cooling rate after a similar anneal. (The FIM samples were given heat treatments at lower temperatures, but the cooling rates were not stated.)

EDS performed on an FEG Philips 400T (Ref 32) could not find evidence of nickel enrichment at the grain boundaries in this material. However, as noted earlier, the nickel enrichment observed at the grain boundaries of stoichiometric Ni_3Al containing boron using a VG HB501 could not be found using the FEG Philips 400T. It seems that using these two techniques, of diffraction contrast on a CTEM and EDS on other than a dedicated STEM, are not ideal when looking for thin, constitutionally disordered thin phases at grain boundaries. Strong contrast is observed from a grain-boundary tilted edge on and viewed under strong diffraction conditions (the best conditions for imaging a phase), making the observation of a narrow phase difficult; however, it is clear that no thick phase, for example, 10 to 20 nm, was present on the grain boundaries of the material examined by Horton and Liu (Ref 32).

Strangely, an alloy similar to that noted above, that is, boron-doped (concentration not given) Ni-24%Al that was melt spun but then given an anneal at 1100 °C for 45 min (cooling rate not given) was examined by Zhu, Cheng, and Zou (Ref 37). Using CBED on a TEM with a probe of approximately 6 nm, they found that there was a (partially) disordered region up to 10 nm from the grain boundary. This result appears to be in agreement with those of Baker and coworkers (Ref 25-27), but some caution should be exercised in that the given anneal may not have completely removed the segregation that occurs upon melt spinning and which is well-documented (Ref 38-40). However, it is the aluminum that is in excess at grain boundaries in rapidly solidified material.

In an interesting experiment, Mackenzie, Vaudin, and Sass (Ref 41), using CTEM, examined the interfaces in two Ni_3Al samples produced by hot pressing thin films of Ni_3Al together. They found that the grain-boundary dislocations in boron-free Ni_3Al were $a\langle 100\rangle$ but that those in boron-doped Ni_3Al were $a/2\langle 110\rangle$. Since these are the shortest translation vectors in $L1_2$ and fcc lattices, respectively, one possible explanation for these observations is that the interface in the latter material was disordered (Ref 41).

Auger Electron Spectroscopy. Liu, White, Koch, and Lee (Ref 42), using Auger electron spectroscopy (AES), first observed the segregation of boron to grain boundaries in boron-doped nickel-rich Ni_3Al. Since then, there have been several AES studies (Ref 8, 25, 43-54) of the grain-boundary chemistry of boron-doped Ni_3Al and related boron-doped $L1_2$ compounds. All of these studies have confirmed segregation of boron to grain boundaries in Ni_3Al, even in stoichiometric and aluminum-rich material (for example, see Fig. 12). Ion sputtering of the grain boundaries has indicated that boron resides only in the top few atomic layers on the grain boundary.

AES for grain-boundary analysis has a significant advantage over other techniques in that a relatively large number of boundaries can be easily examined and sample preparation is normally easy. Furthermore, because the escape depth of Auger electrons is only 1 nm, most of the information comes from the top few atom layers. However, while identification of a segregant is straightforward using AES, orientation effects can result in a 20 to 25% variation in measured compositions, making quantification of nonflat samples difficult (Ref 55).

There are several limitations or problems with AES, some of which are general, some of which may apply only to aluminides, and some of which apply largely to Ni_3Al. In general, AES of *in situ* fractured samples examines only the weakest boundaries. The chemistry of these may not be typical of the chemistry of all boundaries, but it represents either an upper or lower limit (depending on the particular sample) on the concentration of the detrimental or beneficial segregates, respectively. This may be regarded as a problem, but while these boundaries may be atypical, they are also the ones that are most important in fracture.

A particular problem with aluminides is that the fracture surfaces are very reactive and can, in some situations, adsorb oxygen and carbon. This can lead to erroneous quantification of the surface chemistry since the O and C can mask Auger electron peaks, particularly low-energy ones (Ref 56). A further complication in this case is that the adsorbed oxygen can be driven off by increasing the current density in the electron probe, which occurs when the analysis area is reduced for a constant-beam current (changes in the area of analysis occur when looking at surfaces at a fixed magnification at different tilt angles). The result is that the chemical analysis results can depend on the size of the analyzed area even in a homogeneous material (Ref 56).

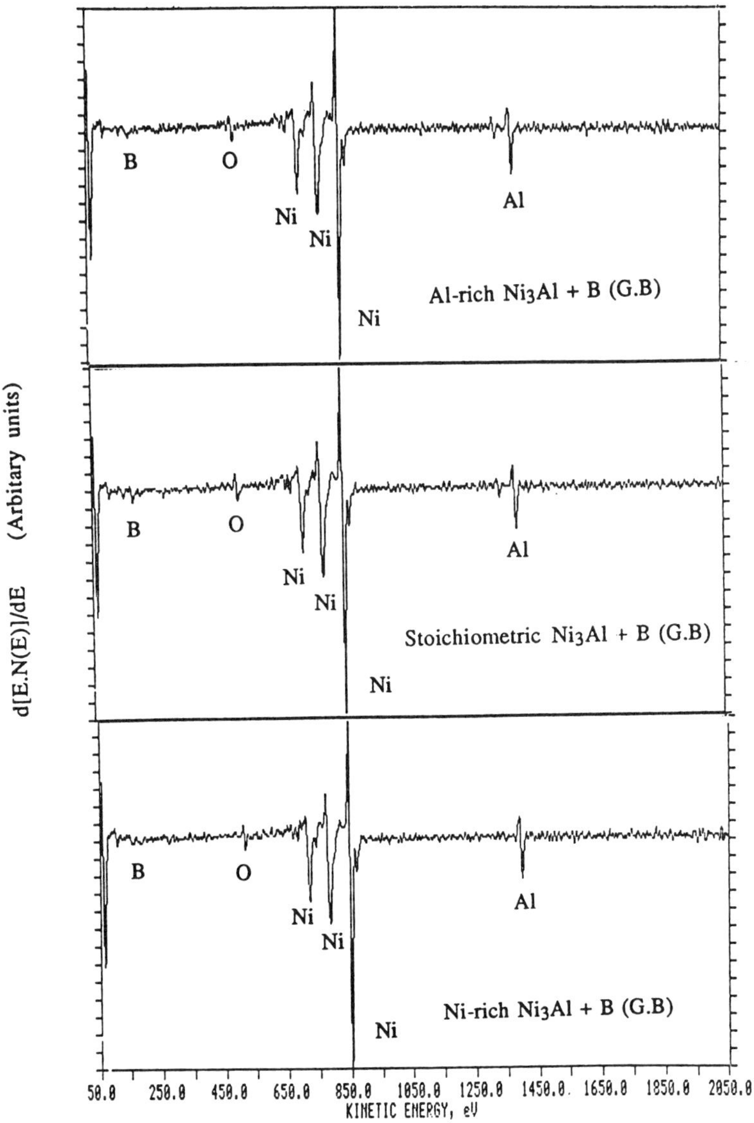

Fig. 12 Auger electron spectra from intergranular facets of B-doped Ni-26%Al, Ni-25%Al, and Ni-24%Al. Source: Ref 25

A technique often used to quantify AES data from grain boundaries is to measure the (differentiated) Auger electron peak height ratios of elements off several transgranular fracture areas where the chemistry is well known and then to use these peak height ratios (PHRs) to quantify the boundary chemistry (Ref 9, 52). It is not clear whether this approach is valid if the matrix is ordered and the grain boundary is disordered because the sensitivity factors used for quantification (Ref 57) may depend on whether an alloy is disordered or ordered.

AES analysis of grain boundaries in boron-doped nickel-rich Ni_3Al has some specific problems. This material is ductile and does not easily fracture intergranularly. Thus hydrogen charging has been employed to produce intergranular fracture (see, for example, Ref 43 and 51). One possible criticism of this technique is that hydrogen charging may cause fracture in boundaries that are not the weakest without hydrogen charging, thereby biasing the results in an unknown way. However, Choudhury *et al.* (Ref 54) have, to some extent, validated this technique by showing that the boron levels on intergranular facets of water-quenched Ni-24%Al containing either 0.14 or 0.24 at.% B were the same with or without hydrogen charging.

Another potential problem in boron-doped nickel-rich Ni_3Al arises from the fact that the scanning Auger microprobe only views surfaces. This is a particular problem when a structure such as that shown in Fig. 7 is present. The question is, does the (hydrogen-induced) "grain-boundary" fracture occur actually along the grain boundary, within the disordered phase, or at the interface between the disordered phase and the ordered matrix? The latter is the fracture plane predicted by Chen *et al.* (Ref 20). In principle this could be answered (at least partly) by examining both sides of a fracture, although in practice this is difficult.

With these problems in mind, some attempts to quantify Auger data in ordered alloys will be considered. An important result by Choudhury, White, and Brooks (Ref 51) was that the extent of boron segregation to grain boundaries in nickel-rich Ni_3Al depended on the heat treatment. More rapid cooling after annealing gave significantly lower boron levels at the grain boundary. This result may explain some of the differences observed in the nickel concentrations and width of the disordered regions after different heat treatments by various researchers, as discussed earlier, if boron and nickel cosegregation occurs. Attempts to measure the degree of boron segregation in Ni_3Al as a function of the Ni:Al ratio have produced some disagreement over whether boron segregation to grain boundaries occurs less strongly (Ref 5) or as strongly in aluminum-rich compositions (Ref 25, 44) as in nickel-rich composi-

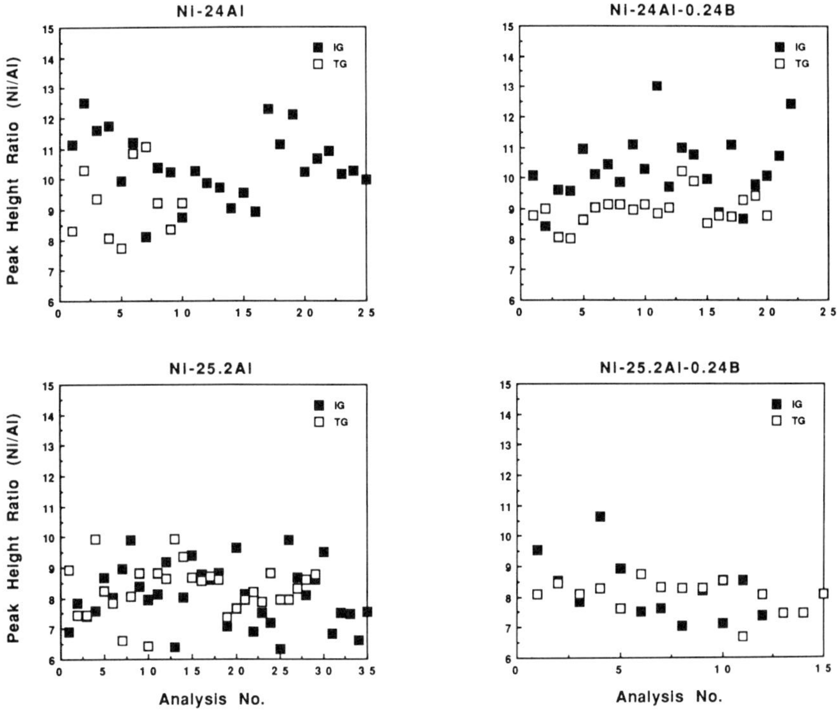

Fig. 13 Ni:Al Auger electron PHRs for various intergranular (IG) and transgranular (TG) areas in *in situ* fractured Ni_3Al alloys. Source: Ref 9

tions. However, it is possible that in the latter study (Ref 44), the material made by rapid solidification was not in equilibrium, even after annealing.

George, Liu, and Padgett (Ref 9) have made the only published attempts to quantify the Ni:Al ratios at grain boundaries as a function of alloy stoichiometry and boron content in Ni_3Al (Fig. 13). The samples, from homogenized ingots that were repeatedly cold forged and annealed, were recrystallized by annealing for 2 h at 1050 °C, after which they were slow cooled (2.4 h) to 500 °C, held at 500 °C for 2 h, and slow cooled (2.1 h) to 300 °C. The experimenters examined from 15 to 35 boundaries, depending on the alloy, and their results, given the limitations in the AES technique outlined above, are broadly in agreement with those of Baker and coworkers (Ref 23-27). The grain boundaries in slightly aluminum-rich (25.2 at.% Al) Ni_3Al with (0.24 at.%) boron and without boron were similar in composition to the matrix. The grain boundaries in boron-free nickel-rich Ni_3Al were nickel enriched (22.4 at.% Al), while those in boron-doped nickel-rich Ni_3Al were on average only slightly more nickel enriched (21.8 at.% Al), with an increase that was much less than the variation in the data. The latter result, they concluded, indicates that boron and nickel cosegregation do not occur. This latter alloy is of the same composition and had a similar heat treatment to one examined by Horton and Liu (Ref 32), using EDS on an FEG Philips EM400T, in which no grain-boundary nickel enrichment was reported.

Averages alone do not convey the whole picture here. For example, the Ni:Al PHRs from grain boundaries ranged from 8 to 12.5 in boron-free nickel-rich Ni_3Al, suggesting initially a large variability in grain-boundary chemistry. However, the peak height ratios from transgranular facets in the same material ranged from 7.7 to 11. It is not clear why the compositions of transgranular areas should be different, but the wide range of PHRs illustrates the difficulty of quantifying AES data.

Similar nickel enrichment (to ~84 at.% Ni) of grain boundaries in nickel-rich (77 at.% Ni) Ni_3Si was reported, with a slight increase in nickel enrichment (to ~87 at.% Ni) of grain boundaries when boron (0.19 at.%) was added (boron also segregated to the grain boundaries) (Ref 52). However, in this case, the data came from far fewer grain boundaries, and there is no other data on the structure of the grain boundaries in this material. The depletion of Si at the grain boundaries in boron-doped $Ni_3(Si, Ti)$ is also evident in the AES data published by Oliver and White (Ref 6).

B2 Compounds

Studies of grain-boundary chemistry/structure in B2 compounds are few. George and Liu (Ref 58) examined 22 grain boundaries in stoichiometric NiAl and found that their average composition was the same as in the transgranularly fractured areas (Fig. 14). Similar AES results were obtained by Baker and Padgett (Ref 59) on stoichiometric NiAl and Ni-49Al (Table 1). In contrast to the AES results on Ni-49Al, Camus *et al.* (Ref 60), using the same material and APFIM, found some grain-boundary regions that were depleted in aluminum (Fig. 15). The width of the depleted regions was less than 0.5 nm. Interestingly, boron added to stoichiometric NiAl segregates to the grain boundaries, as in Ni_3Al, but produces no change in the nickel concentration at the grain boundaries (Ref 58).

Grain-boundary chemistry has been examined in FeAl. Using AES, Liu and George found that boron segregated to the grain boundaries in boron-doped

Table 1 Individual and average Ni:Al and Fe:Al ratios measured using AES on intergranular (IG) and transgranular (TG) fracture surfaces on Ni-49%Al and Fe-45%Al

Fe-45Al				Ni-49Al			
TG		IG		TG		IG	
Fe	Al	Fe	Al	Ni	Al	Ni	Al
55.3	44.7	54.9	45.1			51.0	49.0
56.4	43.6	53.1	46.9	50.8	49.2	50.1	49.9
56.2	43.8	53.9	46.1	50.9	49.1	50.5	49.5
54.4	45.6	54.5	45.5	51.1	48.9	52.2	47.8
52.6	47.4	52.5	47.5	51.2	48.8	52.2	47.5
						50.6	49.4
55.0	45.0	53.8	46.2	51.0	49.0	51.5	48.5
						51.2	48.8

Source: Ref 59

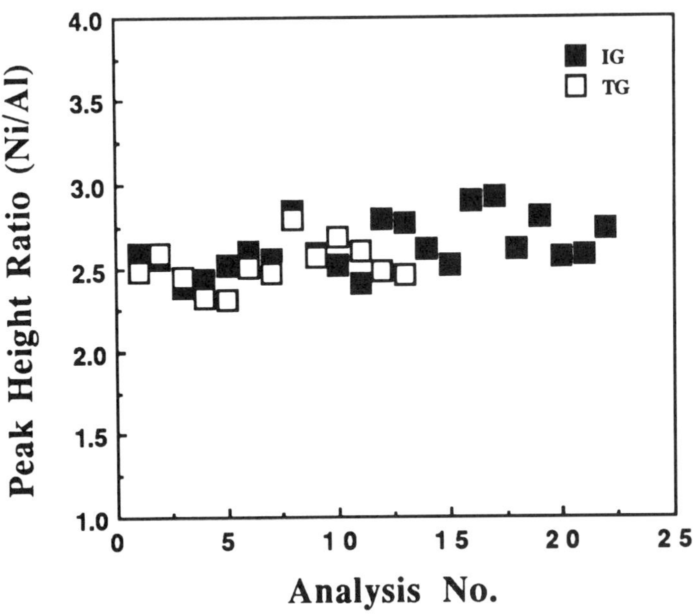

Fig. 14 Ni:Al Auger electron PHRs for various intergranular (IG) and transgranular (TG) areas in stoichiometric NiAl. Source: Ref 58

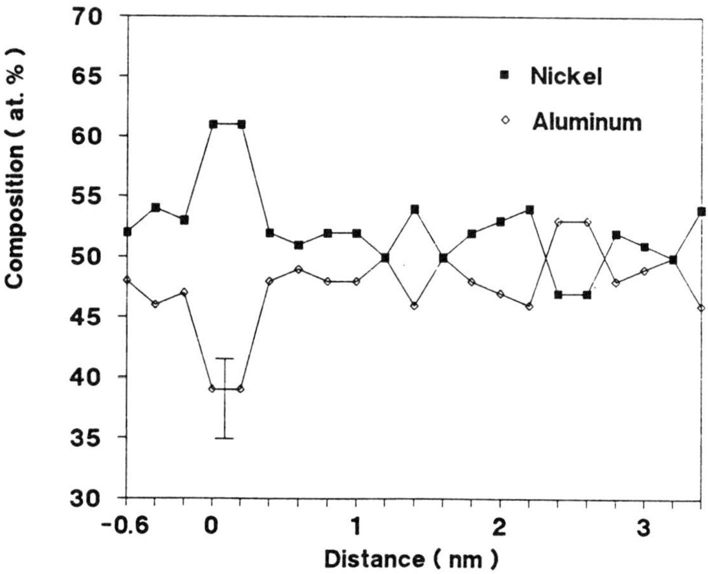

Fig. 15 Atom probe analysis of composition versus distance through a grain boundary in Ni-49%Al. Note the narrow aluminum-depleted region in the boundary. Source: Ref 60

Fe-40%Al (Ref 61) and that the average Fe:Al PHRs were the same for the grain boundaries and matrix in both boron-doped and boron-free material (Ref 62). Similarly, Baker and Padgett (Ref 59), also using AES, found little difference in the Fe:Al PHRs for grain boundaries and transgranular facets in Fe-45%Al (Ref 59) (Fig. 15). Finally, examination of grain boundaries in Fe-40%Al using a VG HB501 found no difference in chemistry from that of the matrix (Ref 63).

Summary

This chapter has reviewed experimental observations of the structure and chemistry of grain boundaries in ordered alloys. Most research has been on Ni_3Al. Although several researchers have noted that the grain boundaries in boron-doped nickel-rich Ni_3Al are enriched in nickel and have found evidence of an fcc phase at the grain boundary, other workers have found no evidence of such a phase. Nickel enrichment and the presence of a phase have been reported at the grain boundaries in boron-free nickel-rich Ni_3Al, and there has been one report of grain-boundary nickel enrichment in boron-doped stoichiometric Ni_3Al. These different results may arise from differences in the Ni:Al ratio, in the levels of boron and other interstitials, and in heat treatment. Because some of the different observations may be due to the examination technique used, the limitations of the various experimental techniques have been discussed.

By comparison, there is evidence that grain boundaries in the B2 compounds NiAl and FeAl have the same composition as the material as a whole.

Acknowledgment

The author would like to thank Profs. S.S. Brenner and S.L. Sass, Drs. A. Choudhury, E.P. George, and M.J. Mills, and Ms. H. Kung for sharing their as yet unpublished results and for providing figures. The author is grateful to Dr. E.P. George and Professor H.J. Frost for commenting on the manuscript. This work was supported by the Division of Materials Sciences, Office of Basic Energy Sciences, U.S. Department of Energy, grant DE-FG02-87ER45311.

References

1. T. Takasugi, E.P. George, D.P. Pope, and O. Izumi, *Scr. Metall.*, Vol 19, 1985, p 155
2. T. Ogura, S. Hanada, T. Masumoto, and O. Izumi, *Metall. Trans. A*, Vol 16A, 1985, p 441
3. K. Aoki and O. Izumi, *Trans. Jpn. Inst. Met.*, Vol 19, 1978, p 203
4. C.T. Liu and C.C. Koch, NBSIR 83-2679-9, National Bureau of Standards, 1983
5. C.T. Liu, C.L. White, and J.A. Horton, *Acta Metall.*, Vol 33, 1985, p 213
6. W.C. Oliver and C.L. White, *Proc. MRS*, Vol 81, 1987, p 241
7. E.M. Schulson, L.J. Briggs, and I. Baker, *Acta Metall.*, Vol 38, 1990, p 207
8. A.I. Taub and C.L. Briant, *Acta Metall.*, Vol 35, 1987, p 1597
9. E.P. George, C.T. Liu, and R.A. Padgett, *Scr. Metall.*, Vol 23, 1989, p 979
10. N. Masahashi, T. Tagasugi, and O. Izumi, *Acta Metall.*, Vol 36, 1988, p 1823
11. F.-S. Chen and A.H. King, *Scr. Metall.*, Vol 20, 1986, p 1401
12. S.P. Chen, A.F. Voter, and D.J. Srolovitz, *Proc. MRS*, Vol 81, 1987, p 45
13. S.P. Chen, D.J. Srolovitz, and A.F. Voter, *J. Mater. Res.*, Vol 4, 1989, p 62
14. S.M. Foiles, *Proc. MRS*, Vol 81, 1987, p 51
15. J.J. Kruisman, V. Vitek, and J.T.M. DeHosson, *Acta Metall.*, Vol 36, 1989, p 2729
16. H.J. Frost, *Acta Metall.*, Vol 36, 1988, p 2199
17. A. King and M.H. Yoo, *Proc. MRS*, Vol 83, 1987, p 99
18. M.H. Yoo and A. King, *J. Mater. Res.*, Vol 3, 1988, p 848
19. H.J. Frost, *Acta Metall.*, Vol 35, 1987, p 519

20. S.P. Chen, A.F. Voter, R.C. Albers, A.M. Boring, and P.J. Hay, *J. Mater. Res.*, Vol 5, 1990, p 955
21. G.J. Ackland and V. Vitek, *Proc. MRS*, Vol 133, 1989, p 105
22. D. Farkas, *Proc. MRS*, Vol 133, 1989, p 137
23. I. Baker, E.M. Schulson, and J.R. Michael, Analytical Electron Microscopy, D.C. Joy, Ed., San Francisco Press, 1987, p 218
24. I. Baker, E.M. Schulson, and J.R. Michael, *Philos. Mag. B*, Vol 57, 1988, p 379
25. I. Baker, E.M. Schulson, J.R. Michael, and R.A. Padgett, *J. Phys.*, Vol 51, 1990, p C1-77
26. I. Baker and E.M. Schulson, *Scr. Metall.*, Vol 23, 1989, p 345
27. I. Baker, E.M. Schulson, J.R. Michael, and S.J. Pennycook, *Philos. Mag. B*, Vol 62, p 659
28. R.A.D. Mackenzie and S.L. Sass, *Scr. Metall.*, Vol 22, 1988, p 1807
29. H. Kung and S.L. Sass, Paper presented at the TMS Fall Meeting, Indianapolis, IN, The Minerals, Metals & Materials Society, Oct 1989
30. H. Kung and S.L. Sass, unpublished research
31. M.J. Mills, *Scr. Metall.*, Vol 23, 1989, p 2061
32. J.A. Horton and C.T. Liu, *Scr. Metall.*, Vol 24, 1990, p 1251
33. M.J. Mills, S.H. Good, and J.R. Whetstone, unpublished research
34. D.D. Sieloff, S.S. Brenner, and M.G. Burke, *Proc. MRS*, Vol 81, 1987, p 87
35. D.D. Sieloff, S.S. Brenner, and H. Ming-Jian, *Proc. MRS*, Vol 133, 1989, p 155
36. S.S. Brenner and Hua Ming-Jian, *Scr. Metall.*, Vol 24, 1990, p 667, 671
37. J. Zhu, Z.Y. Cheng, and D.X. Zou, *Scr. Metall.*, Vol 24, 1990, p 439
38. J.A. Horton, C.C. Koch, and C.T. Liu, in *Proceedings of the 41st Electron Microscope Society of America*, 1983, p 248
39. M.K. Miller and J.A. Horton, *Scr. Metall.*, Vol 20, 1986, p 789
40. J.A. Horton and M.K. Miller, *Acta Metall.*, Vol 35, 1987, p 133
41. R.A.D. Mackenzie, M.D. Vaudin, and S.L. Sass, *Proc. MRS*, Vol 122, 1988, p 461
42. C.T. Liu, C.L. White, C.C. Koch, and J. Lee, *Proc. Electrochem. Soc.*, Vol 83 (No. 7), 1983, p 32
43. E.P. George, C.T. Liu, and R.A. Padgett, *Scr. Metall.*, Vol 23, 1989, p 979
44. C.L. Briant and A.I. Taub, *Acta Metall.*, Vol 36, 1988, p 2761
45. S.C. Huang, A.I. Taub, and K.-M. Chang, *Acta Metall.*, Vol 32, 1984, p 1703
46. C.L. White, R.A. Padgett, C.T. Liu, and S.M. Yalisgrove, *Scr. Metall.*, Vol 18, 1984, p 1417
47. C.L. White and A. Choudhury, *Proc. MRS*, Vol 81, 1987, p 427
48. A.I. Taub, S.C. Huang, and K.-M. Chang, *Metall. Trans. A*, Vol 15A, 1984, p 399
49. T. Tagasugi, O. Izumi, and N. Masahashi, *Acta Metall.*, Vol 33, 1985, p 1259
50. A.I. Taub, C.L. Briant, S.C. Huang, K.-M. Chang, and M.R. Jackson, *Scr. Metall.*, Vol 20, 1986, p 129
51. A. Choudhury, C.L. White, and C.R. Brooks, *Scr. Metall.*, Vol 20, 1986, p 1061
52. I. Baker, E.M. Schulson, and R.A. Padgett, *Scr. Metall.*, Vol 23, 1989, p 1969
53. C.L. White, C.T. Liu, and R.A. Padgett, *Acta Metall.*, Vol 36, 1988, p 2229
54. A. Choudhury, C.L. White, and C.R. Brooks, *J. Vac. Sci. Technol.*, in press
55. B. Rothman and M. Menyhard, *Scr. Metall.*, Vol 23, 1989, p 1169

56. I. Baker, E.P. George, and R.A. Padgett, *Scr. Metall.*, in press
57. L.E. Davis, N.C. McDonald, P.W. Palmberg, G.E. Riachand, and R.E. Weber, *Handbook of Auger Electron Spectroscopy*, Physical Electronic Industries, 1976
58. E.P. George and C.T. Liu, *J. Mater. Res.*, Vol 5, 1990, p 754
59. I. Baker and R.A. Padgett, unpublished research
60. P.P. Camus, I. Baker, J.A. Horton, and M.K. Miller, *J. Phys.*, Vol 49, 1988, p C6-329
61. C.T. Liu and E.P. George, *Scr. Metall.*, Vol 24, 1990, p 1285
62. E.P. George, private communication, Oak Ridge National Laboratory, Aug 1990
63. I. Baker and J.R. Michael, unpublished research

5

TEM Observations of Tilt Boundaries in B_2-Ordered Fe_3Al

A. Buis, F.D. Tichelaar, and F.W. Schapink
Laboratory of Metallurgy, Delft University of Technology
Rotterdamseweg 137, 2628 AL Delft, The Netherlands

The structure of two low-angle tilt boundaries and a near $\Sigma = 3$ tilt boundary in B_2-ordered Fe_3Al have been investigated using transmission electron microscopy (TEM).

A symmetrical low-angle tilt boundary was found, consisting of two dislocation configurations: a regular array of $a_0\langle 111\rangle$-type superdislocations located in the boundary plane and an occasional set of irregular ordinary dislocations associated with an antiphase boundary (APB) from the bulk. The separation of the ordinary dislocations comprising each superdislocation in the boundary was found to be larger than the calculated value of the equilibrium separation using anisotropic elasticity theory. Possible reasons for this discrepancy are indicated.

Part of the second low-angle boundary could be described as an asymmetrical tilt boundary, which consisted of two types of $a_0\langle 110\rangle$ dislocations. These dislocations were dissociated into ordinary dislocations, forming a mixture of ordinary dislocations bounding APBs from the bulk and $a_0\langle 110\rangle$-type superdislocations with the connecting APB in the $\{112\}$ boundary plane. It is the authors' belief that $a_0\langle 110\rangle$-type superdislocations have not been observed before in the B_2 structure. The origins of the dislocation configurations found in both low-angle boundaries are discussed.

The structure of the asymmetrical $\{111\}_1\{115\}_2$ $\Sigma = 3$ tilt boundary was analyzed geometrically. Nearest-neighbor distances at the boundary are for most of the atoms comparable to the nearest-neighbor distances in the bulk if the continuity of the atomic density at the boundary is taken into account. The $\{111\}_1\{115\}_2$ near $\Sigma = 3$ bicrystal found in one of the specimens showed a 4° deviation from the coincident orientation. The hexagonal contrast in this boundary was interpreted as a dislocation network accommodating the extra misorientation. However, Burgers vectors consistent with the misorientation were found to be lattice vectors instead of the expected smaller displacement shift complete (dsc) lattice vectors belonging to the $\Sigma = 3$ boundary.

Alloys based on the DO_3-ordered Fe_3Al and B_2-ordered FeAl intermetallics show a number of properties that make them suitable for aerospace materials, especially at elevated temperatures (Mendiratta, Ehlers, Dimiduk, Kerr, Mazdiyasni, and Lipsitt, Ref 1; Vedula and Stephens, Ref 2). These properties are a high oxidation and sulfidation resistance due to the formation of a protective and adherent Al_2O_3 scale; a low density compared to Ni-based superalloys; low material cost; and a sufficient yield strength, with a value between 230 and 700 MPa in the temperature range from 0 to 550 °C. The yield strength of powder-processed Fe_3Al drops from 700 MPa at 0 to 200 °C to approximately 230 MPa at 500 °C (Mendiratta *et al.*, Ref 3), which is comparable to the values of DO_3-ordered Fe_3Al made from a melt (Morgand, Mouturat, and Sainfort, Ref 4). The yield strength of iron-rich B_2-ordered Fe-Al (~40%) was found to be constant up to approximately 550 °C. Morgand, Mouturat, and Sainfort (Ref 4) found a value of approximately 250 MPa, and Vedula and Stephens (Ref 2), approximately 500 MPa. Stoichiometric FeAl has a higher yield strength at room temperature (Crimp, Vedula, and Gaydosh, Ref 5; Baker and Gaydosh, Ref 6).

The major problem encountered in the development and application of polycrystalline aluminides is low room-temperature ductility, caused by grain-boundary embrittlement, also known in $L1_2$-ordered alloys. Stoichiometric B_2-ordered FeAl shows a negligible room-temperature elongation (Baker and Gaydosh, Ref 6). Fe_3Al (DO_3 structure) and iron-rich Fe-Al (B_2 structure) show a small elongation of 3 to 7% (Mendiratta *et al.*, Ref 3; Baker and Gaydosh, Ref 6; Morgand *et al.*, Ref 4; Sainfort, Mouturat, Pepin, Petit, Cabane, and Salesse, Ref 7), while Marcinkowski, Taylor, and Kayser (Ref 8) found no elongation at all in carbon-rich (100 ppm) Fe_3Al. The amount of elongation before fracture decreases with a higher carbon content. Carbon can form inclusions of the type Fe_3AlC at grain boundaries (Kerr, Ref 9), which might, for instance, affect the cohesive strength at the boundary.

Grain-boundary failure in ordered alloys is often attributed to a lack of cohesive strength at the boundary. Crimp, Vedula, and Gaydosh (Ref 5) found an increase in ductility and transgranular fracture in B_2-ordered Fe-40% Al after the addition of boron, which presumably improves grain-boundary cohesion by segregation. Marcinkowski, Taylor, and Kayser (Ref 8), Marcinkowski and Larsen (Ref 10), and Marcinkowski (Ref 11) attributed the intrinsic weakness of grain boundaries to the higher degree of disorder at boundaries, which makes the boundary a potential crack propagation path, leading to intergranular fracture. A second reason for grain-boundary embrittlement in ordered alloys might be associated with a limited possibility of strain transfer across grain boundaries, in contrast to the disordered case. When strain transfer across a grain boundary is restricted, stress accumulates at the boundary, which might lead to intergranular fracture. This was examined in detail for $L1_2$ alloys by King and Yoo (Ref 12). The reader is referred also to Schapink and Tichelaar (Ref 13) and Tichelaar and Schapink (Ref 14, 15).

Since both cohesive strength and plastic deformation at grain boundaries are associated with grain-boundary structure, the investigation of the phenomenon of grain-boundary embrittlement in ordered alloys requires a detailed knowledge of the grain-boundary structure. In the past this has led to a number of theoretical investigations of both B_2- and $L1_2$-ordered alloys. In the B_2 structure, Takasugi and Izumi (Ref 16, 17) investigated the structure of high-angle CSL (coincident site lattice) boundaries. The structure of low-angle tilt and twist boundaries was analyzed by Marcinkowski (Ref 11, 18), and Czernichow, Gudas, Marcinkowski, and Tseng (Ref 19). Experimental observations of grain boundaries in the B_2 structure are limited to low-angle boundaries in CuZn (Rinnovatore and Brown, Ref 20), FeCo (Czernichow *et al.*, Ref 19), and Fe-39.9%Al (Morgand, Mouturat, and Sainfort, Ref 4). However, no details about the boundary structure were given in these studies.

This chapter examines in detail a symmetrical and an asymmetrical low-angle tilt boundary, as well as the structure of a near {111} {115} $\Sigma = 3$ tilt boundary. The structure of the symmetrical low-angle boundary has been described in more detail previously (Buis, Tichelaar, and Schapink, Ref 21).

Theoretical Considerations

The phase diagram of the Fe-Al system has been discussed in detail by Kuentzler (Ref 22). Two types of long-range order are found in Fe_3Al, based on the body-centered cubic cell. Below 550 °C the DO_3 structure is stable. This structure consists of four interpenetrating fcc sublattices, one of which is occupied by Al atoms. Both first and second nearest-neighbor atom pairs are ordered (no Al-Al pairs). Above 550 °C the second nearest-neighbor pairs are disordered, and an imperfect B_2 structure is formed, stable up to 776 °C (Swann, Duff, and Fisher, Ref 23). The B_2 structure consists of two interpenetrating simple cubic sublattices, one of which is occupied by Fe atoms, while Fe and Al are distributed at random on the other sublattice, thereby creating an imperfectly ordered B_2 structure.

Antiphase boundaries (APBs) between different domains in the B_2 structure are characterized by a translation of type $p = (1/2)a_0 \langle 111 \rangle$. Hence APBs may be bounded by ordinary dislocations with Burgers vector $(1/2)a_0\langle 111 \rangle$. Perfect dislocations in the B_2 structure can be of the type $a_0\langle 100 \rangle$, $a_0\langle 110 \rangle$, or $a_0\langle 111 \rangle$. The energy of these perfect dislocations (or superdislocations) can be decreased by a dissociation into ordinary dislocations connected by a ribbon of APB:

$$a0[100] \rightarrow \tfrac{1}{2}a_0[111] + \tfrac{1}{2}a_0[1\bar{1}\bar{1}] + \text{APB} \qquad \text{(Eq 1)}$$

$$a_0[110] \rightarrow \tfrac{1}{2}a_0[111] + \tfrac{1}{2}a_0[11\bar{1}] + \text{APB} \qquad \text{(Eq 2)}$$

$$A_0[111] \rightarrow \tfrac{1}{2}a_0[111] + \tfrac{1}{2}a_0[111] + \text{APB} \qquad \text{(Eq 3)}$$

According to Frank's criterion (low Σb^2), dissociation Eq 1 is energetically unfavorable. However, the total energy of the superdislocation as given by Eq 4 should be taken into account, which makes this dissociation possible in some materials as calculated from anisotropic elasticity theory by Potter (Ref 24). The total energy of the superdislocation is given by:

$$E = 2E_s + E_{12} + \gamma \cdot r \qquad \text{(Eq 4)}$$

where E_s is the self energy of an ordinary dislocation, E_{12} is the interaction energy of the ordinary dislocations, γ is the energy of the connecting APB, and r is the equilibrium separation of the superpartials. The value of r is calculated from the balance between the repulsive forces of the dislocations against the tension (γ) of the connecting APB as:

$$r_{eq} = Kb^2/2\pi\gamma \qquad \text{(Eq 5)}$$

where K is the energy factor, depending on the dislocation line direction and the elastic constants, and b is the size of the Burgers vector of the ordinary dislocations.

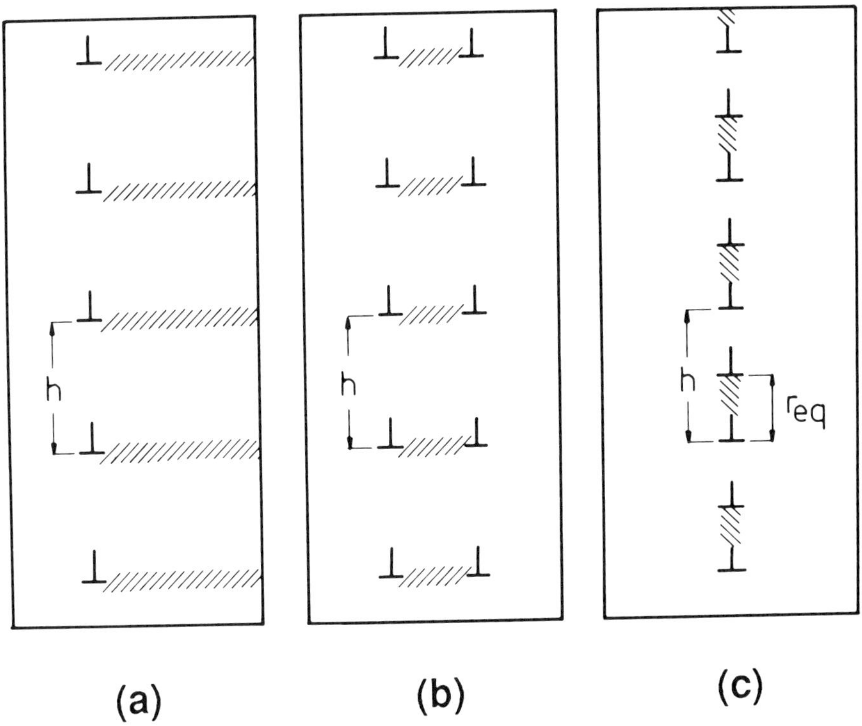

Fig. 1 Possible configurations of a symmetrical tilt boundary in an ordered alloy. Antiphase boundaries (APBs) are indicated by the hatched areas. Source: Ref 11

Marcinkowski and Brown (Ref 25) did not observe superdislocations in a deformed B_2-ordered Fe_3Al film. They observed that deformation occurred by the movement of single ordinary dislocations leaving behind a ribbon of APB, which was attributed to the low APB energy on the {110} glide plane. Superdislocations of type $a_0\langle 111\rangle$ were found by Ray, Crawford, and Cockayne (Ref 26) in bulk deformed B2-ordered Fe-35%Al, which has a larger APB energy than B_2-ordered Fe_3Al.

Marcinkowski (Ref 11) considered three possible structures for a low-angle symmetrical tilt boundary in ordered alloys (Fig. 1). When ordinary edge dislocations are introduced into the ordered material, APBs are produced along the slip planes of the dislocations. A wall of parallel ordinary edge dislocations is formed, each dislocation leaving behind a ribbon of APB in the bulk (Fig. 1a). The structure shown in Fig. 1(b) is obtained when the energy of each APB in Fig. 1(a) is reduced by moving a second wall of ordinary dislocations along the same slip traces. This results in a wall of regularly spaced superdislocations, of the type shown in Eq 3, perpendicular to the boundary plane. A third configuration (Fig. 1c) is obtained when the superdislocations are located in the boundary plane. In this case the extension of each superdislocation is deduced from the anisotropic elasticity equations given by Stroh (Ref 27) and Buis, Tichelaar, and Schapink (Ref 21):

$$r_{eq} = \frac{h}{\pi} + \mathrm{an}^{-1}\left(\frac{Kb^2}{2h\gamma}\right) \quad \text{(Eq 6)}$$

where h is the separation of the superdislocations. A symmetrical tilt boundary can also be comprised of superdislocations of the type shown in Eq 1 and 2 if the superdislocations are present on (100) and (110) planes, respectively. In these cases each superdislocation is considered to be one dislocation with the Burgers vector perpendicular to the boundary plane.

The APB energy γ for the imperfectly B_2-ordered Fe_3Al alloy is expressed in terms of the ordering energies V and W for the first and second nearest-neighbor atoms, respectively, (Buis, Tichelaar, and Schapink, Ref 21; Leamy, Schwellinger, and Warlimont, Ref 28) as:

$$\gamma = \frac{l}{2a_0^2\sqrt{N}}[-2hV + (h+k+l)W],\ h \geq k+l \quad \text{(Eq 7)}$$

$$\gamma = \frac{(h+k+l)}{2a_0^2\sqrt{N}}(-V+W),\ h < k+l \quad \text{(Eq 8)}$$

where $N = h^2 + k^2 + l^2$ and $\{hkl\}$ is part of the unit stereographic triangle: $h \geq k \geq l \geq 0$. It was argued by Buis, Tichelaar, and Schapink (Ref 21) that Eq 7 and 8 hold for any APB plane if the average composition at the APB is stoichiometric. For the case $h + k + l$ = odd, when two successive $\{hkl\}$-APB planes have different compositions, this can be achieved by alternating the APB on the two types of $\{hkl\}$ planes.

The ordering energies are defined as:

$$V = v^1(\mathrm{FeAl}) - \tfrac{1}{2}\{v^1(\mathrm{FeFe}) + v^1(\mathrm{AlAl})\} \quad \text{(Eq 9)}$$

$$W = v^2(\mathrm{FeAl}) - \tfrac{1}{2}\{v^2(\mathrm{FeFe}) + v^2(\mathrm{AlAl})\} \quad \text{(Eq 10)}$$

where $v^1(XY)$ is the bonding energies between the ith nearest-neighbor atoms X and Y. The ordering energy (V) for the imperfect B_2-ordered alloy Fe_3Al

can be roughly estimated from the ordering temperature (T_c) using the Bragg-Williams approximation involving nearest neighbors only (Muto and Takagi, Ref 29):

$$V = \frac{-kT_c}{3} \qquad \text{(Eq 11)}$$

With T_c = 1049 °K (Swann, Duff, and Fisher, Ref 23), we find $V = -4.8 \times 10^{-21}$ J. The fraction W/V has been estimated by comparing results predicted from the Bragg-Williams approximation with experimental data from x-ray analysis. Rudman (Ref 30) and Schapink (Ref 31, 32) found $W/V \approx 0.50$. Krzanowski and Allen (Ref 33) deduced from a comparison of the calculated and experimental Fe-Al phase diagram that $V = -7.3 \times 10^{-21}$ J and $W = -3.8 \times 10^{-21}$ J. Alternatively, values for V and W can be deduced independently from the APB energies as measured by Crawford, Ray, and Cockayne (Ref 34) from fourfold dissociated superdislocations on a {110}-plane in DO_3-ordered Fe-26%Al. This results in $V = -9.2 \times 10^{-21}$ J and $W = -5.1 \times 10^{-21}$ J.

The structure of high-angle boundaries in the B_2-ordered structure was examined by Takasugi and Izumi (Ref 16, 17). $\Sigma = 3$ twin boundaries were also considered by Cahn and Coll (Ref 35) and Marcinkowski (Ref 18). It was found that {112}-type twin boundaries have a low energy because first-neighbor distances are the same at the boundary as in the single crystal, contrary to the case of the {111} twin boundary. Twin boundary planes of type {112} were observed at large annealing twins (~300 μm) in B_2-ordered Fe-23.5%Al (Cahn and Coll, Ref 35). In this discussion, a different type of $\Sigma = 3$ boundary is considered in the B_2 structure, the asymmetrical $(\bar{1}11)_1$ $(\bar{1}\bar{1}5)_2$ tilt boundary (Fig. 2).

First, the density of the underlying bcc lattice sites at the boundary will be considered. The bicrystal can be constructed from the dsc lattice as follows: a divisional plane is constructed through coincident sites on $(\bar{1}11)_1 = (\bar{1}\bar{1}5)_2$ and the lattice sites belonging to lattice 1 above the plane and the lattice sites belonging to lattice 2 on and below this plane are subsequently removed. Coincident lattice sites at this boundary in the $[211]_1 = [552]_2$ direction occur at a period that is a third of the corresponding period of the $(\bar{1}11)$ twin boundary. The lattice site density in both lattices can be written as $(d_1 \cdot p_1 \cdot c)^{-1} = (d_2 \cdot p_2 \cdot c)^{-1}$, where d_1 is the interplanar spacing in lattice 1, p_1 is the period of the lattice sites on the corresponding planes in the $[211]_1$ direction, and c is the period in the $[0\bar{1}1]_1 = [1\bar{1}0]_2$ direction (the bicrystal tilt axis). At the boundary plane the lattice site density will be too high if the plane containing the coincident sites is considered to belong to the lattice with the smallest value of p (p_1), while the interplanar spacing d changes abruptly from d_1 in lattice 1 to d_2 in lattice 2. The lattice site density will be maintained at the grain boundary in this case if the interplanar spacing d_{GB} at the boundary in lattice 2 is equal to:

$$d_{GB} = \frac{(d_1+d_2)}{2}$$

The number of surplus atoms per periodic boundary area $p_2 \cdot c$ is $(d_{GB}-d_2) \cdot p_2 \cdot c/d_2 \cdot p_2 \cdot c$. In the case with $p_1 = a_0/2\sqrt{27}$ and $p_2 = a_0/2\sqrt{3}$, this number is one atom per period p_2. This corresponds to exactly one atom layer in lattice 2, which has been left out in Fig. 2 in order to maintain the bulk lattice density at the boundary. It should be noted that the lattice site density at the boundary will be too small if the plane containing the coincident sites is considered to belong to the lattice with the largest value of p (p_2), while the interplanar spacing d changes abruptly from d_2 in lattice 2 to d_1 in lattice 1.

Figure 2 shows four different structures of the boundary, which arise by a different occupation of the two sublattices on both sides of the boundary plane. Different boundary structures can be separated by terminating APBs in the boundary plane. Two of the structures are off stoichiometric, which, however, does not result in a larger number of wrong first nearest-neighbor bonds, in contrast to the stoichiometric boundary structures. The wrong bonds were drawn under the assumption that each atom would have eight nearest-neighbor bonds after an expected relaxation of some of the atoms occurred. It should be noted that only a small relaxation is necessary to accommodate the misfit at some of the lattice sites. Therefore, a significant rigid-body translation is not expected at this type of boundary. Also, a faceting of the boundary plane on $\{112\}_c$ and $\{111\}_c$ is possible, but the atomic misfits at the boundary are not avoided in this way.

Specimen Preparation and Experiment Details

B_2-ordered Fe_3Al specimens for TEM were prepared as follows: a melt was made from pure Fe (99.96%, MRC Europe) and Al (99.999%, Highways International) by arc melting. The alloy was homogenized at 1050 °C in an Ar-filled quartz ampul for 41 h and cut into 600 m slices. Recrystallization and grain growth following 15% deformation by rolling at room temperature and annealing at 1050 °C for 1.5 h resulted in an average grain size of about 600 μm, with some grains containing subgrain boundaries. Following the annealing treatment, the specimen was lowered from the vertical oven and cooled to room temperature in a stream of Ar. It is suspected that small B_2-type domains

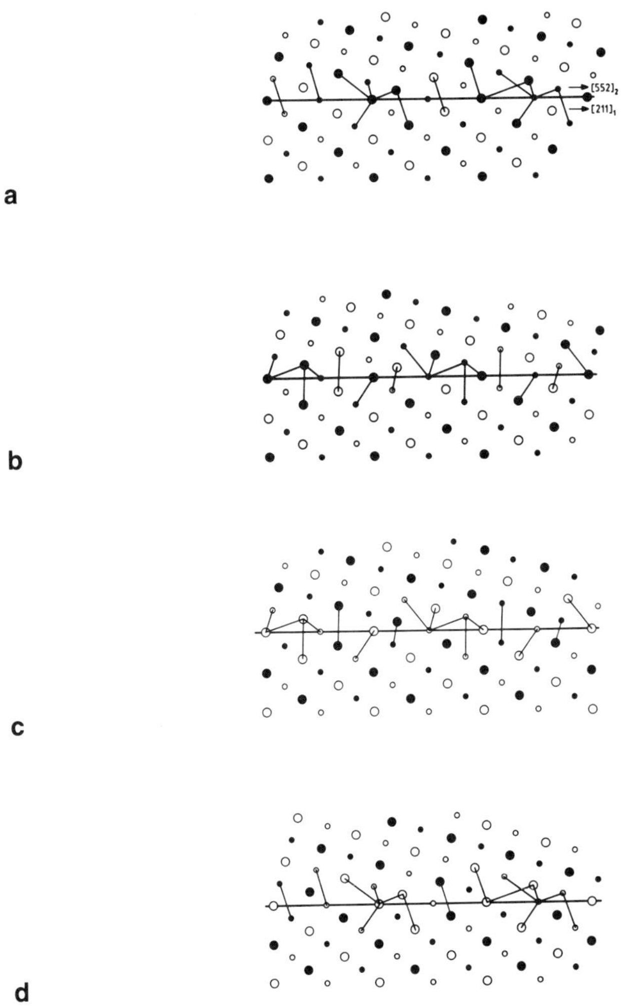

Fig. 2 The $(\bar{1}11)_1(\bar{1}\bar{1}5)_2$ asymmetrical $\Sigma = 3$ tilt boundary in the B_2 structure projected along the $[01\bar{1}]_1 = [\bar{1}10]_2$ tilt axis. The different structures correspond to different occupations of the two sublattices in both crystals. Filled and empty symbols represent different occupations of the sublattices; small and large circles represent lattice sites at different heights. The atomic compositions in the boundary structures (b) and (c) are stoichiometric. In (a) and (d), a surplus of black and white atoms, respectively, exists at the boundaries. The boundary structures are in the unrelaxed state. Wrong nearest-neighbor bonds after a likely relaxation are indicated.

were present in the material at this stage since even a rapid quench (50,000 °C/s) to room temperature was not sufficient to suppress the B_2-type order (Swann, Duff, and Fisher, Ref 23). In order to obtain the ordered metastable B_2 structure at room temperature with an average domain size of about 20 μm, the alloy was further annealed at 600 °C for 1 h and water quenched. All annealing treatments were performed under Ar pressure (10^5 N/m^2).

The B_2-ordered slices were subsequently polished to 60 μm sheet. From this sheet disks were punched and electrochemically thinned for TEM by jet polishing using an electrolyte containing 20% perchloric acid and 80% acetic acid at 11 °C and 20 V. Chemical analysis of the 60 μm sheet showed the Al concentration to be 25.3%, close to the stoichiometric composition.

The lattice parameter of the quenched B_2-ordered alloy was determined by x-ray diffraction to be 2.901 ± 0.001 Å, which is 0.003 Å higher than the value

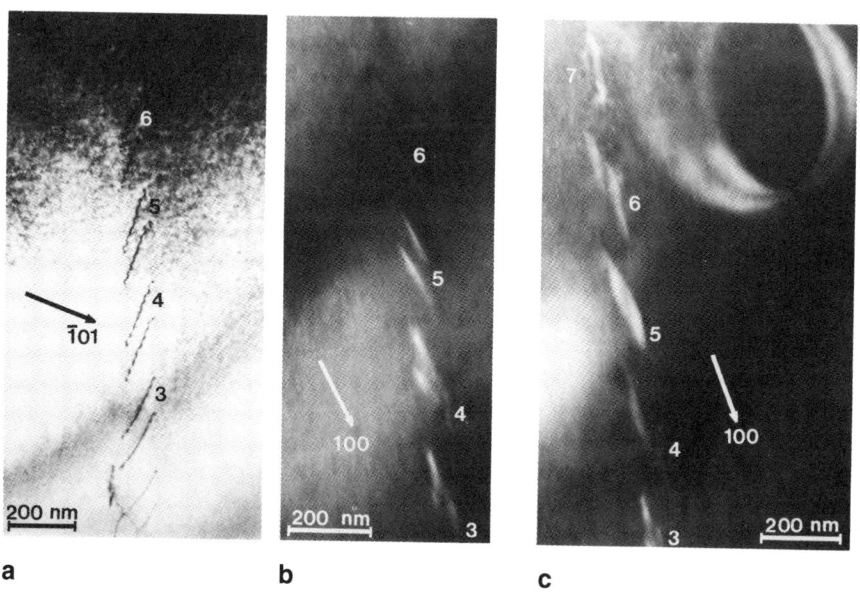

Fig. 3 Electron micrographs of the same area in B_2-ordered Fe_3Al showing a low-angle symmetrical tilt boundary consisting of superdislocations, using (a) a fundamental reflection (bright field, $B \approx [121]$), (b) the weak-beam method (g/3g, centered dark field, $B \approx [025]$), and (c) the superreflection $g = [100]$, $B \approx [025]$ (centered dark field). In (a) and (b) the ordinary dislocations bounding the APBs visible in (c) are imaged. The foil normal is $FN \approx [133]$.

measured by Okamoto and Beck (Ref 36) in quenched powder specimens with the same composition. The origin of this difference is unclear.

The specimens were examined in a Philips 400T electron microscope, operating at 120 kV. In order to study grain-boundary structure, bicrystals with a coincidence site lattice (CSL) or CL (a CSL with a rigid-body translation at the boundary; see Delavignette *et al.*, Ref 37) misorientation are favorable because common reflections can be used. In order to identify CL bicrystals, the orientation relationship between two crystals was determined by solving the rotation matrix R from $B_1^i = R \cdot B_2^i$ (i = 1,2,3,) in which B_1^i and B_2^i are the common electron beam directions expressed in the coordinates of crystals 1 and 2, respectively, for tilt i. The electron beam directions are determined accurately (~0.1°) from experimentally obtained Kikuchi patterns using convergent beam electron diffraction (CBED). A small deviation from a CL misorientation can be found by calculating the deviation matrix A from:

$$A = R \cdot R_c^{-1} \qquad \text{(Eq 12)}$$

in which R and R_c^{-1} are the actual rotation matrix and the CL misorientation matrix, respectively.

Forty high-angle boundaries were investigated in this manner. Three bicrystals with a small deviation from a CL misorientation were found, a near $\Sigma = 3$ boundary which is discussed in this chapter. In order to investigate this boundary and two of the low-angle boundaries, both fundamental reflections ($h + k + l$ = even) and superreflections ($h + k + l$ = odd) were used, for which APBs are either invisible and visible respectively (Marcinkowski, Ref 38). It was not always possible to correct the astigmatism caused by the magnetic Fe_3Al specimen.

Experimental Results

The Symmetrical Low-Angle Tilt Boundary. Part of low-angle boundary 1 is shown in Fig. 3. This part consists of a regular array of dislocation pairs, imaged in Fig. 3(a) using a fundamental reflection and in Fig. 3(b) using the weak-beam method, employing a superlattice reflection. In Fig. 3(c) APB contrast is visible between the dislocations of dislocation pairs 4, 5, and 6 (APB in the bulk). It should be noted that the beam directions in Fig. 3(b) and Fig. 3(c) are the same. Consequently these dislocations must be ordinary dislocations with Burgers vectors of the type $b = (1/2)a_0\langle 111\rangle$, each dislocation pair forming a superdislocation. To determine b, the $g \cdot b = 0$ invisibility criterion was applied to eight fundamental reflections (Buis, Tichelaar, and Schapink,

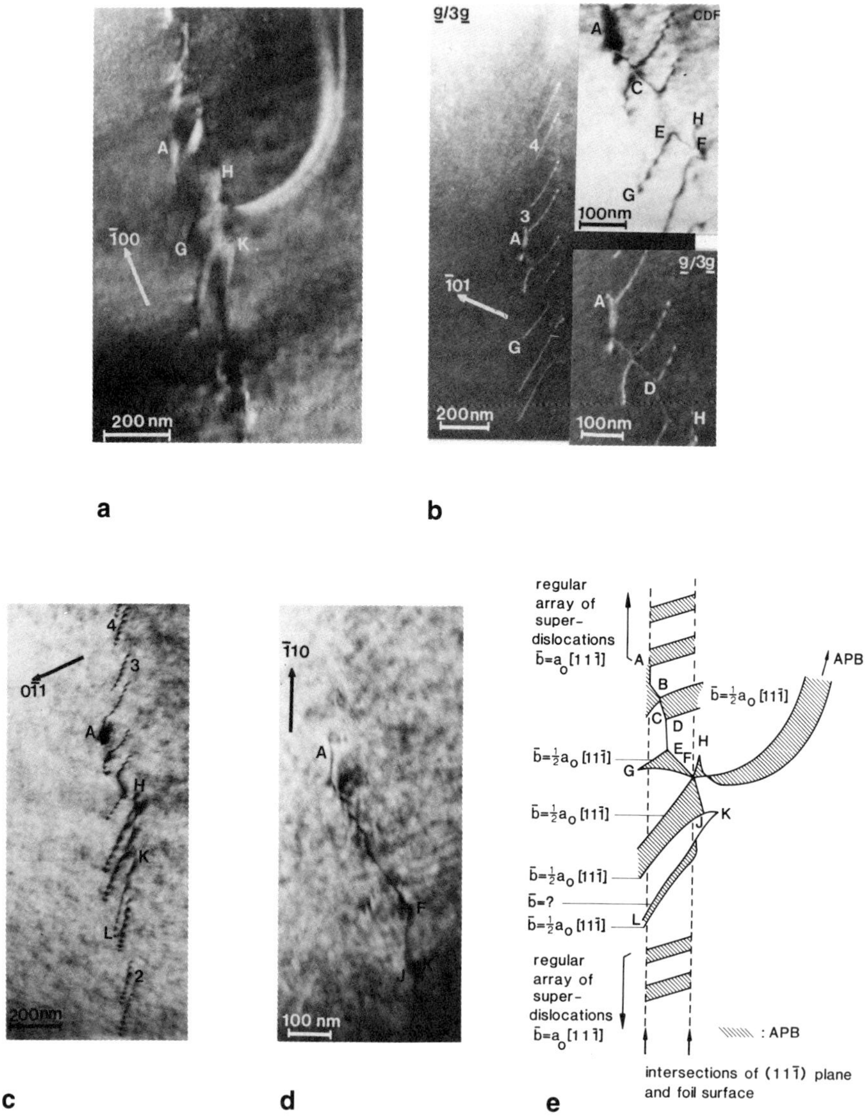

Fig. 4 Centered dark-field electron micrographs of approximately the same area in B_2-ordered Fe_3Al showing the interaction of an APB from the bulk (a) with the low-angle boundary (b) containing an extrinsic dislocation (D). In (b) a weak-beam image shows the dislocation network, also illustrated schematically in (e). A fundamental reflection is used in (c) and (d). Corresponding positions are indicated.

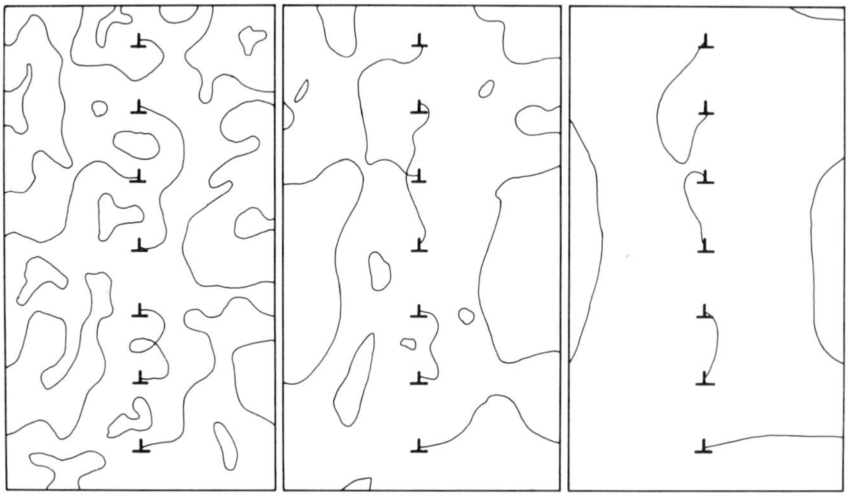

Fig. 5 The growth of ordered domains in a B_2-ordered alloy containing a wall of ordinary dislocations with Burgers vector $(1/2)a_0[11\bar{1}]$ is illustrated at different moments during annealing. APBs bounded by the dislocations tend to align with the boundary plane when the average domain size is large compared to the dislocation separation.

Ref 21) (see for example Fig. 3a, and Fig. 4c and d). From the analysis of all four possible $(1/2)a_0\langle 111\rangle$-type Burgers vectors, only $b = (1/2)a_0[11\bar{1}]$, which is parallel to the boundary normal (n), is consistent with the observations. The latter was deduced from the edge-on position of the boundary. The line direction of the ordinary dislocations was determined to be u = [101], perpendicular to b. The angle of tilt was determined to be $\theta = 0.16°$ from $\theta = b/h$, where $b = \sqrt{3}\,a_0$ and $h = 175$ nm, the separation of the superdislocations. It is concluded that this part of the boundary is a low-angle symmetrical tilt boundary comprised of a regular array of superdislocations; each superdislocation consists of two regular dislocations lying in the boundary plane and separated by a ribbon of APB (Fig. 1c). A symmetrical tilt boundary consisting of a regular array of superdislocations has also been observed in the B_2-ordered FeCo alloy by Marcinkowski (see Amelinckx, Ref 39), but no APB contrast was detected in this material.

APBs interacting with the low-angle symmetrical tilt boundary have always been observed to disarrange the regular array of superdislocations (Buis,

Tichelaar, and Schapink, Ref 21). The particular boundary area shown in Fig. 4 consists of an irregular dislocation network (Fig. 4b) in which an APB from the bulk terminates (Fig. 4a). This network is noncoplanar and contains an extrinsic dislocation (ABCDEFJKL in Fig. 4e), imaged in Fig. 4(d) using a fundamental reflection for which the intrinsic dislocations are invisible. Burgers vector analysis using the $g \cdot b = 0$ method of part of the dislocation network reveals that the dislocation reaction GE + EF → DE (Fig. 4e) can be described by:

$$\tfrac{1}{2}a_0\,[11\bar{1}] + \tfrac{1}{2}a_0\,[\bar{1}11] \rightarrow a_0\,[010]$$

which is the reverse of the reaction shown in Eq 1. The same dislocation reaction has been observed in twist boundaries in B_2-ordered FeCo (Czernichow *et al.*, Ref 19). Dislocation section GEFH (Fig. 4e) is not associated with a pair of dislocations and is therefore assumed to bound the APB from the bulk.

Part of the low-angle symmetrical tilt boundary investigated consisted of superdislocations of type $a_0\langle 111\rangle$ located in the boundary plane. Also, ordinary dislocations bounding APBs from the bulk were found in some areas. The third possible configuration proposed by Marcinkowski (Ref 11) (Fig. 1b), consisting of $a_0\langle 111\rangle$-type superdislocations perpendicular to the boundary plane, was not observed. This configuration is conceivable only if polygonization has taken place in the ordered material. However, in the specimen investigated, it is thought that polygonization had already occurred in the disordered state, forming a regular array of perfect dislocations in the bcc lattice with Burgers vector $(1/2)a_0\langle 111\rangle$.

Ordering sets in when the specimen is cooled below T_c after recrystallization. It is suggested that a particular sequence occurs. The disorder-order transition in a B_2-ordered alloy is believed to be a higher-order phase transition (Swann, Duff, and Fisher, Ref 40) in which the homogeneous nucleation of ordered domains takes place. Domains that have an identical occupation of the same sublattice coalesce, and thermal APBs are formed where different domains meet. Also, APBs will nucleate on the ordinary dislocations in the low-angle boundary. When ordering is complete, a network of APBs is present (Fig. 5a). All APBs are either closed surfaces or end in a dislocation or at the specimen surface. During the annealing at 600 °C, domain growth occurs by the coalescence of domains and by the disappearance of small, isolated domains (Fig. 5a to c). This process of the annihilation of migrating APBs is discussed in more detail by Marcinkowski (Ref 38).

The energy of APBs connecting different dislocations can be reduced by minimizing the area of the APBs. In this process the APBs will tend to align with the boundary plane, and when the average domain size is large compared to the separation of the dislocations, most neighboring dislocations will be connected by an APB (Fig. 5c). During further annealing, the energy of these APBs can be reduced by forming a superdislocation in the boundary plane (Fig. 1c). The distance between the ordinary edge dislocations constituting the superdislocation is then necessarily decreased by a climb process. The relative number of dislocations bounding APBs from the bulk (Fig. 3) is determined by the average domain size in the bulk. The observed disarrangement of dislocations bounding APBs from the bulk is possibly a result of the asymmetric configuration with respect to the boundary plane in this case: the energy of the APB can be reduced by rearrangement of the bounding dislocation, depending on the interaction forces between the dislocations.

The equilibrium separation (r_{eq}) of the dislocations forming each superdislocation in the boundary is found from Eq 6 with $K = 1.085 \times 10^{11}$ N/m^2 (Buis, Tichelaar, and Schapink, Ref 21). The APB energy γ for a {111} plane is calculated from Eq 8. Since V and W can be estimated only roughly (see the section "Theoretical Considerations" earlier in the chapter), both low and high values are considered. From Eq 11 and $W/V = 0.5$, it is found that $\gamma\{111\} = 0.025$ J/m^2, and from Eq 6, $r_{eq} = 37$ nm. On the other hand, it is considered more realistic to deduce V and W from Crawford's measurements (Ref 34) (see the same earlier section), yielding $\gamma\{111\} = 0.042$ J/m^2 and $r_{eq} = 24$ nm. From the experimental value $r_{eq} = 64$ nm it is found that $\gamma\{111\} = 0.009$ J/m^2, which is considerably lower than the calculated values. Both calculated values of r_{eq} are smaller than the measured 64 nm, which suggests that equilibrium has not been reached. Buis, Tichelaar, and Schapink (Ref 21) pointed out, first, that the difference between the calculated and the measured value of r_{eq} can also be explained by Fe segregation to the APB plane, which lowers the APB energy, and, second, that this difference is not explained by a thin film effect.

The Asymmetrical Low-Angle Tilt Boundary. The second low-angle boundary studied in detail is partly shown in Fig. 6 and 7. Corresponding dislocations in these figures are indicated by numbers. In Fig. 6 the dislocations comprising the boundary are imaged using fundamental reflections. Dislocations 2 through 13 (2...13) form a regular array of dislocation pairs lying in a slightly curved plane with an approximate boundary normal of $n = [12\bar{1}] \pm 10°$, determined from the edge-on position. Below dislocation 1 this array is disarranged by the interaction of a dislocation from one of the crystals with the low-angle boundary. The dislocations indicated by 14...21 have a boundary plane different from dislocations 2...13. In Fig. 7 the boundary is imaged by applying both a fundamental reflection (Fig. 7a) and a superreflection (Fig. 7b).

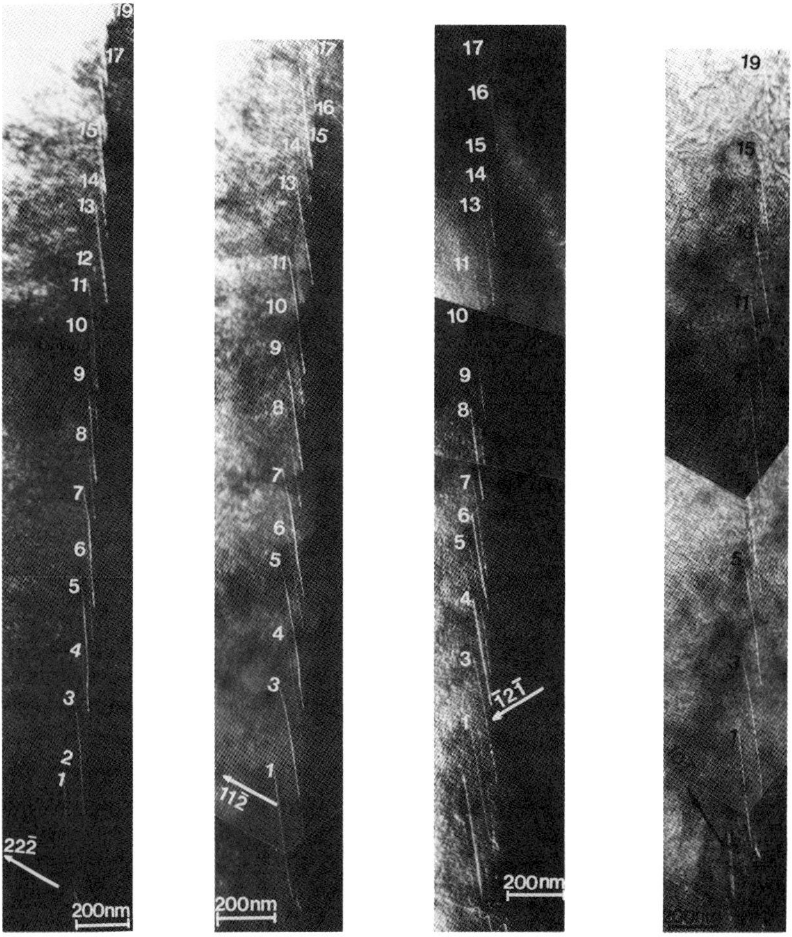

a b c d

Fig. 6 Weak-beam electron micrographs (~$g/2g$, centered dark field) of the same area in B_2-ordered Fe_3Al showing part of the low-angle boundary 2 consisting of an array of ordinary dislocations with different Burgers vectors (see Fig. 7). Foil normal, [313]. (a) $B = [134]$, all dislocations are visible. (b) $B = [375]$, dislocations 2 and 12 are invisible. (c) $B = [332]$. Dislocations 2 and 12 are invisible. (d) $B = [111]$, the dislocations indicated by the numbers 2,4,...,14,16, 17,18 are invisible.

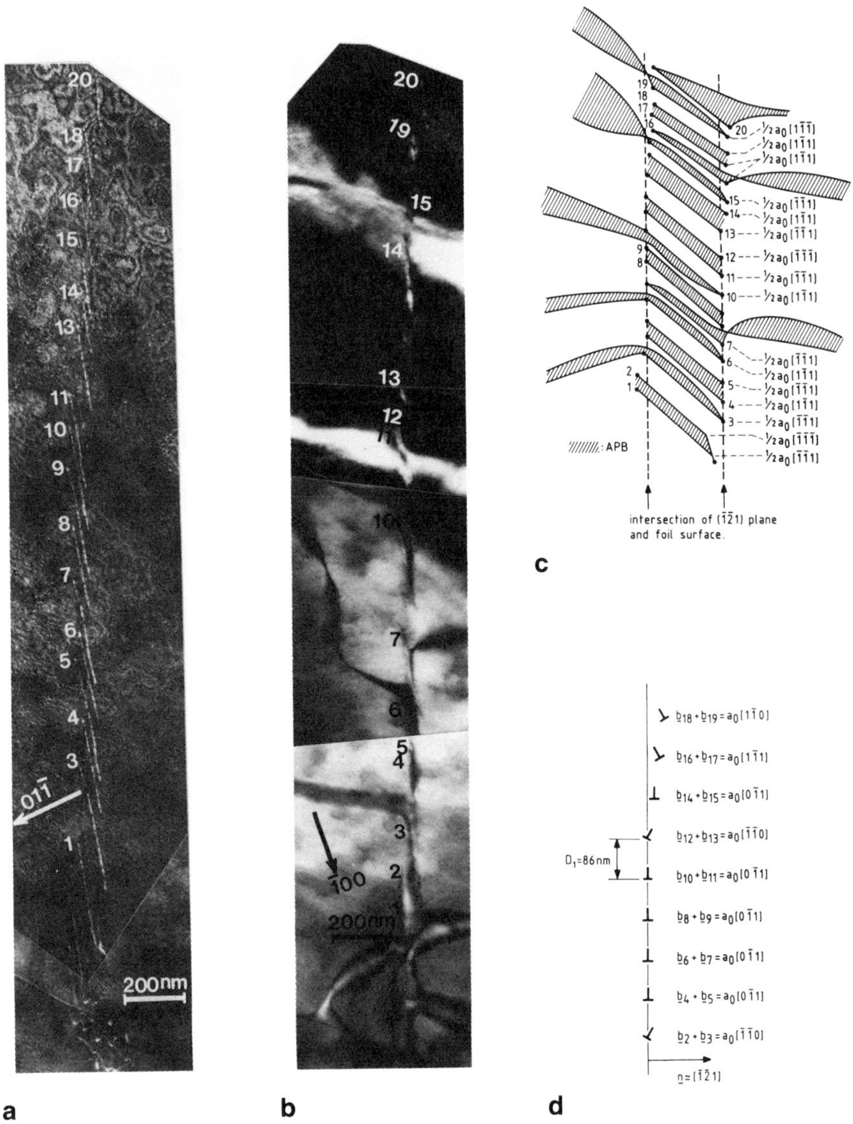

Fig. 7 Electron micrographs of the same area in B_2-ordered Fe_3Al showing part of the low-angle boundary 2 (Fig. 6) using a fundamental reflection in (a) ($B \approx [011]$) and a superreflection in (b) ($B \approx [011]$). Dislocations 2, 12, and 19 show weak contrast in (a). The dislocations imaged in (a) are bounded by an APB from the bulk or by an APB in the boundary plane, as shown in (c). The dislocations may be added by pairs, as shown in (d). Part of the boundary then forms an asymmetrical tilt boundary comprised of two sets of ⟨110⟩-type dislocations.

APBs from both crystals terminate in the dislocations 3, 6, 7, 10, 15, 16, 19, and 20. Because of a variation of the foil thickness on a small scale, the APB contrast is not always clear. However, APB contrast can be observed between dislocations 1 and 2, and 11 and 12, and to some extent between 4 and 5. These observations suggest that the boundary is comprised of both ordinary dislocations, with $b = a_0\langle 111\rangle/2$, and superdislocations.

To determine the Burgers vector, the $g \cdot b = 0$ invisibility criterion was applied to 11 fundamental reflections. For example, the even-numbered dislocations are invisible in Fig. 6(d). All four $(1/2)a_0\langle 111\rangle$-type Burgers vectors were considered in the analysis, the result of which is given in Fig. 7(c). The regular array indicated by 1...13 is shown to consist of different dislocation types, which partly alternate. The odd-numbered dislocations have $b = \pm(1/2)a_0[\bar{1}\bar{1}1]$, whereas the even-numbered dislocations have $b = \pm(1/2)a_0[1\bar{1}1]$, except dislocations 2 and 12, which have $b = \pm(1/2)a_0[\bar{1}\bar{1}\bar{1}]$. The higher contrast of the odd-numbered dislocations compared to the even-numbered dislocations in Fig. 6(a) and (b), and vice versa in Fig. 6(c), is consistent with the higher value of $g \cdot b$ in these cases. The line direction of the dislocations 1...14 was determined from the projected line directions at different electron beam directions to be $u = [\bar{1}11]$. The average separation (D_1) between the even-numbered (or between the odd-numbered) dislocations in the regular array was determined from the projected distances on the electron micrograph at different beam directions to be 86 ± 7 nm.

The ordinary dislocations indicated by 2...13 can be interpreted as an array of two types of edge dislocations with Burgers vectors $a_0[0\bar{1}1]$ and $a_0[\bar{1}\bar{1}0]$ ($u = [\bar{1}11]$) by adding the Burgers vectors of the ordinary dislocations, as indicated in Fig. 7(d). An average Burgers vector (b_{av}) can be assigned to this part of the low-angle boundary. The value of b_{av} can be calculated as $b_{av} = 4a_0[0\bar{1}1] + 2a_0[\bar{1}\bar{1}0] = 2a_0[\bar{1}\bar{3}2]$, which is almost parallel to the observed boundary normal. This indicates that this part of the low-angle boundary can be regarded as an asymmetrical tilt boundary, as described, for instance, by Read (Ref 41). The tilt of the boundary can be calculated from Frank's formula (Hirth and Lothe, Ref 42), which results in $\theta = b_{av}/6D_1$. With $D_1 = 86$ nm, it is found that $\theta = 0.24°$.

It is concluded that the regular array 2...13, which is part of the low-angle boundary 2, is a low-angle asymmetrical tilt boundary comprised of two sets of $a_0\langle 110\rangle$-type edge dislocations. These dislocations are dissociated into ordinary dislocations, which are either bounded by an APB from one of the crystals or separated by a ribbon of APB in the boundary plane. The latter dislocations form $a_0\langle 110\rangle$-type superdislocations of the type shown in Eq 2. The APB-bounded ordinary dislocations are reminiscent of the structure proposed by

Marcinkowski (Fig. 1a). The dislocations 14...20 can be regarded as an arbitrary low-angle boundary comprised of ordinary dislocations bounded by APBs from the bulk, superdislocations of type $a_0\langle 110\rangle$, and one superdislocation of type $a_0\langle 111\rangle$. The origin of the APB structure at the boundary must be a process similar to that outlined in the section above, "The Symmetrical Low-Angle Tilt Boundary." The origin of the dissociated $a_0\langle 110\rangle$-type edge dislocations is more difficult to assess. During polygonization in the disordered state, ordinary dislocations with different Burgers vectors of type $a_0\langle 111\rangle$ must have arranged their positions in the observed alternating manner in the region 3...11 in order to form a low-energy configuration. The energy (E) (see Eq 4) of $a_0\langle 110\rangle$-type superdislocations can be expected to be lower than that of $a_0\langle 111\rangle$-type superdislocations because the interaction energy, and consequently the separation distance of the superpartials, is lower in the $a_0\langle 110\rangle$-type superdislocation. Since the APB configuration at the asymmetrical part of the boundary was highly irregular, a calculation of the superpartial separation in the boundary was not attempted.

The Asymmetrical Near $\Sigma = 3$ Tilt Boundary. In one of the specimens, the misorientation of the $\Sigma = 3$ bicrystal is described by a 60° clockwise rotation of crystal 2 relative to crystal 1 around the common [111] axis, indicated as [111]c. The boundary normal was estimated from the edge-on position at three different beam directions as $n = [\bar{6}65]_1 \pm 5°$ (subscripts 1 and 2 refer to the coordinate systems 1 and 2), which is close to $[\bar{1}11]_1 = [\bar{1}\bar{1}5]_2$. Figure 8(a) shows a micrograph of the boundary using a noncommon superreflection for which APBs are visible. The APBs terminate abruptly in the boundary and can therefore separate different boundary structures (Fig. 2). The deviation (θ) and the associated rotation axis (c) from the $\Sigma = 3$ orientation relationship were determined from the shift and rotation of the Kikuchi lines at the near $[111]_c$-axis and from the calculation of matrix A (Eq 12) as $\theta = (4.1 \pm 0.6)°$ and $c = [\bar{1}20]_1 \pm 18°$. Because of this relatively large deviation from the coincident lattice, almost no common diffraction vectors were available, Therefore a possible rigid-body translation at the boundary could not be determined with the method of α-fringes (Pond, Ref 43). Figure 8(b) shows a dark-field electron micrograph of the boundary, with a hexagonal network of lines and a spacing of about 50 Å. For small spacings, the distinction between a moiré pattern and a dislocation network is often difficult to make (Thölén, Ref 44). A moiré pattern perpendicular to the employed fundamental reflection with a spacing of 80 Å was expected but was not observed. Also, the observed line contrast did not show the regularity expected in a moiré pattern. Because the hexagonal contrast cannot be explained by a moiré effect, it might have been caused by a

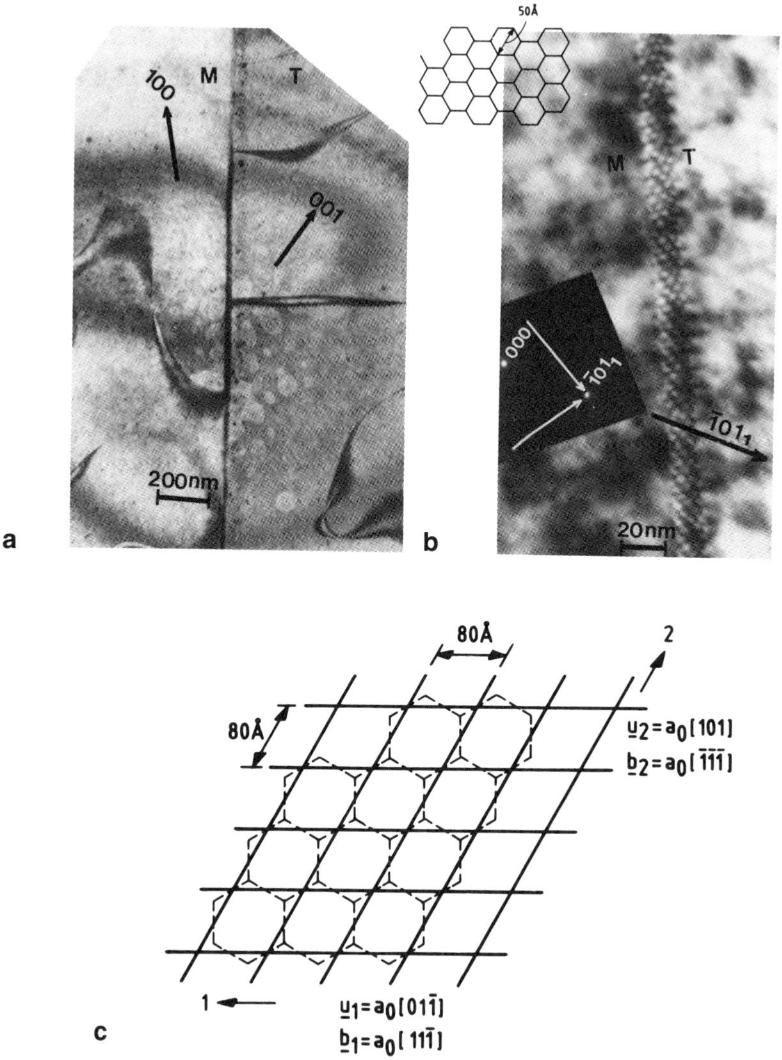

Fig. 8 (a) A bright-field electron micrograph employing a noncommon ⟨100⟩-type superreflection shows APBs terminating at a near $(\bar{1}11)_1(\bar{1}15)_2\ \Sigma = 3$ grain boundary in B_2-ordered Fe_3Al. (b) The almost common fundamental reflection used in a dark-field electron micrograph reveals a hexagonal dislocation network in the boundary plane. The projection of this network is shown schematically in the inset. (c) A hypothetical network of two sets of superdislocations in the boundary plane is shown by the drawn lines. This network can explain part of the deviation from the coincident misorientation (see text). Reactions between dislocations of different sets can produce a hexagon-shaped network, indicated by the dotted lines.

network of dislocations in the boundary, which accommodates the observed deviation from the CL. The observed deviation from the CL misorientation was small enough to be accommodated by a dislocation network, according to most criteria, as discussed by Déchamps, Baribier, and Marrouche (Ref 44). Hexagonal dislocation networks have been observed before (in low-angle boundaries) in B_2-ordered alloys (Czernichow *et al.*, Ref 19; Morgand, Mouturat, and Sainfort, Ref 4).

Figure 8(c) shows a hypothetical lozenge-shaped dislocation network in the $(\bar{1}11)_1$ boundary plane, consisting of two sets of superdislocations with Burgers vectors $b_1 = a_0[11\bar{1}]$ and $b_2 = a_0[\overline{111}]$ and a spacing of 80 Å. (Note that the Burgers vectors of these grain-boundary dislocations can be decomposed into smaller vectors of the dsc lattice of the $\Sigma = 3$ boundary. The smallest dsc vectors of the ordered lattice are $[111]_c/3$ and vectors of type $\langle 11\bar{2}\rangle/3$. However, such a decomposition was not observed, possibly due to the limited resolution.) The application of Frank's formula for low-angle boundaries (Hirth and Lothe, Ref 42) to this network yields an (extra) misorientation angle of 4.1° around $[\bar{1}10]_1$, which is in reasonable agreement with the experimental values. A distorted hexagonal network (depicted by the dotted lines in Fig. 8c) can arise from the lozenge-shaped network by the dislocation reaction $a_0[11\bar{1}]$ + $a_0[\overline{111}] = a_0[002]$. Such a hexagonal network was proposed by Marcinkowski (Ref 18). Unfortunately, this model could not be verified by a Burgers vector analysis of the dislocations because of the narrow spacing of the dislocations, the problems with astigmatism, and the lack of common reflections.

Summary and Conclusions

This chapter has examined the structures of two types of low-angle boundaries and the structure of a near $\Sigma = 3$ boundary. Low-angle bou ndary 1, a symmetrical tilt boundary, consists of two dislocation configurations: a regular array of $a_0\langle 111\rangle$-type superdislocations located in large areas of the boundary plane and occasional ordinary dislocations with an irregular line direction bounding APBs from the bulk. The extension r_{eq} of the superdislocations comprising the boundary is larger than the equilibrium value calculated from the anisotropic elasticity theory. Presumably either the origin of this discrepancy is associated with a nonequilibrium extension of the superdislocations in the specimen or the APB energy is lower than assumed, probably due to segregation to the APBs.

Low-angle boundary 2 was found to have a more irregular structure. However, part of the boundary could be described as an asymmetrical tilt boundary comprised of two types of $a_0\langle 110\rangle$ dislocations. These dislocations were dissociated into ordinary dislocations, forming a mixture of ordinary dislocations bounding APBs from the bulk and $a_0\langle 110\rangle$-type superdislocations with the APB parallel to the {112} boundary plane. It is believed that $a_0\langle 110\rangle$-type superdislocations have not been observed before in the B_2 structure.

Superdislocations perpendicular to the boundary plane, as considered by Marcinkowski (Ref 11), were not found in either of the two low-angle boundaries, which is in agreement with the idea that polygonization takes place in the disordered state.

The structure of the asymmetrical $(\bar{1}11)_1$ $(\bar{1}\bar{1}5)_2$ $\Sigma = 3$ boundary was analyzed geometrically. Nearest-neighbor distances at the boundary are for most of the atoms comparable to the nearest-neighbor distances in the bulk if the continuity of the atomic density at the boundary is taken into account. Therefore, a major rigid-body translation at this type of boundary is not expected. Experimentally, a near $\Sigma = 3$ bicrystal with a 4° extra misorientation was found. The hexagonal contrast in this boundary was interpreted as a dislocation network accommodating the deviation from the CSL misorientation. However, Burgers vectors consistent with the misorientation were found to be lattice vectors, not the expected smaller dsc lattice vectors belonging to the $\Sigma = 3$ boundary.

References

1. M.G. Mendiratta, S.K. Ehlers, D.M. Dimiduk, W.R. Kerr, S. Mazdiyasni, and H.A. Lipsitt, *Mater. Res. Soc. Symp. Proc.*, Vol 81, 1987, p 393
2. K. Vedula and J.R. Stephens, *Mater. Res. Soc. Symp. Proc.*, Vol 81, 1987, p 387
3. M.G. Mendiratta, S.K. Ehlers, D.K. Chatterjee, and H.A. Lipsitt, *Metall. Trans. A*, Vol 18, 1987, p 283
4. P. Morgand, P. Mouturat, and G. Sainfort, *Acta Metall.*, Vol 16, 1968, p 867
5. M.A. Crimp, K.M. Vedula, and D.J. Gaydosh, *Mater. Res. Soc. Symp. Proc.*, Vol 81, 1987, p 499
6. I. Baker and D.J. Gaydosh, *Mater. Sci. Eng.*, Vol 96, 1987, p 147
7. G. Sainfort, P. Mouturat, P. Pépin, J. Petit, G. Cabane, and M. Salesse, *Mém. Sci. Rev. Métall.*, Vol LX, 1963, p 125
8. M.J. Marcinkowski, M.E. Taylor, and F.X. Kayser, *J. Mater. Sci.*, Vol 10, 1975, p 406
9. W.R. Kerr, *Metall. Trans. A*, Vol 17, 1986, p 2298
10. M.J. Marcinkowski and J. Larsen, *Metall. Trans.*, Vol 1, 1970, p 1034
11. M.J. Marcinkowski, *Philos. Mag.*, Vol 17, 1968, p 159
12. A.H. King and M.H. Yoo, *Mater. Res. Soc. Symp. Proc.*, Vol 81, 1987, p 99
13. F.W. Schapink and F.D. Tichelaar, *Phys. Status Solidi (a)*, Vol 106, 1988, p 433
14. F.D. Tichelaar and F.W. Schapink, *Philos. Mag. A*, Vol 62, 1990a, p 53
15. F.D. Tichelaar and F.W. Schapink, *Philos. Mag. A*, in press, 1990b

16. T. Takasugi and O. Izumi, *Acta Metall.*, Vol 31, 1983, p 1187
17. T. Takasugi and O. Izumi, *Acta Metall.*, Vol 35, 1987, p 823
18. M.J. Marcinkowski, *Mém. Sci. Rev. Métall.*, Vol LXIX, 1972, p 579
19. J. Czernichow, J.P. Gudas, M.J. Marcinkowski, and W.F. Tseng, *Metall. Trans.*, Vol 2, 1971, p 2185
20. J.V. Rinnovatore and N. Brown, *Philos. Mag.*, Vol 14, 1966, p 629
21. A. Buis, F.D. Tichelaar, and F.W. Schapink, *Philos. Mag.*, Vol A59, 1989, p 861
22. R. Kuentzler, *J. Phys.*, Vol 44, 1983, p 1167
23. P.R. Swann, W.R. Duff, and R.M. Fisher, *Trans. Metall. Soc. AIME*, Vol 245, 1969, p 851
24. D.I. Potter, *Mater. Sci. Eng.*, Vol 5, 1969-1970, p 201
25. M.J. Marcinkowski and N. Brown, *Acta Metall.*, Vol 9, 1961, p 764
26. I.L.F. Ray, R.C. Crawford, and D.J.H. Cockayne, *Philos. Mag.*, Vol 21, 1970, p 1027
27. A.N. Stroh, *Philos. Mag.*, Vol 3, 1958, p 625
28. H.J. Leamy, P. Schwellinger, and H. Warlimont, *Acta Metall.*, Vol 18, 1970, p 31
29. T. Muto and Y. Takagi, *Solid State Phys.*, Vol 1, 1955, p 193
30. P.S. Rudman, *Acta Metall.*, Vol 8, 1960, p 321
31. F.W. Schapink, *Scr. Metall.*, Vol 2, 1968, p 635
32. F.W. Schapink, Ph.D. thesis, Technical University of Delft, 1969
33. J.E. Krzanowski and S.M. Allen, *Surf. Sci.*, Vol 144, 1984, p 153
34. A.C. Crawford, I.L.F. Ray, and D.J.H. Cockayne, *Philos. Mag.*, Vol 27, 1973, p 1
35. R.W. Cahn and J.A. Coll, *Acta Metall.*, Vol 9, 1961, p 138
36. H. Okamoto and P.A. Beck, *Metall. Trans.*, Vol 2, 1971, p 569
37. P. Delavignette, T. Karakostas, G. Nouet, and F.W. Schapink, *Phys. Status Solidi (a)*, Vol 107, 1988, p 551
38. M.J. Marcinkowski, *Electron Microscopy and Strength of Crystals*, G. Thomas and I. Washburn, Ed., Interscience, 1963
39. S. Amelinckx, in *Dislocations in Solids*, Vol 2, F.R.N. Nabarro, Ed., North Holland, 1979, p 279
40. P.R. Swann, W.R. Duff, and R.M. Fisher, *Metall. Trans.*, Vol 3, 1972, p 409
41. W.T. Read, *Dislocations in Crystals*, McGraw-Hill, 1953
42. J.P. Hirth and J. Lothe, *Theory of Dislocations*, McGraw-Hill, 1968
43. R.C. Pond, *Proc. R. Soc. (London) A*, Vol 357, 1977, p 471
44. M. Déchamps, F. Baribier, and A. Marrouche, *Acta Metall.*, Vol 35, 1987, p 101
45. A.R. Thölén, *Phys. Status Solidi (a)*, Vol 2, 1970, p 537

6

Characterization of Grain Boundaries in Ordered Ni_3Al

D. Farkas and H. Jang
Department of Materials Engineering
Virginia Polytechnic Institute and State University, Blacksburg, Virginia

Grain boundaries can be completely characterized by nine crystallographic parameters. Three of these refer to the misorientation of one grain with respect to the other and are usually addressed first in characterizing grain boundaries. This chapter addresses the question of grain-boundary-plane orientation. Conventional transmission electron microscopy (TEM) has been used to study the relative misorientation and grain-boundary-plane orientation in boron-doped Ni_3Al. The results show that for misorientations corresponding to a high density of coincident sites the boundary plane tends to be located in planes that take maximum advantage of the coincident site structure. These results are discussed in the light of a symmetry analysis of grain boundaries and are related to computer simulation studies. The implications of these results for the mechanical behavior of Ni_3Al are also discussed.

It has been known for several years that grain boundaries in ordered Ni_3Al are intrinsically brittle and that microalloying with boron can ductilize nickel-rich polycrystalline material by strengthening the grain boundaries. Despite the large effort that has been devoted to the problem, no clear understanding has emerged of either the reason for the intrinsic grain-boundary brittleness or the mechanism of boron strengthening. The few clearly established experimental facts are (Ref 1-3):

- Most grain boundaries in the polycrystalline material are brittle, with the exception of some $\Sigma = 3$ boundaries
- Boron segregates to the grain boundary, apparently also producing some nickel cosegregation
- Boron ductilizes only nickel-rich material

Theoretical studies of grain-boundary structure have been largely restricted to symmetrical tilt or twist boundaries and, in most cases, to low Σ values. These results clearly show a trend of higher grain-boundary energy for a higher aluminum content of the boundary (Ref 4-6). There is also evidence that the grain-boundary structures tend to be more open than in fcc materials (Ref 4-7).

A number of other features of the grain boundary have been studied, with no unequivocal conclusion; for example, the possible presence of a disordered layer in the grain-boundary region. According to theoretical estimates of the disordering effect near grain boundaries, this should not be a significant effect up to about 80% of the order-disorder temperature for the bulk (Ref 8). Some experimental observations of disordered structures near the boundary region have been reported (Ref 9). These observations have not been widely confirmed.

It is possible that grain-boundary features depend strongly on the geometry of the particular grain boundary. It is therefore of extreme importance to have a characterization of the types of grain boundaries present in ductile and brittle Ni_3Al. Some work along these lines has been done on the Σ distribution of boundaries. However, very little characterization work that includes the orientation of the grain-boundary plane has been done. In the present work, experimental results regarding the statistical distribution of Σ values and grain-boundary-plane location are discussed.

Geometric Characterization

Macroscopic Grain-Boundary Parameters. Five macroscopic parameters characterize grain-boundary geometry: three are related to the misorientation (two for the misorientation axis and one for the misorientation angle); the other two are related to grain-boundary-plane orientation.

The first three parameters determine the value of Σ, the reciprocal volumetric density of coincident sites. These parameters completely determine the coincident site lattice (CSL). The structure, energetics, and behavior of the boundary are strongly dependent on the grain-boundary-plane orientation.

One might think that there are some preferred grain-boundary planes for a given CSL misorientation. This would be expected because special planes of the CSL have particularly high planar density of coincident sites. It is not clear whether this corresponds to an actual driving force for the location of the grain-boundary plane at these particular sites. Furthermore, even if this driving force exists, it is not clear whether it is sufficient for establishing a relationship in the material among misorientation and grain-boundary-plane orientation. It is therefore of interest to see whether there is experimental evidence for a correlation between grain-boundary-plane location and the bicrystal misorientation.

Microscopic Grain-Boundary Parameters. There are four microscopic degrees of freedom to the grain-boundary structure. These correspond to the values of the rigid body translation of one crystal with respect to the other (three degrees of freedom) and to the movement of the grain-boundary plane parallel to itself. It should be noted that these degrees of freedom are much more important in ordered alloys than in pure metals since they may induce different grain-boundary stoichiometries and have a significant effect on the number of "wrong bonds" present in the grain-boundary structure.

Microscopic parameters are not easily measured. These measurements involve complex TEM work (Ref 10). An understanding of the microscopic nature of G · B has largely been obtained from computer simulation studies. The present work concentrates on macroscopic characterization.

Experimental Characterization

Several techniques have been used in the past to measure Σ distribution in polycrystalline Ni_3Al. Each technique involves obtaining the crystal orientation of the electron beam in the two grains in order to obtain the misorientation between the two grains. This process has to be done twice so that the rotation axis and angle can be obtained in order to determine the rotation matrix. This matrix then must be compared to the 24 possible rotation matrices for each CSL misorientation. The indexing can be performed using scanning electron microscopy (SEM) or TEM techniques. Hanada *et al.* (Ref 11) used an SEM technique to obtain Σ distributions in brittle Ni_3Al. These results showed a distribution different from that found for ductile fcc metals (see, for example, the work of Lim and Raj, Ref 12). Farkas *et al.* (Ref 13) used similar techniques to study the Σ distribution in both ductile and brittle Ni_3Al. The results show agreement with Hanada *et al.* (Ref 11) for brittle samples and a distribution similar to that of pure recrystallized Al for the ductile material. Mackenzie *et al.* (Ref 14), using an automatized SEM technique, found no difference in the distributions for ductile and brittle materials. More recent work by Lin and Pope also includes results for Σ distributions in brittle Ni_3Al (Ref 15). The lack of agreement between available measurements reflects the effect of a number of factors related to sample history, measurement technique, and data analysis on the Σ distribution observed. The most important of these factors are discussed below.

Effects of Sample Processing History. The distribution of misorientations certainly does not reflect an equilibrium situation resulting from different grain-boundary energies, but rather is a result of the kinetic process of grain growth, or recrystallization. These processes depend on the sample's stress and thermal history. A dramatic example of this influence is the number of $\Sigma = 3$ (twins) found. These boundaries are simply annealing twins, and their number is mostly determined by sample processing.

Effects of Measurement Techniques. Different techniques can give different results for Σ distribution because of a possible bias in the operator's choice of boundaries to be studied. For example, if the technique includes a low-magnification metallography, annealing twins ($\Sigma = 3$) can be easily recognized because they occur within a parent grain and can then be avoided by the operator. Other techniques may depend on observing a change in backscattering patterns to avoid low-angle boundaries.

Data Analysis. Finally, the data on misorientation axis and angles should be processed such that a Σ value can be assigned to each boundary. Two main questions arising from this procedure are, what is the allowed deviation from the exact CSL misorientation, and which is the value of Σ above which boundaries should be considered random? The first question is discussed in detail below. Regarding the second question, the cutoff values used in the past have usually been $\Sigma = 29$, $\Sigma = 49$, and $\Sigma = 19$. However, there is no physical basis for the choice of any of these values.

Criteria for Deviations from CSL Misorientations. Considerable experimental evidence has shown that a special boundary maintains its properties of low-energy interfacial structures when it deviates slightly from the exact CSL misorientation. Several criteria have been suggested for the maximum allowable deviation at which the boundary sustains its special properties. According to Read and Shockley (Ref 16), the maximum deviation of a boundary from a special misorientation is given by:

$$\Delta\theta_c = \frac{b}{d_{\min}} \qquad \text{(Eq 1)}$$

where b is the average Burgers vector in the displacement shift complete (dsc) lattice and d_{min} is the closest allowed spacing of the dislocations that is of the order of the boundary periodicity (p). The boundary periodicity varies with the boundary plane, and the Burgers vector depends on the grain-boundary-plane orientations. Warrington and Grimmer (Ref 17) proposed that the mean value of p varies as a function of $\Sigma^{1/3}$ and the mean value of b varies as a function of $\Sigma^{-1/3}$ since the volume of the CSL unit cell varies linearly with Σ and the volume of the unit cell of the Burgers vectors varies with $1/\Sigma$. Then the maximum allowable deviation angle from exact CSL misorientations are:

$$\Delta\theta_c = \theta_0\Sigma^{-2/3} \quad \text{(Eq 2)}$$

where θ_0 is constant for all CSLs and is the maximum deviation angle for low-angle grain boundaries. Considering the special case of dsc Burgers vectors lying in a boundary plane perpendicular to a ⟨100⟩ misorientation axis, p varies as $\Sigma^{1/2}$, and b varies as $\Sigma^{-1/2}$. Then the maximum deviation is (Ref 18):

$$\Delta\theta_c = \theta_0\Sigma^{-1} \quad \text{(Eq 3)}$$

On the other hand, Brandon's criterion (Ref 19), which has been used by many researchers, considered the boundary periodicity only in the special case of a boundary plane perpendicular to a ⟨100⟩ misorientation axis. The Brandon's criterion is:

$$\Delta\theta_c = \theta_0\Sigma^{-1/2} \quad \text{(Eq 4)}$$

Therefore, the criterion can be adjusted to specific grain boundaries depending on the geometry of the grain boundary, as considered by Ishida and Mclean (Ref 18) for ⟨100⟩ twist grain boundaries, or based on the periodicity of the grain boundary alone, as considered by Brandon (Ref 19), and there seems to be no criterion that is applicable to all grain-boundary geometries. The choice of criterion can influence the statistical distribution of Σ in a polycrystal. Different criteria may significantly affect the obtained distribution of Σ values.

The estimation of the maximum deviation angle for low-angle grain boundaries (θ_0) is another factor to be considered in determining the maximum tolerable deviation angle from exact CSL misorientation. First, Read and Shockley (Ref 16) suggested that a valid limit of θ_0 is 15° from their dislocation model of the grain-boundary structure. On the other hand, TEM observation of the maximum deviation angle of observable discrete grain-boundary dislocations showed that the maximum allowable deviation angle of a low-angle boundary was 8° using a conventional TEM technique in a two-beam dynamic condition. In the mid 1980s, Ichinose and Ishida (Ref 20) showed an evidence

Table 1 Comparison of Σ distributions obtained using different criteria for the maximum allowed deviation from the exact CSL misorientation

	Σ distribution, %												
Criterion	3	5	7	9	11	13	17	19	21	25	27	29	>29
Brandon (a)	15.1	1.4	2.7	1.4	4.1	1.4	1.4	1.4	1.4	1.4	1.4	1.4	65.5
Warrington (b)	15.1	1.4	2.7	1.4	2.7	1.4	0	1.4	1.4	0	0	0	72.5
Ishida (c)	15.1	1.4	1.4	0	1.4	0	0	0	0	0	0	0	80.7

(**a**) Ref 19. (**b**) Ref 17. (**c**) Ref 18

of discrete dislocations maintained up to 15° using high-resolution electron microscopy.

Experiments

Experiments were performed to investigate grain-boundary-type distribution using different criteria. Moreover, the statistical distribution of the grain-boundary plane normal was also studied for possible relationships between misorientation and grain-boundary-plane orientation. The material investigated was ordered Ni_3Al having a chemical composition of 24 at.% Al-76 at.% Ni with 500 ppm B. The samples were vacuum annealed at 1050 °C for 6 h to reduce the defect density and control the average grain size of the material. After annealing, the average grain size of the material was about 5.4 μm in diameter. A final thinning procedure was done with a twin-jet electrochemical polisher and a chemical solution of 70 vol% CH_3OH + 30 vol% HNO_3. Experimental work was carried out using a Japan Electron Optics Limited (JEOL) 200CX transmission electron microscope (Ref 21). The misorientation was determined first in the usual way. The grain-boundary plane was obtained from trace analysis and the measurement of the sample thickness.

Statistical Distribution of Σ Values. From the geometric analysis of grain boundaries using TEM, a statistical distribution of grain-boundary types was obtained from polycrystalline Ni_3Al. Grain-boundary misorientations were measured from 73 boundaries. In order to determine the proportion of special boundaries from randomly oriented polycrystals, three different criteria for maximum deviation were examined. Table 1 shows the influence of different criteria on the statistical distribution of grain-boundary types. It should be noted that the influence of criteria used to obtain the frequency of a given Σ increases as Σ increases. Table 2 shows a summary of the measurements of the present work and that of other researchers on the Σ distribution in ductile and brittle Ni_3Al. In this data, wide variations in the number of $\Sigma = 3$ and low-angle

Table 2 Comparison of Σ distributions obtained for Ni_3Al, Ni, and Al

Material	Σ distribution, %			
	Low angle	Σ = 3	Σ = 5-29	Random
Cast Ni_3Al(a)	8.7	0	5.8	85.5
Recrystallized Ni_3Al, 0.3 mm grain size(a)	8	12.4	4.4	75.1
Recrystallized Ni_3Al, 1 mm grain size(a)	3	14.5	3	79.5
B-doped Ni_3Al-24.8% Al(b)	0	15.1	19.4	65.5
B-doped Ni_3Al-24.8 %Al, 0.06 mm grain size(c)	26	0	25	49
B-doped Ni_3Al-24.8% Al, 0.16 mm grain size(c)	26	0	23	51
B-doped Ni_3Al-25%Al, 0.15 mm grain size(c)	17	4.5	11.5	67
B-doped Ni_3Al-25%Al, 0.28 mm grain size(c)	18	5	12	65
Annealed pure Ni_3Al(d)	2.2	25.1	3.8	68.9
Annealed B-doped Ni_3Al(d)	1.8	20.1	5.3	72.8
Ni_3Al(e)	3.4	...	16.4	83.6

(a) Hanada *et al.* (Ref 11). **(b)** Present work. **(c)** Farkas *et al.* (Ref. 23). **(d)** Mackenzie *et al.* (Ref 14). Authors include in the random boundaries those with Σ > 19. **(e)** Lin and Pope (Ref. 15). Authors include in the low-Σ boundaries those with Σ = 3 to 49.

boundaries are observed, probably due to the different techniques used. There are additional differences in the percentage of random and low-Σ boundaries related to the criterion used for acceptable directions from exact CSL misorientations. In addition, some investigators consider low Σ for Σ = 5 to 19, and others take Σ = 5 to 29 or Σ = 5 to 49. If the latter criteria are used, some differences between ductile and brittle material can be identified. These differences are not significant if Σ > 19 boundaries are taken to be random.

Statistical Distribution of Grain-Boundary Planes. It has been known that the grain-boundary energy of a certain Σ grain boundary varies as a function of the boundary-plane orientation. Thus not only the value of Σ but also grain-boundary-plane orientations should be considered. This has been suggested by grain-boundary-energy calculations using atomistic simulation (Ref 22). A review of symmetry arguments for low interfacial energy proposed that the energetically stable grain-boundary structure can be related to the high planar density of coincident sites. The planar CSL site density depends on both Σ and the grain-boundary-plane orientation. Thus the statistical distribution of the grain-boundary plane normal is important in order to investigate the energetically stable orientation of grain-boundary planes and to determine the critical value of Σ to be considered as a random boundary regardless of the special characteristics of the boundary.

The statistical distribution of grain-boundary-plane orientations from 73 grain boundaries was determined. The Σ was determined using traditional

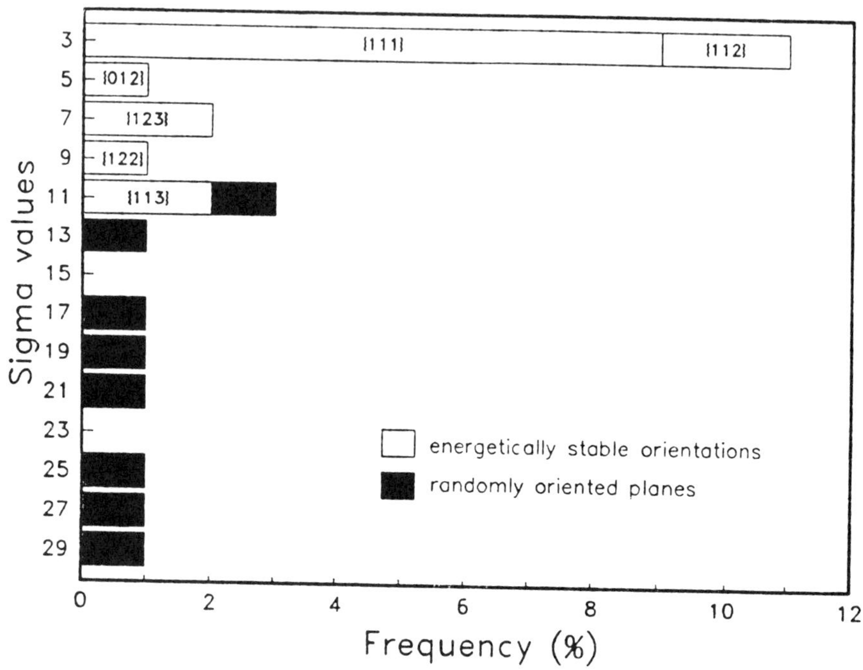

Fig. 1 The statistical distribution of grain-boundary-plane orientations from 73 grain boundaries

Brandon criteria (Ref 19). Figure 1 shows the result on grain-boundary-plane orientations adopted in each Σ in Ni_3Al. In the case of Σ = 3 grain boundaries, all planes showed {111} or {112} plane orientation. These orientations are the lowest energy, {111}, and second lowest energy, {112}, configurations for Σ = 3 grain boundaries. The low-energy configurations are to be expected for grain-boundary planes having indexes (*hkl*), which follow a rule of $n\,\Sigma = h^2 + k^2 + l^2$, where *n* is 1 or 2, since these are the planes of high coincident site density. The trend of having a high coincident site density plane was maintained on low-Σ boundaries, up to Σ = 9, and random orientations of grain-boundary planes became dominant after Σ = 11. The trend of deviation from the energetically stable geometries after Σ = 11 is believed to come from the fact that the driving force for choosing energetically stable orientations of boundary planes during material processing is not sufficiently strong in high-Σ boundaries. The occurrence of particular grain-boundary planes for a given misorientation is expected from the symmetry of the coincidence lattice. For example, for $\Sigma = 5$ the coincident site lattice is tetragonal and is formed by the axis [2 1 0], [1 $\bar{2}$ 0], [0 0 1]. The planes that are high in coincident sites are then (2 1 0) (1 $\bar{2}$ 0), (3 $\bar{1}$ 0), (1 $\bar{3}$ 0), and (001). One expects these planes to be preferred orientations for the grain-boundary plane. In general for a given Σ these planes are $(h\,k\,l)$ such that $h^2 + k^2 + l^2 = \Sigma$ or $h^2 + k^2 + l^2 = 2\Sigma$. If $l = 0$, one may also expect the 001 plane, and if $h = k$, the {110} type planes will be expected. Finally if $h = k + l$ or $k = h + l$ or $l = h + k$, {111} planes may appear.

Grain-Boundary Steps. Corresponding to energetic driving forces for the choice of particular grain-boundary planes for a given misorientation, one expects to observe grain-boundary steps. Figure 2 shows an example of $\Sigma = 3$ in which the grain boundary is clearly oriented along very particular planes. In this case there is a clear energetic advantage for the grain boundary to be located in {111} planes. For a higher-value Σ, the driving force is probably lower, and this type of phenomenon was not widely observed.

Discussion

A large number of theoretical and experimental studies on the grain-boundary structure have been reported during past several decades. The main effort of these studies has been focused on the possible relationship between grain-boundary structure and properties such as grain-boundary strength, creep, corrosion behavior, etc. There is much evidence on this relationship, and it is now generally accepted that low interfacial energy in grain boundaries plays a major role in grain-boundary properties and behavior.

Special efforts have been devoted to the relationship between the Σ distribution of grain boundaries and the ductile-brittle behavior of materials. However, there are discrepancies in results (Ref 11-15). The present authors' results show that these may be due to the different experimental methods used to obtain the grain-boundary geometry and the criteria for the special characteristics of the grain boundary, or they may be due to the differences in material processing.

The criteria for the allowable deviation from exact CSL misorientations were examined. The criterion proposed by Brandon (Ref 19) has been developed in the special case of a boundary plane perpendicular to a ⟨100⟩ misorientation axis by considering the boundary periodicity only. Warrington and Grimmer suggested a criterion by considering the mean value of boundary periodicity and dsc Burgers vector with respect to the reciprocal volume density of coincident sites. For the estimation of the maximum deviation angle for low-angle boundaries (θ_0), an evidence from high-resolution electron

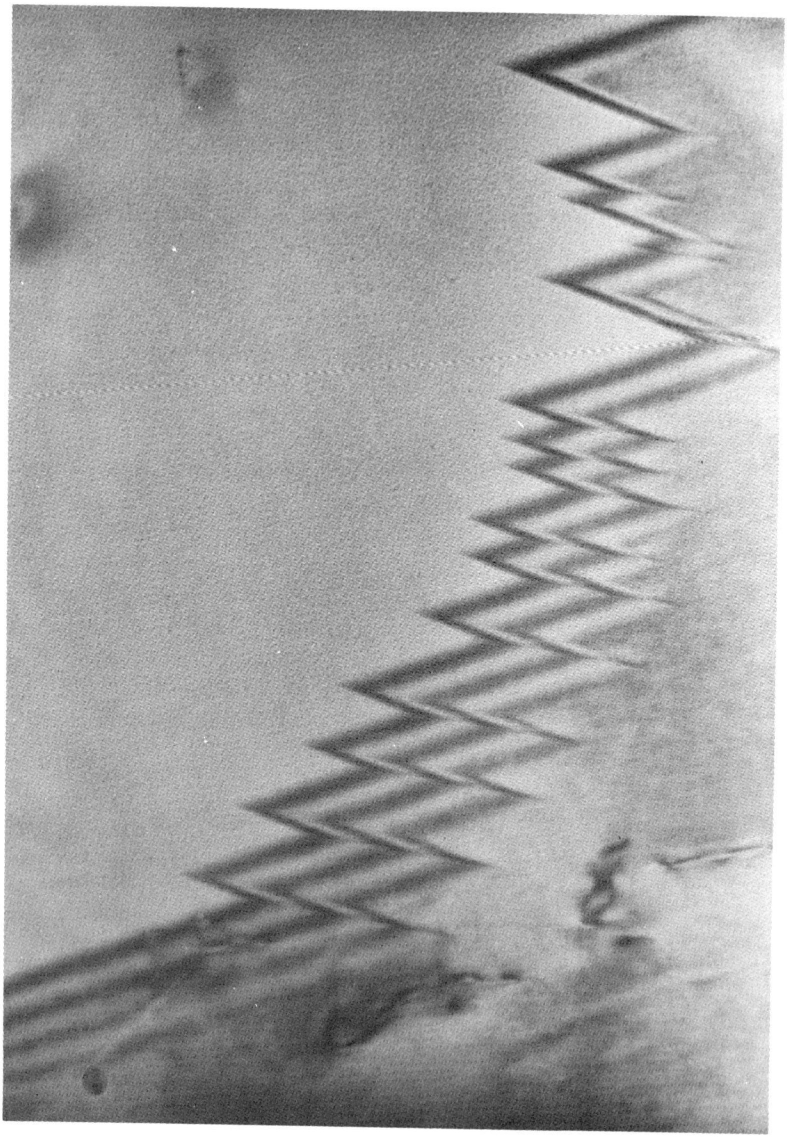

Fig. 2 A $\Sigma = 3$ grain boundary showing steps. 34,000×

microscopy showed the same value of 15° as for θ_0, which was suggested by Read and Shockley (Ref 16).

In most studies the orientation of the grain-boundary plane is not determined. Boundaries with the same Σ but different grain-boundary-plane orientation may behave differently. In the present work the authors tried to analyze possible favored locations for the grain-boundary plane for a given Σ.

In the authors' study of the distribution of grain-boundary-plane orientation, boundary normals were oriented in the plane of high density of coincident sites for low-Σ boundaries up to $\Sigma = 9$, and they deviated from planes with a high density of coincident sites for $\Sigma = 11$ or higher. This result implies that most of the grain-boundary-plane with low Σ values would be oriented in special planes since the driving force for energetically stable geometries of grain boundaries is high. However, the driving force for high-Σ boundaries is small during polycrystal thermal processing such as solidification, recrystallization, or grain growth, and the grain-boundary plane remains of low planar density of coincident sites.

The authors' work suggests the value $\Sigma = 11$ as a cutoff for the grain-boundary-plane orientation along a favored plane in the CSL. This cutoff value is possibly dependent on the details of materials processing and should not be taken as generally valid. The grain-boundary-plane orientation is an extremely important parameter in determining the grain-boundary behavior. It will, therefore, be very interesting to test the validity of the cutoff value suggested here for different processing conditions and in different materials. Particularly, it will be interesting if this cutoff changes upon the microalloying of Ni_3Al with boron. Most studies on the statistical distribution of grain-boundary types in terms of Σ from polycrystals have considered grain boundaries with relatively high Σ boundaries to be random boundaries after $\Sigma = 29$ or $\Sigma = 49$ without detailed investigation of grain-boundary-plane orientations. On the other hand, the authors' investigation of grain-boundary-plane orientation showed that the threshold value for random boundaries is about $\Sigma = 11$, which is considerably lower than 29 or 49. However, as mentioned above, the statistical distribution of grain-boundary types is sensitive to the thermal history of the sample, and one should be very careful when speculating about the possible relationship between the statistical distribution of grain-boundary types and mechanical properties.

Conclusions

A statistical distribution of grain-boundary types for Ni_3Al was obtained according to three different criteria for the maximum allowable deviation from the exact CSL misorientation. The result showed that the choice of criteria changed the Σ distribution of the material. There seemed to be no criterion applicable to all possible geometries of grain boundaries. From the statistical investigation of grain-boundary normals, the trend of having high coincident site density planes was maintained up to $\Sigma = 9$, and grain boundaries above $\Sigma = 11$ were oriented randomly regardless of Σ values.

Acknowledgment

This work was partly supported by the Department of Energy, Energy Conversion and Utilization Technologies program, under subcontract 19X-89768V with Martin Marietta Energy Systems Inc. and was monitored by Oak Ridge National Laboratories.

References

1. K. Aoki and O. Izumi, *Nippon Kinzaku Gakkaishi*, Vol 43, 1979, p 1190
2. C.T. Liu, C.L. White, and J.A. Horton, *Acta Metall.*, Vol 33, 1985, p 213
3. E.M. Schulson, T.P. Weihs, I. Baker, H.J. Frost, and J.A. Horton, *Acta Metall.*, Vol 34, 1986, p 1935
4. S.P. Chen, D.J. Srolovitz, and A.F. Voter, *J. Mater. Res.*, Vol 4 (No. 1), 1989, p 62
5. S.P. Chen, A.F. Voter, and D.J. Srolovitz, *Scr. Metall.*, Vol 20, 1986, p 1389
6. J.J. Kruisman, V. Vitek, and T.M. De Hosson, *Acta Metall.*, Vol 36, 1988, p 2729
7. D. Farkas and V. Rangarajan, *Acta Metall.*, Vol 35, 1987, p 353
8. D. Farkas and H. Jang, *Phys. Rev. B*, Vol 39, 1989, p 11769
9. R.A.D. Mackenzie, M.D. Vaudin, and S.L. Sass, *Proc. MRS Symp.*, Vol 122, 1988, p 461
10. F.D. Tichelaar and F.W. Schapink, *Philos. Mag.*, Vol A54, 1986, p L55
11. S. Hanada, T. Ogura, S. Watanabe, O. Izumi, and T. Masumoto, *Acta Metall.*, Vol 34, 1986, p 13
12. L.C. Lim and R. Raj, *Acta Metall.*, Vol 32, 1984, p 1177
13. D. Farkas, M.O. Lewus, and V. Rangarajan, *Scr. Metall.*, Vol 22, 1988, p 1195
14. R.A.D. Mackenzie, M.D. Vaudin, and S.L. Sass, *Proc. MRS Symp.*, Vol 122, 1988, p 461
15. H. Lin and D.P. Pope, Paper presented at MRS Spring Meeting, San Francisco, CA, Materials Research Society, 1990
16. W.T. Read and W. Shockley, *Phys. Rev.*, Vol 78, 1950, p 275
17. D.H. Warrington and H. Grimmer, *Philos. Mag.*, Vol 30, 1974, p 461
18. Y. Ishida and M. Mclean, *Philos. Mag.*, Vol 27, 1973, p 1125
19. D.G. Brandon, *Acta Metall.*, Vol 14, 1966, p 1479
20. H. Ichinose and Y. Ishida, *J. Phys.*, Vol C4, 1985, p 39
21. H. Jang, Ph.D. thesis, Virginia Polytechnic Institute and State University, 1990
22. D. Farkas, *Proc. MRS Symp.*, Vol 133, 1989, p 137

Section III:

Interfaces in Oxides and Ceramics

7

Control of the Structure and Mechanical Properties of Metal-Ceramic Interfaces through Interfacial Reactions

F.-S. Shieu
The Dow Chemical Company
Midland, MI 48667
R. Raj and S.L. Sass
Department of Materials Science and Engineering
Cornell University, Ithaca, NY 14850-1501

The use of interfacial reactions to control the structure and shear strength of metal-ceramic interfaces was studied in the NiO-Pt system. Interfaces were formed by hot pressing together thin NiO single crystals and thick Pt polycrystalline films. The suitable choice of annealing temperature, time, and oxygen partial pressure allowed the introduction at the interface of a layer of either an intermetallic compound NiPt with thickness between 1 and 65 nm or a Ni-Pt solid solution. The shear strength of the NiO-Pt interface with and without the different interlayers present was measured by the periodic-cracking method. Compared to the originally hot-pressed state, the shear strength of the NiO-Pt interface was increased by a factor of at least four by the presence of the NiPt layer or by the solid solution. The use of interfacial reactions to control interfacial strength may also be applicable to other metal-ceramic systems in which the metal and the cation form intermetallic compounds and in which the oxidation potentials of the metal and the cation are significantly different.

The interest in the structure and mechanical properties of metal-ceramic interfaces is reflected in several papers on this topic in recent conference proceedings (Ref 1-3). The issues being studied are four in number: the dislocation and/or atomic structure of the interface, for example, Ruhle, Burger, and Mader (Ref 4), Fecht and Gleiter (Ref 5), and Shieu and Sass (Ref 6); the mechanical properties of the interface itself, for example, Marshall, Cox, and Evans (Ref 7), and Agrawal and Raj (Ref 8); the thermodynamics of interfacial reactions, for example, Klomp (Ref 9); and the influence of interfacial reactions on mechanical strength, for example, Pask (Ref 10).

In addition to the work on the interfaces themselves, theoretical and experimental papers have reported on the relationship between interface strength and the toughness of metal-ceramic composites (Ref 11, 12). Ashby, Blunt, and Bannister have recently shown that the strength of the interface can significantly change the local stress-to-displacement curve for deformation in the metal, thereby having a marked influence on the toughness of the composite (Ref 13). These studies make it clear that the control of interface strength is a key element in the design of metal-ceramic composites.

This chapter gives examples of the effect of reactions at a NiO-Pt interface on the shear strength of the interface. These reactions, which produce either interlayers of the intermetallic compound NiPt with different thicknesses or a Ni-Pt solid solution, are controlled through the choice of annealing temperature, time, and oxygen activity. It is found that both reactions significantly enhance the shear strength of the original interface produced by hot pressing at high temperature in air.

Experimental Procedure

Thin single crystals of NiO with various thicknesses were grown on the (001) surface of cleaved MgO crystals by the thermal decomposition of $NiBr_2$ in the presence of water vapor (Ref 6, 14). The metal-ceramic interface was produced by hot pressing together the thin NiO film, still mounted on its MgO substrate, and a polycrystalline Pt plate approximately 10 × 30 × 1 mm. The bonding was carried out in air at 1200 °C for 10 h at a pressure of 0.5 to 1 MPa. The specimen geometry is shown in Fig. 1, where the Pt plate is sandwiched by two MgO substrates 1 mm thick, one of which is covered with the thin NiO film. This configuration was used to eliminate bending that can result during

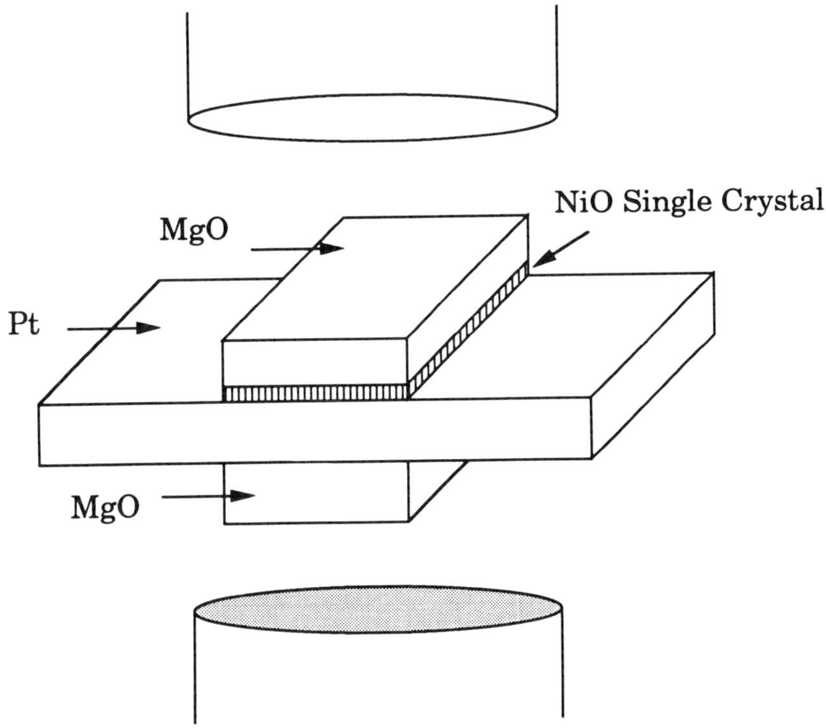

Fig. 1 Hot-pressing configuration of NiO-Pt specimen for mechanical testing

cooling due to a difference in the thermal-expansion coefficients of NiO and Pt. Pressure was applied after reaching the hot-pressing temperature and released 30 min before cool-down. The MgO was then dissolved in a 15% H_2SO_4-H_2O solution at a temperature of 70 °C, leaving the Pt plate and a layer of NiO with a thickness between 3 and 7 μm. The dissolution process usually resulted in the separation of some part of the thin NiO film from the Pt plate. The separated NiO film was used to advantage to measure the film's thickness. It should be pointed out that during hot pressing, MgO dissolves in the NiO, and a NiO-MgO solid solution may be present.

The Pt plate with the thin NiO film on it was then deformed in tension using the approach of Agrawal and Raj (Ref 8) to determine the ultimate shear strength of the NiO-Pt interfaces. The specimen was intermittently removed from the Instron machine and examined in a scanning electron microscope (SEM) to measure the development of cracks in the NiO film.

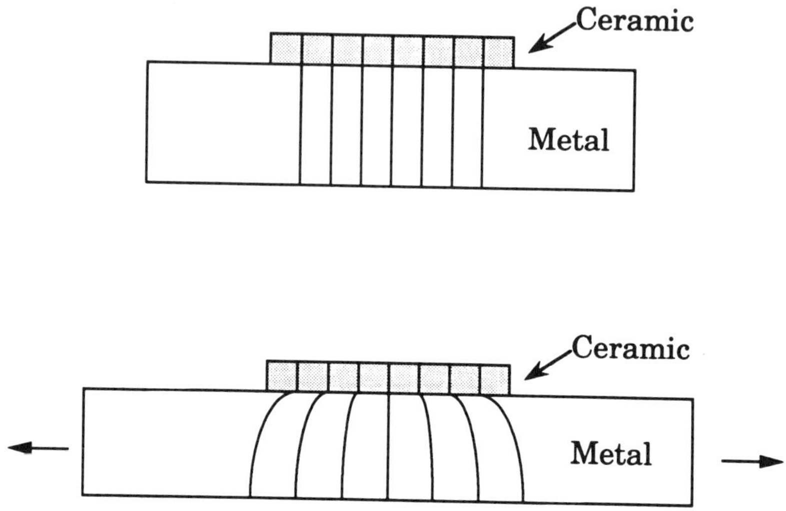

Fig. 2 Load transfer between metal and ceramic via shear lag of axial displacement in loading direction

According to the Ni-Pt phase diagram (Ref 15), an intermetallic compound, NiPt, forms at temperatures below 645 °C. In the case of the NiO-Pt specimens, NiPt might form at the interface between the metal and the ceramic. After the initial hot pressing to form the metal-ceramic interface, some of the specimens were annealed at 500 °C, in either air or a 0.1% CO-CO_2 mixture, which corresponds to an oxygen partial pressure of 10^{-23} atm. In addition, one specimen was annealed at 420 °C in a 0.1% CO-CO_2 mixture, which corresponds to an oxygen partial pressure of 10^{-28} atm, to examine the feasibility of lowering the annealing temperature; a lower temperature might be very useful for processing either composites or metal layers on ceramic substrates in electronic packaging applications. In all cases of annealing at 500 °C and below, the formation of a thin layer of the NiPt intermetallic compound at the interface was observed. According to the Ni-Pt-O_2 phase diagram (Ref 16), annealing the NiO-Pt specimen at 1200 °C in an oxygen partial pressure of 10^{-5} atm is expected to produce a 20 at.% solid solution of Ni in Pt in the vicinity of the interface. All interface microstructures were studied by transmission electron microscopy (TEM).

Analysis for Maximum Shear Stress at or near the Interface

With regard to the specimen geometry in this study, when a load is applied to the Pt plate, shear stresses develop in the vicinity of the interface due to the

Table 1 The integration constant, k, in Eq 2 for different shear stress distributions

Reference	k
19	6
20	2
8	π
Present work	π

difference in axial displacement between NiO and Pt (Fig. 2). As the strain in the Pt plate increases, periodic cracks develop in the thin NiO film, dividing it into many parallel strips. When the cracking ceases, the maximum in-plane stress, which occurs at the center of the strip, is equal to the fracture stress of the NiO film. The in-plane stress is assumed to be uniform through the thickness of the NiO film. Since the NiO single crystal is brittle, it can be assumed to deform elastically. The fracture stress can then be calculated from Hooke's law using Young's modulus of NiO and the fracture strain obtained from the crack density-versus-strain curve (see the section "Experimental Results" in this chapter). If the normal tensile stress at the interface near the edge of the strip does not cause delamination, the form of the in-plane stress that can cause fracture in the NiO film is coupled to the shear stress at or near the interface through an integral equation:

$$\sigma(x) = \frac{1}{\delta}\int_0^x \tau(x)dx \qquad \text{(Eq 1)}$$

where σ is the in-plane tensile stress in the NiO film; δ is the thickness of the NiO film; τ is the shear stress at the NiO-Pt interface, and x is the interface coordinate that originates at the crack. From this force equilibrium equation, the maximum shear stress τ_{max}, at the NiO-Pt interface is written as:

$$\tau_{max} = \frac{k\,\delta\,\sigma_f}{\lambda_{max}} \qquad \text{(Eq 2)}$$

In Eq 2, $\lambda_{max}/2$ is the distance from the edge of a crack in the ceramic film over which the in-plane stress in the film and the shear stress at the interface are constrained by Eq 1, such that the in-plane stress, $\sigma(x)$, increases monotonically up to the fracture stress of the film. Therefore, $\lambda_{max}/2$ is the closest possible spacing of the cracks if it is assumed that the strength of the film is single valued. Similarly, the greatest spacing between any two cracks will be less than λ_{max} because the maximum in-plane stress between two cracks that

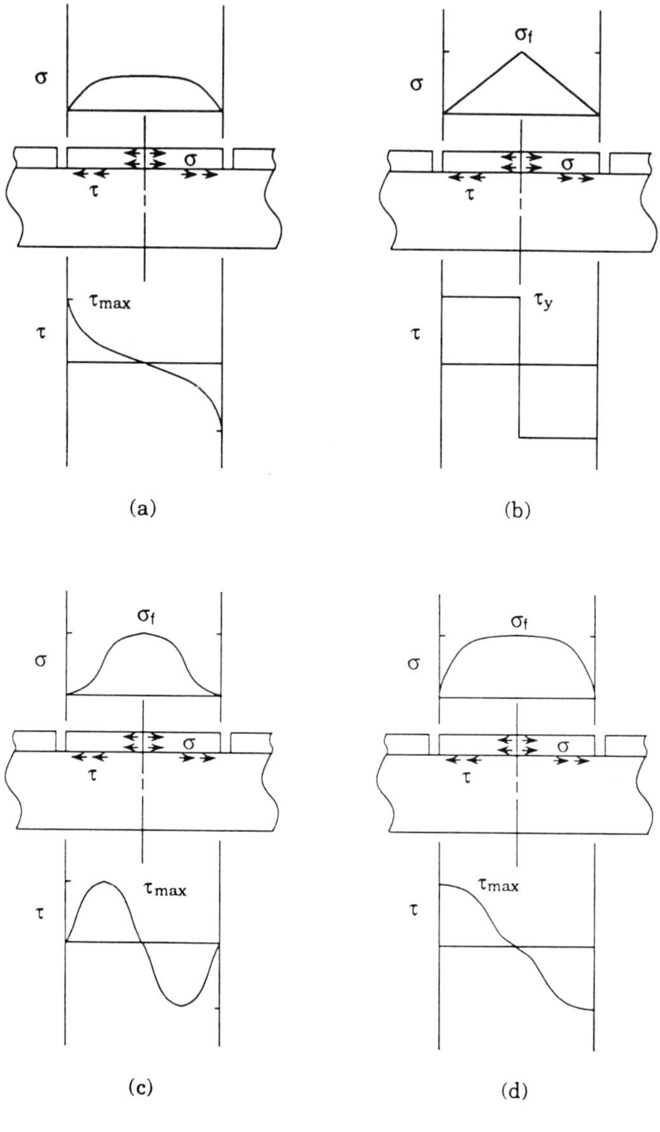

Fig. 3 Various models of shear stress distribution along the metal-ceramic interface and the resultant tensile stress in the ceramic strip. Source: (a) Ref 19. (b) Ref 20. (c) Ref 8. (d) Present work

are less than λ_{max} apart will fall below the fracture stress. Thus the maximum and minimum spacing between the cracks will vary from $\lambda_{max}/2$ to λ_{max}.

The factor k in Eq 2 depends on the shape of the shear stress distribution at the interface adjacent to a crack in the film. Cox (Ref 17) and Dow (Ref 18) used a proportional shear lag analysis to calculate the shear stress at the interface and obtained a much smaller stress, especially near the strip edge, than that measured experimentally using the photoelastic method employed by Tyson and Davies (Ref 19). Kelly (Ref 20) discussed the situation in which the matrix can undergo plastic flow, as in a metal, and assumed that the maximum value of shear stress at the interface is limited to the shear yield stress of the metal (τ_y).

Figure 3 summarizes and compares the various shear stress distributions and the resultant in-plane stresses in the ceramic film. Table 1 gives the values of k calculated for the different distributions in Fig. 3, where, for the sake of simplicity, a parabolic form is assumed for the case of Tyson and Davies (Ref 19). The choice of different k values can change the measured τ_{max} by as much as a factor of three. In the present work, a shear stress distribution in the form of a sine wave, which gives τ_{max} near the strip edge, was chosen, with the singularity at the free surface of the edge of the crack being neglected. This led to $k = \pi$, the same as the result of Agrawal and Raj (Ref 8). Since the aim of the present study was to explore the feasibility of tailoring the mechanical properties of metal-ceramic interfaces by interfacial reactions, the absolute value of the ultimate shear stress at the interface was not important. Rather, the point of interest was the change in the ultimate shear stress produced by the introduction of different interlayers.

Experimental Results

Determination of the Fracture Stress of NiO. The periodic cracks obtained in the specimen hot pressed at 1200 °C in air at 3.5% strain are shown in Fig. 4(a). A plot of crack density versus strain, given in Fig. 4(b), can be considered the superposition of two different curves. In stage I, which shows sigmoidal behavior, no crack was observed for strains of less than 1%, but crack density increased rapidly for strains between 1 and 2%, as predicted by Agrawal and Raj (Ref 8). In stage II, crack density increased gradually with strain, a behavior that could be due to the triaxial state of stress in the NiO thin film resulting from the surface steps (slip bands or lines) in the plastically deformed Pt plate.

The fracture strain (ε_f) for NiO is taken to be 1.5 ± 0.5%, which gives a fracture stress $\sigma_f = E_{NiO}\,\varepsilon_f = 1.44$ GPa, where E_{NiO}, the Young's modulus of NiO, is 95.8 GPa (Ref 21). This value of the fracture strain of NiO is slightly higher

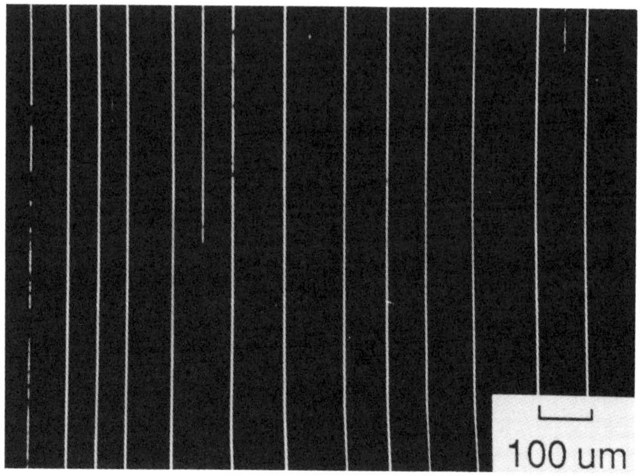

a

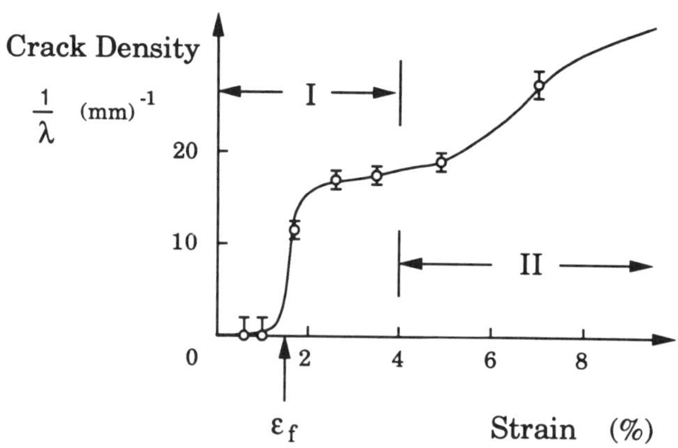

b

Fig. 4 (a) Periodic cracks observed in a "nonreactive" NiO-Pt interface for 3.5% strain in the Pt plate. (b) The variation in the crack density in the NiO film with the strain applied to the Pt plate

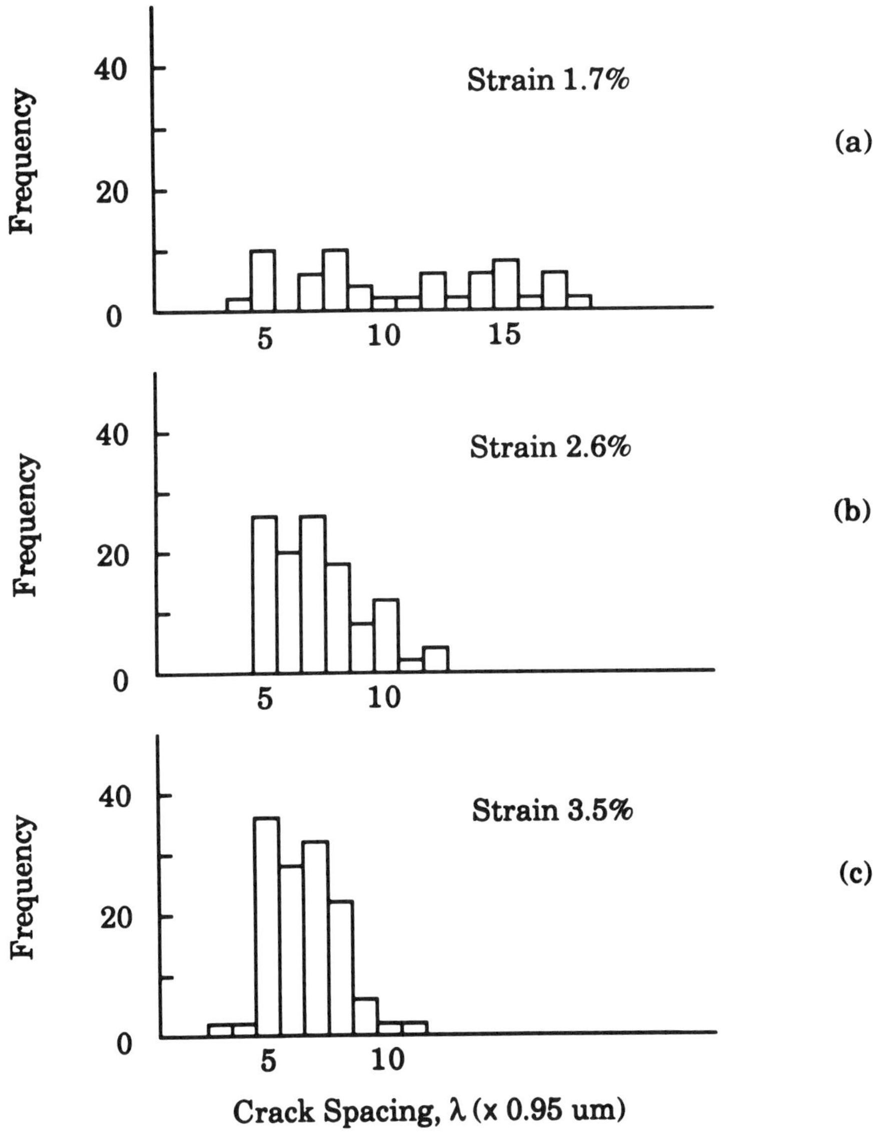

Fig. 5 The distributions of crack spacings (λ) obtained in the Pt plate for tensile strains of (a) 1.7%, (b) 2.6%, and (c) 3.5%

than that obtained by Evans, Rajdev, and Douglass (Ref 22). This is reasonable because the measurements of interest were made on thin films of NiO, which are expected to have a higher fracture stress and strain than do thick specimens.

The distributions of the crack spacings for strains near ε_f are given in Fig. 5. At ε_f the crack spacing spans a large range, from 38 to 171 μm. For strains above ε_f, the average crack spacing decreases, as does the range, which is 48 to 105 μm, for a strain of 3.5%.

Ultimate Shear Stress of the Original Hot-Pressed NiO-Pt Interfaces. The maximum crack spacing in Fig. 4(a) and 5(c), $\lambda_{max} = 105 \pm 5$ mm, is approximately twice the minimum spacing, $\lambda_{min} = 48 \pm 5$ μm, as predicted by Agrawal and Raj (Ref 8). By substituting $\sigma_f = 1.44$ GPa, $\lambda_{max} = 105$ μm, and $\delta = 5.5$ mm in Eq 2 with $k = \pi$, the ultimate shear stress was found to be 236 MPa. For three other specimens having the same processing conditions, the ultimate shear stress was determined to be between 224 and 446 MPa. The average value of all determinations is 335 MPa.

Ultimate Shear Stress of NiO-Pt Interfaces with Interlayers. Annealing at a temperature of 500 °C and below, in either air or a 0.1% CO-CO_2 mixture, gives a layer of NiPt intermetallic compound at the interface. Figure 6(a) shows a specimen that was annealed in a 0.1% CO-CO_2 mixture at 500 °C for 36 h, giving a NiPt layer approximately 20 nm thick. Close examination shows subtle contrast features near the Pt-NiPt and NiPt-NiO interfaces. Lines that are parallel to these interfaces are displaced approximately 5 nm into both the NiO and Pt (indicated by arrows). The contrast of these lines changes with defocus, suggesting that they are interfaces with a mean inner potential different from that of the neighboring regions (Ref 23, 24). Figure 6(b) shows the selected-area diffraction pattern of the region containing the NiPt layer, where weak reflections from the NiPt compound can be identified. Figure 6(c) is a diagram of the diffraction pattern. The superlattice reflections lack the expected twofold symmetry of the tetragonal structure of NiPt (Ref 25), probably because some of them are too weak to be detected.

The interlayer produced by annealing for 6 h at 500 °C in air seen in Fig. 7(a) is 1 nm thick. Identifying the atomic structure of such a thin layer is difficult. Further annealing under these conditions produced a layer that was clearly NiPt, and it is believed that the 1-nm-thick layer in Fig. 7(a) has the same structure. An approximately 2-nm-thick NiPt layer was produced at the NiO-Pt interface for the specimen annealed in 0.1% CO-CO_2 mixture at 420 °C for 6 h. For a specimen annealed in a 0.1% CO-CO_2 mixture at 1200 °C for 4 h, in which the formation of a solid solution of Ni in Pt is expected near the a

b

c

Fig. 6 (a) The bright-field image of the approximately 20 nm-thick NiPt layer at the NiO-Pt interface after annealing in a 0.1% CO-CO_2 mixture at 500 °C for 36 h. (b) The selected-area diffraction pattern obtained from the interface region, showing reflections from the NiPt compound. (c) Schematic diagram of (b). Solid diamond symbol represents reflection from NiPt interlayer; solid circle, Pt crystal; open circle, (001) NiO crystal; open circle with ×, transmitted beam.

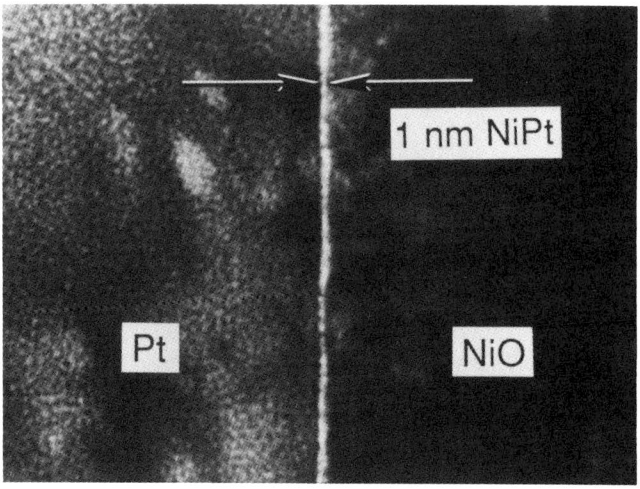

a

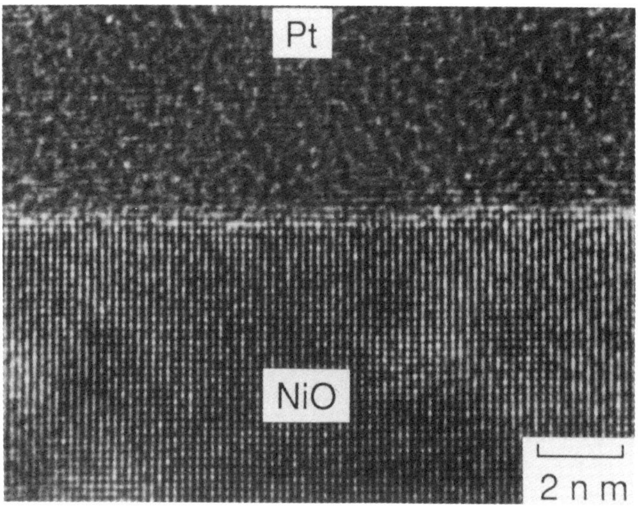

b

Fig. 7 (a) A bright-field image showing a NiPt layer approximately 1 nm thick produced by annealing 6 h at 500 °C in air. The image contrast changes with defocus. (b) The lattice image of the NiO-Pt(Ni) solid solution interface showing a "clean" interface structure, without an interlayer

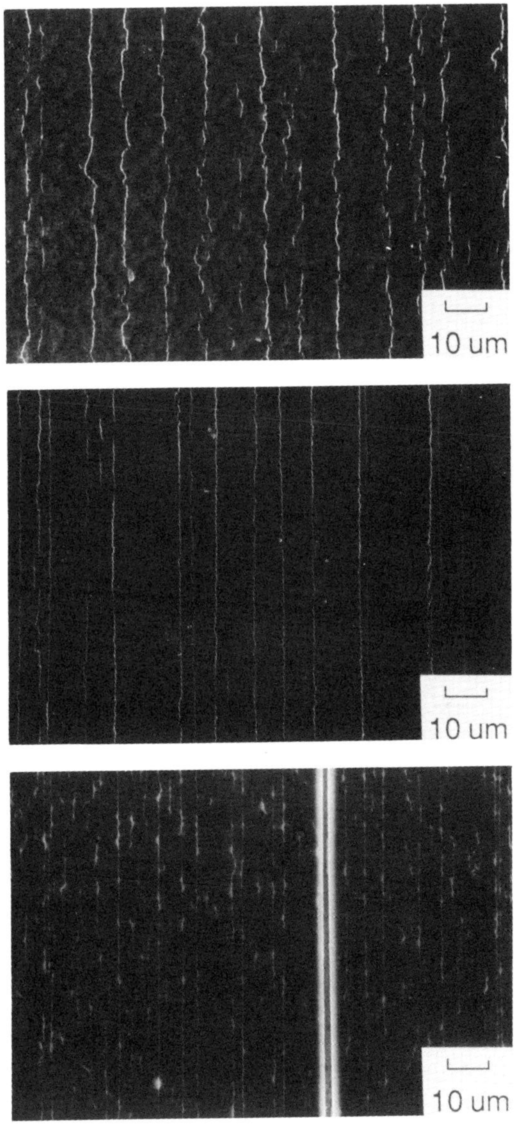

a

b

c

Fig. 8 Periodic cracks observed in the NiO films for different annealing conditions. (a) 500 °C in air, giving a NiPt interlayer approximately 1 nm thick. (b) 500 °C in a 0.1% CO-CO_2 mixture, giving a NiPt interlayer 10 to 65 nm thick. (c) 1200 °C in a 0.1% CO-CO_2 mixture, giving a Ni-Pt solid solution near the interface. Annealing at 420 °C in a 0.1% CO-CO_2 mixture gives results similar to those in (b).

Table 2 Ultimate shear stress of NiO-Pt interfaces

Processing	Number of specimens	Annealing time, h	NiPt thickness, nm	τ_{max}, MPa
Hot pressing				
1200 °C	4	...	...	224-446
500 °C in air	3	6	1.0	1080-1740
500 °C, $P_{O_2} = 10^{-23}$ atm	2	12-36	10-65(a)	1360-2040
420 °C, $P_{O_2} = 10^{-28}$ atm	1	6	2.0	1450-1910
1200 °C, $P_{O_2} = 10^{-5}$ atm	1	4	...	2100-4460

(a) These thicknesses correspond to the 36-h anneal.

interface, a slightly wavy interface structure was observed (Fig. 7b), with no evidence of a NiPt interlayer.

Periodic cracks in specimens with a NiPt layer at the interface are shown in Fig. 8(a and b) for different annealing conditions. In contrast to the cracks in Fig. 4(a) for the specimen only hot pressed at 1200 °C in air, the average crack spacing is smaller when NiPt is present, and two types of cracks were observed, wavy fine and straight coarse. The coarse cracks are similar both in spacing and morphology to those in Fig. 4(a). Their origin is discussed in the section "Comparison with SiO_2-Cu Results" in this chapter. Using the measured maximum spacing of the wavy cracks and the NiO film thickness, the ultimate shear stress for specimens annealed in air at 500 °C was calculated by Eq 2, with $k = \pi$, to be between 1.08 and 1.74 GPa, with an average value of 1.41 GPa. This is approximately four times larger than the value for the original NiO-Pt interface without NiPt. For specimens annealed in a 0.1% CO-CO_2 mixture at 500 or 420 °C, the ultimate shear stress ranged from 1.36 to 2.04 GPa and 1.45 to 1.91 GPa, with an average value of 1.70 and 1.68 GPa, respectively. For a specimen annealed in a 0.1% CO-CO_2 mixture at 1200 °C (Fig. 8c), conditions that are expected to produce a solid solution of 20 at.% Ni in Pt (Ref 16), the ultimate shear stress was found to be between 2.10 and 4.46 GPa. Table 2 summarizes all ultimate shear stresses for the different heat treatments.

Discussion

The Magnitude of τ_{max}. The ultimate shear stress that is estimated from the periodic-cracking method is most likely a measure of the maximum shear stress supported by the interface when the thin ceramic film fractures. Rigorously, this maximum stress (τ_{max}) is a lower bound for the ultimate shear stress of the interface, which may actually be greater than τ_{max} if yielding occurs away from the interface. It was found that with the NiPt interlayer present, τ_{max} increased by a factor of approximately four. This is believed to reflect a true increase in the yield strength of the interface because if yielding had taken place in the metal away from the interface, the intermetallic layer would not have produced a fourfold increase in τ_{max}.

The bulk yield stress (τ_y) of Pt is 6.9 MPa (Ref 26), and the ultimate tensile strength is 145 MPa (Ref 26). For the unreacted NiO-Pt interface, when the Pt plate was deformed by 3.5%, τ_{max} was greater than the tensile strength of Pt. When there was a Ni-Pt solid solution at the interface, the local stress (τ_{max}) even reached the fracture stress of the NiO, 1.44 GPa, as measured by the authors. It is still not known why the measured τ_{max} of the interface is so much greater than the ultimate tensile strength of the bulk Pt. This behavior may be related, at least in part, to the constraint in the Pt near the interface (Ref 13). At present little is known about deformation mechanisms in the vicinity of metal-ceramic interfaces. It is interesting to note that a similar difference between the bulk yield stress and ultimate shear stress of the interface was reported for single-filament tests on graphite fibers in an epoxy matrix (Ref 27). The measured τ_{max} may reflect the theoretical strength of the interface, as determined by the strength of the Ni-O, Pt-Pt, or Ni-Pt bonds, in the same sense that the ideal shear strength of crystals depends upon intrinsic bonding between atoms.

Comparison with SiO_2-Cu Results. It is worthwhile to compare some of the characteristics of NiO-Pt and SiO_2-Cu (Ref 8) interfaces. The single crystal of NiO was bonded to the Pt plate in an orientation such that the loading direction was always along the ⟨100⟩ direction of NiO. The cleavage plane in NiO is {100}. The crack direction in the thin NiO film is, therefore, along the ⟨100⟩ direction orthogonal to the loading direction. Slip lines or bands were observed in the Pt plate and in some areas of the thin NiO film when the strain was greater than 4%. The slip lines or bands were always along the ⟨110⟩ direction of NiO, which is 45° away from the crack direction, a result of the slip system in NaCl-type structures, which is {110} ⟨110⟩. In contrast, although the Cu underneath was deformed to 25%, no slip lines or bands were observed in the SiO_2 since SiO_2 is amorphous. Large amounts of deformation produced shear bands in the Cu substrate along a direction 45° to the tensile axis, which caused some spalling of the SiO_2 film from the substrate. The difference in this behavior for the two systems most likely explains the differences in the crack

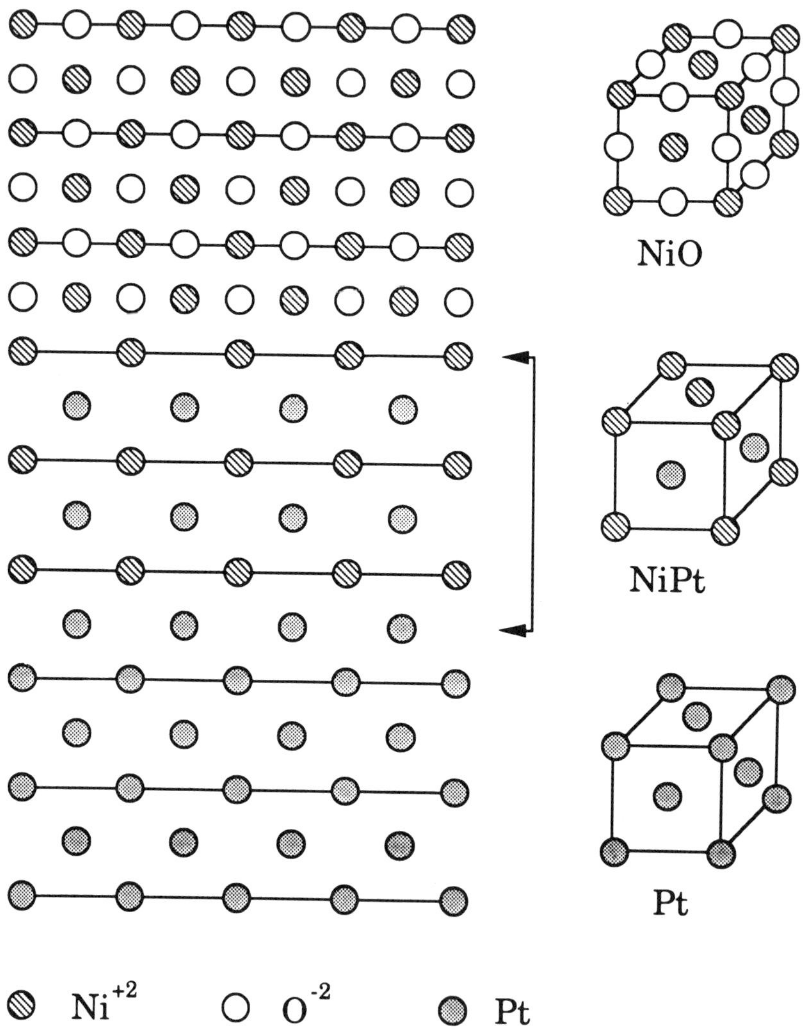

Fig. 9 A cross-sectional view of the atomic structure of an exact (001) epitaxial boundary between NiO and Pt with a layer of NiPt intermetallic compound at the interface. The atomic spacings in the drawing are not to scale.

density-versus-elongation curves for strains much above ε_f. The crack density in the thin NiO film was influenced by the heavily plastically deformed metal substrate, whereas this was not the case with the SiO_2 film.

Cracking Due to Thermal Stresses. Two types of cracks occurred when the NiPt compound was present. Probably the straight cracks initiated in the NiO single crystal because of the presence of surface defects such as etch pits and edge irregularities. When the specimen was loaded in tension, it is likely that stress concentrations from defects in the NiO film gave premature cracks, instead of cracks that originated at the interface, where the maximum shear stress is present. Some coarse cracks may also have come from thermal stresses (Ref 28) since this type of crack was seen after the MgO substrate was dissolved, leaving the thin NiO film bonded to the Pt plate, and before mechanical testing. The wavy cracks were evidently due to the presence of the NiPt layer. NiPt is believed to be brittle, like most intermetallic compounds. Its fracture behavior is expected to be different from that of single-crystal NiO, which exhibits well-defined {100} cleavage planes, resulting in sharp, straight cracks.

Effect of the Thickness of the Intermetallic Layer. The observations reported here on the NiO-Pt system demonstrate that the shear strength of metal-ceramic interfaces can be controlled by interfacial reactions to produce layers with intermediate compositions. Annealing the NiO-Pt system in air at 500 °C for 6 h gives a layer approximately 1 nm thick that is believed to be the NiPt compound. When the annealing time was increased, a thicker layer of NiPt was obtained (unpublished research). Annealing in a 0.1% CO-CO_2 mixture for 36 h at 500 °C produced a thicker layer of NiPt. It is evident that decreasing the oxygen partial pressure increases the growth rate of the NiPt layer, probably because the activity of the Ni in the NiO is increased by the reducing conditions of the low P_{O_2}. The annealing temperature can, therefore, be lowered by reducing the partial pressure of oxygen, as illustrated by the formation of the NiPt layer at 420 °C.

Studies on steel wire/aluminum composites by Friedrich, Pompe, and Kopjov (Ref 29) showed that a particular thickness of intermetallic layer was needed to produce optimal mechanical properties. Although the thickness in the study was typically several microns, the optimum value is expected to vary from system to system. Therefore, different thicknesses of NiPt would be expected to give different increases in the ultimate shear stress. This is consistent with the observations in Table 2, although the change is not marked. It will be interesting to examine in more detail the influence of the thickness of the NiPt layer on τ_{max}.

A Simple Model for the Atomic Structure of NiPt Interfaces. Arguments based on the bonding energy between ions or atoms can help to explain the increase in the shear strength of the NiO-Pt interface with the introduction of a NiPt interlayer. From a high-resolution electron microscopy study of CdO-Ag interfaces produced by internal oxidation, Necker and Mader (Ref 30) showed that, for an exact (111) epitaxial boundary, a terminating layer of O^{-2} ions is present at the interface between CdO and Ag. It is reasonable to speculate, therefore, that, for the exact (001) epitaxial boundary of the NiO-Pt interface, Pt atoms at the interface sit above O^{-2} sites. Since the bonding between Pt and O^{-2} involves van de Waals forces, the shear strength of the original NiO-Pt interface is expected to be relatively weak.

The crystal structure of the NiPt compound consists of alternating (002) planes of Ni and Pt atoms, with an ABAB stacking sequence (Ref 25). It is expected that the Ni-O, Pt-Pt, and Ni-Pt bonds that result from the formation of the NiPt compound are stronger than the Pt-O bond. The structures of the interfaces may be like those shown in Fig. 10, where on one side of the NiPt layer there is an atomic plane of Ni facing the NiO crystal, and on the other side, an atomic plane of Pt facing the Pt crystal. This would be favorable since all the Pt-O bonds in the original NiO-Pt interface have now been replaced by either Ni-O or Pt-Pt bonds. It should be noted, however, that this is a simplified explanation of the influence of the NiPt compound on the shear strength of the interface, since other factors, such as the crystallographic orientation and the lattice mismatch (i.e., strain energy), are also expected to play a role. The number of stronger bonds can also be increased by forming a Ni-Pt solid solution at the metal-ceramic interface. Further work is needed to understand the large effect of a layer of NiPt as thin as 1 nm on the ultimate shear strength of the NiO-Pt interface.

Importance of Oxygen Activity. Based on the phase diagram at 1200 °C for the NiO-Pt system (Ref 16), it was predicted that a solid solution of 20 at.% Ni in Pt would form when the specimen was annealed in a 10^{-5} atm partial pressure of oxygen. A tenfold increase in strength was observed for this specimen over that of the nonreactive NiO-Pt interface. Since solid solution hardening of Pt near the interface could change the shear stress distribution along the interface, modifying the k in Eq 2 may be necessary to calculate the ultimate shear stress in this case. Thus the authors hesitate to state that the solid solution gives a stronger interface than the NiPt compound because the increase is only a factor of two. It will be interesting to examine the consequences for τ_{max} of varying the concentration of Ni in Pt by using different oxygen partial pressures.

Remarks and Conclusion

Initially it seems remarkable that the introduction of a 1-nm-thick layer of NiPt at the NiO-Pt interface increases its ultimate shear strength by a factor of at least four. There have been cases, however, in which drastic changes in the mechanical properties were produced by interface modifications at the nanometer level, for example, the ductile-to-brittle transition of grain boundaries in Fe due to the segregation of a fraction of a monolayer of S, P, Sn, or Sb at these boundaries (for example, Ref 31), and the brittle-to-ductile transition of grain boundaries in Ni_3Al due to the segregation of a fraction of a monolayer of boron and changes in stoichiometry at these boundaries (for example, Ref 32 and 33). The role of NiPt in modifying interface properties may, therefore, be another example of interface nanometer-level changes in the local structure and chemistry that have a large influence on properties. Because of the availability of the relatively simple periodic-cracking method for measuring the ultimate shear strength of a particular interface, the NiO-Pt system may represent a useful model for studying the fundamentals of interface property control by the introduction of 1-nm-thick layers.

It is believed that interfacial reactions can also be used to control the mechanical properties of other metal-ceramic systems in which the metal and the cation form intermetallic compounds. A critical requirement for this would seem to be that the metal have a relatively unstable oxide or no stable oxide, that is, that it be noble or nearly so. Thus, the oxidation potentials of the metal and the cation should be very different. In order to obtain an intermetallic compound in a metal-ceramic system, the proper choice of annealing temperature, time, and partial pressure of oxygen must be made. This requires information on phase equilibria in that system, information that is rarely available at present.

Since the thermal-expansion coefficient of NiPt is expected to be between those of NiO and Pt, the intermetallic compound may not only strengthen the interface but also enhance its resistance to thermal shock.

The ultimate shear strength of the NiO-Pt interface can be controlled by reactions that produce layers of either a NiPt intermetallic compound or a solid solution at the interface. The formation and thickness of the intermetallic-compound layer and the thickness and composition of the metallic solid solution layer can be varied by the choice of annealing temperature, time, and oxygen partial pressure.

In order to tailor the mechanical properties of metal-ceramic interfaces, phase equilibrium data are needed to provide guidance in the selection of the appropriate annealing conditions to produce the desired interfacial structure.

The use of interfacial reactions to control interfacial strength may also be applicable to other metal-ceramic systems in which the metal and the cation form intermetallic compounds and in which the oxidation potentials of the metal and the cation are significantly different.

Acknowledgment

This research was supported by the U.S. Office of Naval Research under Contract N00014-88-K-0331. Partial support was received from the Air Force Office of Scientific Research under Contract AFOSR-86-0321. The use of the Electron Microscopy, Materials Preparation and Mechanical Testing Facilities of the Materials Science Center at Cornell University, which is supported by the National Science Foundation, is acknowledged. The phase equilibria information of the NiO-Pt system was kindly provided by Prof. R. Dieckmann. The assistance of B.F. Addis is gratefully appreciated. The early ideas underlying this work arose out of discussions with Prof. J.T. Klomp.

References

1. C.G. Pantano and E.J.H. Chen, Ed., *Tailored Interfaces in Composite Materials*, Materials Research Society, 1989
2. R. Raj and S.L. Sass, Ed., Interface Science and Engineering '87, *J. Phys.*, Tome 49, 1988
3. J.A. Pask and A.G. Evans, Ed., *Surfaces and Interfaces in Ceramic and Ceramic-Metal Systems*, Plenum Press, 1981
4. M. Ruhle, K. Burger, and W. Mader, *J. Microsc. Electron.*, Vol 11, 1986, p 163
5. H.J. Fecht and H. Gleiter, *Acta Metall.*, Vol 33, 1985, p 557
6. F.-S. Shieu and S.L. Sass, *Acta Metall. Mater.*, Vol 38, 1990, p 1653
7. D.B. Marshall, B.N. Cox, and A.G. Evans, *Acta Metall.*, Vol 33, 1985, p 2013
8. D.C. Agrawal and R. Raj, *Acta Metall.*, Vol 37, 1989, p 1265
9. J.T. Klomp, *Fundamentals of Diffusion Bonding*, Y. Ishida, Ed., Elsevier, 1987
10. J.A. Pask, *Am. Ceram. Soc. Bull.*, Vol 66, 1987, p 1587
11. L.S. Sigl, P.A. Mataga, B.J. Dalgleish, R.M. McMeeking, and A.G. Evans, *Acta Metall.*, Vol 36, 1988, p 945
12. M.S. Hu and A.G. Evans, *Acta Metall.*, Vol 37, 1989, p 917
13. M.F. Ashby, F.J. Blunt, and M. Bannister, *Acta Metall.*, Vol 37, 1989, p 1847
14. R.E. Cech and E.I. Alessandrini, *Trans. ASM*, Vol 51, 1959, p 150
15. M. Hansen, *Constitution of Binary Alloys*, McGraw-Hill, 1958
16. R. Dieckmann, *Solid State Ionics,* Vol 45, 1991, p 271
17. H.L. Cox, *Br. J. Appl. Phys.*, Vol 3, 1952, p 72
18. N.F. Dow, Report R 63SD61, General Electric Company, 1963
19. W.R. Tyson and G.J. Davies, *Br. J. Appl. Phys.*, Vol 16, 1965, p 199
20. A. Kelly, *Strong Solids*, Clarendon Press, 1966
21. M.R. Notis, R.M. Spriggs, and W.C. Hahn, Jr., *J. Appl. Phys.*, Vol 44, 1973, p 4165

22. A.G. Evans, D. Rajdev, and D.L. Douglass, *Oxid. Met.* Vol 4, 1972, p 151
23. M. Ruhle and M. Wilkens, *Proc. Fifth Eur. Cong. Electron Microsc.*, Vol 1, 1972, p 416
24. M. Ruhle and S.L. Sass, *Philos. Mag.*, Vol 49, 1984, p 759
25. A. Schneider and U. Esch, *Z. Elektrochem.*, Vol 50, 1944, p 268
26. *Properties and Selection: Nonferrous Alloys and Pure Metals, Metals Handbook*, Vol 2, 9th ed., American Society for Metals, 1979
27. A.N. Netravali, R.B. Heustenburg, S.L. Phoenix, and P. Schwartz, *Polym. Compos.*, Vol 10, 1989, p 226
28. F.-S. Shieu and S.L. Sass, *Acta Metall. Mater.*, Vol 39, 1991, p 539
29. E. Friedrich, W. Pompe, and I.M. Kopjov, *J Mater. Sci.*, Vol 9, 1974, p 1911
30. G. Necker and W. Mader, *Proc. MRS Symp.*, Vol 122, 1988, p 43
31. H.L. Marcus and P.L. Palmberg, *Trans. AIME*, Vol 245, 1969, p 1644
32. K. Aoki and O. Izumi, *Nippon Kinzaku Gakkaishi*, Vol 43, 1979, p 1190
33. C.T. Liu, C.L. White, and J.A. Horton, *Acta Metall.*, Vol 33, 1984, p 213

8

Interfacial Chemistry and Structure in Ceramic Composites

Russell H. Jones, Natalio T. Saenz, and C.H. Schilling
Pacific Northwest Laboratory*
Richland, Washington 99352

The interfacial chemistry and structure of ceramic matrix composites (CMCs) play a major role in the properties of these materials. Fiber-matrix interface chemistries are vitally important in the fracture strength, fracture toughness, and fracture resistance of ceramic composites because they influence fiber loading and fiber pullout. Elevated-temperature properties are also linked to the interfacial characteristics through the chemical stability of the interface in corrosive environments and the creep/pullout behavior of the interface. Physical properties such as electrical and thermal conductivity are also dependent on the interface. Fiber-matrix interfaces ranging in thickness from 1-μm-thick multilayers of amorphous and graphitic carbon to a 1-nm-thick layer of SiO_2 can result from sintering operations for some composite systems. Fibers coated with C, BN, C/B_4C/BN, and Si are also used to produce controlled interface chemistries and structures. Growth interfaces within the matrix resulting from the processing of CMCs can also be crucial to the behavior of these materials. The evaluation of the interfacial chemistry and structure of CMCs requires the

*Operated for the U.S. Department of Energy by Battelle Memorial Institute under Contract DE-AC06-76RLO 1830.

use of a variety of analytical tools, including optical microscopy (reflected and transmission), scanning electron microscopy (SEM), Auger electron spectroscopy (AES), and transmission electron microscopy (TEM) coupled with energy-dispersive x-ray (EDS) analysis.

This chapter reviews the interfacial chemistry and structure of SiC-whisker- and fiber-reinforced Si_3N_4 and SiC/SiC materials. Where possible, correlations with fracture properties and high-temperature stability are made.

Ceramic matrix composites contain a wide range of interfaces, such as reinforcement-matrix and matrix-matrix interfaces. Matrix-matrix interfaces range from grain boundaries in hot-pressed material to growth interfaces in material produced by chemical vapor infiltration (CVI) methods. The chemistry and structure of these interfaces may be tailored for specific properties or may result from processing conditions. Mechanical properties such as fracture and creep and physical properties such as electrical and thermal conductivity can be dependent on the chemistry and structure of these interfaces.

The chemistry and structure of the interface between the matrix and the fibers or whiskers in ceramic composites must assume many functions. The interface must have properties that optimize fracture strength, toughness, and resistance, as well as the slow crack growth, of the composite. In addition, it may be necessary to coat the fibers or whiskers to minimize reaction with the matrix during processing and with the environment during service. The reinforcement of ceramic bodies with high-strength, high-moduli ceramic whiskers or fibers as a means to toughen these materials is under intensive investigation. Current theories suggest that microcracking (Ref 1), crack deflection (Ref 2), and crack bridging (Ref 3, 4) are the predominant toughening mechanisms, although the relative contributions of each mechanism to toughening have not been clearly delineated. However, there is general agreement that the interfaces between the whiskers and the matrix are of prime importance in enhancing toughness. Weakly bonded interfaces significantly enhance the toughness of composites by facilitating crack bridging and whisker pullout during crack propagation, whereas strongly bonded interfaces facilitate whisker fracture, yielding little or no increase in toughness.

Coating the whiskers with an inert substance may be necessary to prevent interfacial reactions between whiskers and matrix during composite fabrication because these reactions are likely to cause strong interfacial bonding. The type of coating material and its thickness may be varied to yield interfacial phases that, in addition to preventing interfacial reactions, optimize the toughness of the composite. For elevated-temperature applications in severe environments, interfacial phases with good high-temperature creep strength and resistance to oxidation and corrosion will be required to minimize slow crack growth and creep.

Matrix-matrix interfaces such as grain boundaries in Si-based ceramic composites generally contain a Si-rich glass phase (Ref 5-11). This glass phase can affect high-temperature slow crack growth and creep properties and the material's interaction with the environment. Growth interfaces such as those produced by CVI have not had their chemistry or structure well characterized, but it is expected that they have nonequilibrium structures and may have trapped impurities from the CVI process. The effect of growth interfaces on properties is also relatively unknown.

Chemistry and Structure of Grain Boundaries in Ceramics

A common feature of many polycrystalline ceramic materials is the presence of a thin, 1- to 10-nm-thick glass phase in the grain boundaries. Evidence of grain-boundary glass phases has been observed in a variety of polyphase, engineering structural ceramics such as silicon nitride, silicon carbide, transformation-toughened zirconia, alumina, Sialon, and multicomponent substrate ceramics (Ref 5-11). Grain-boundary glass phases have been shown to affect a variety of high-temperature properties such as creep deformation, subcritical crack growth, oxidation, and corrosion (Ref 12-15). Similar grain-boundary phases have been observed in the interfaces of SiC-whisker-reinforced Si_3N_4 by Nutt and Phillips (Ref 16) and Schoenlein *et al.* (Ref 17), as shown in Fig. 1. Because of the prevalence of intergranular and interfacial glass phases in ceramic materials, considerable effort has been devoted to understanding the processes that produce these phases.

Clarke (Ref 18, 19) has reviewed both the processes that produce intergranular and interfacial glass phases and the processes that control their distribution. Grain-boundary glass phases form in many ceramics as a result of liquid-phase sintering. Incomplete crystallization of the glass phase in glass ceramics and the formation of glass phases from impurities in other ceramics are also processes by which intergranular and interfacial glass phases form. High-resolution electron microscopy has been used to demonstrate clearly that glass phases are most prevalent at grain-boundary triple points with 1- to 10-nm-thick layers present between individual grains. The noncrystallinity of these phases has been verified by microdiffraction and diffuse dark-field imaging electron microscopy techniques.

Clarke (Ref 18, 19) observed that the thickness of the grain-boundary glass phase is relatively constant and independent of orientation for a given

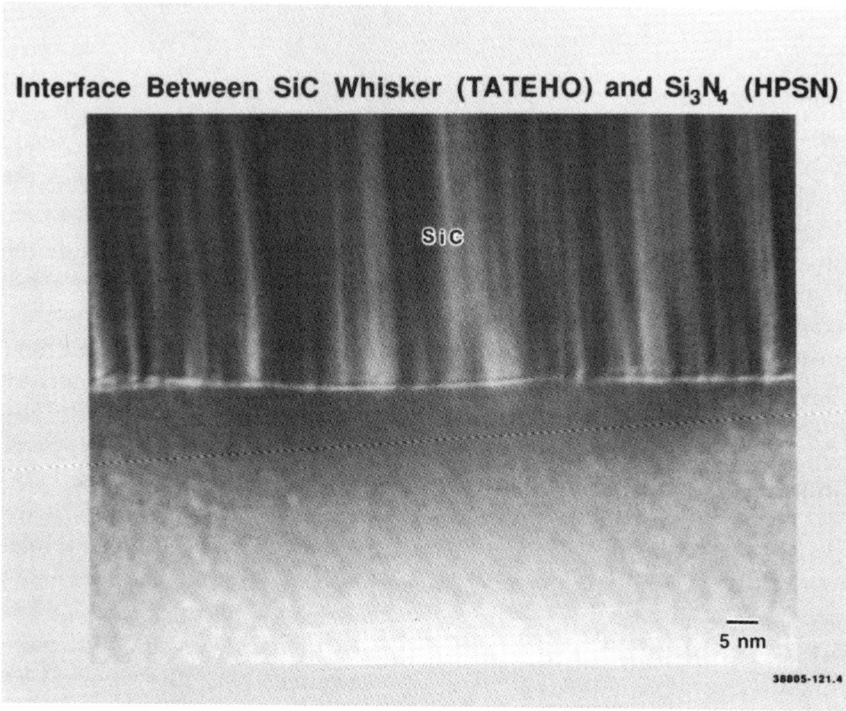

Fig. 1 Interface between SiC whisker (Tateho) and Si_3N_4 (HPSN) matrix

material. Exceptions to this finding include "special" boundaries such as coherent twin boundaries and low-angle boundaries where there is evidence that a glass phase is not present. Also, the thickness of the interfacial phase varies for interfaces between different materials, such as silicon nitride and yttrium-silicon-oxynitride, as opposed to Si_3N_4-Si_3N_4 interfaces. Clarke (Ref 19) calculated the equilibrium thickness of the intergranular glass phase using an interfacial energy and a force balance approach, which predicts that intergranular films with a thickness of 1 to 2 nm are stable. This equilibrium thickness results from the competing interactions of an attractive van der Waals dispersion interaction between adjacent grains and a repulsive force resulting from the intergranular liquid. These calculations also suggested that film thickness is dependent on the dielectric constant of the adjacent grains, thereby causing the different thicknesses of intergranular phase in various materials and at interfaces between different phases in polyphase material (Ref 6).

More progress has been made in determining the thickness and structure (amorphous) of the intergranular phases in ceramics than in establishing their chemical characterization. The thinness of the intergranular phases is primarily responsible for the lack of reliable chemical information since beam spreading within the thin foil during EDS analysis increases the effective volume of material under examination. Therefore, a 1- to 2-nm layer produces a small fraction of the total x-ray emission used for EDS analysis. For this reason, analysis has largely been performed on intergranular phases at triple points rather than on the interface between two grains. However, the consensus is that intergranular phases in silicon nitride materials are highly siliceous (Ref 10). Impurities and elements from the sintering aid collect in the intergranular glass phase, although, for the few reported analyses, the composition is close to that of silica. Lou *et al.* (Ref 6) concluded that the intergranular glass phase in commercial silicon nitride contained Mg but was not a Mg silicate phase. For silicon nitride sintered with Y_2O_3, Bradley and Karasek (Ref 20) determined that the yttria concentration in the glass phase was as high as 40%. They also concluded that the K-phase ($YSiO_2N$) was present in the grain boundaries.

Grain-boundary glass phases significantly affect the high-temperature strength and fracture properties of polycrystalline ceramic materials. Creep and viscous flow of the grain-boundary glassy phase were identified as a limitation to high-temperature creep in the early stages of silicon nitride development. Lange *et al.* (Ref 21) evaluated the compressive creep behavior of silicon nitride with variable Al_2O_3 contents and found that cavitation creep could be interpreted in terms of the volume fraction of the viscous phase. Direct observation of wedge-shaped cavities in silicon nitride by TEM and the effect of a glass phase on the high-temperature properties of Si_3N_4 has focused attention on modification of the properties of the grain-boundary glass phase. Karunaratne and Lewis (Ref 14) reported on efforts to crystallize the grain-boundary phase in Si-Al-O-N ceramics by the controlled removal of metallic impurities such as Mg, Mn, and Ca and the consequent crystallization of the grain-boundary glass phase into $\beta\ Si_3N_4$. Govila *et al.* (Ref 22, 23) also reported that Y_2O_3 additions increase the softening temperature of the grain-boundary glass phase. Optimal properties have been obtained by adding a small percentage of Y_2O_3 and Al_2O_3 to aid in sintering, thereby minimizing the volume fraction of the glass phase and increasing the glass-softening temperature.

The segregation of impurities to the grain boundaries of metals can have a dramatic effect on fracture properties. The primary result of this segregation, which can occur by equilibrium or nonequilibrium processes, is the reduction in intergranular fracture strength. However, there have been very few reports of impurity segregation to grain boundaries of ceramics or of potential effect on properties. This lack of information may stem from the fact that the presence of amorphous glass phases in grain boundaries has been the dominant concern for intergranular properties of ceramics, while low-temperature brittle fracture generally occurs by transgranular cleavage. Therefore, monolayer levels of intergranular impurity segregation have not been a primary concern, although the effect of second-phase precipitation or agglomeration at grain boundaries on magnetic, electrical, and mechanical properties has been well characterized.

Yan, Cannon, and Bowen presented a theoretical framework for solute segregation to ceramic interfaces (Ref 24). They analyzed three components of the interaction potentials between solutes and interfaces in ceramics: an electrostatic interaction, an elastic interaction, and a dipole interaction. The electrostatic interaction is caused by the space charge distribution adjacent to interfaces. Space charges arise from differences in the formation energies of cation and anion vacancies and the consequent differences in their concentration adjacent to interfaces. The elastic interaction is caused by the size misfit between solute and matrix; the dipole interaction results from the dipole moment associated with solute-vacancy complexes. The analysis of Yan, Cannon, and Bowen has demonstrated that the noncoulombic elastic interaction alters the electrostatic potential such that enhanced solute segregation can be predicted. Examples of solute segregation at ceramic interfaces that substantiate the predictions of Yan, Cannon, and Bowen have subsequently been presented.

Impurity segregation to the free surface of sapphire and Al_2O_3 has been reported by Mukhopadhyay *et al.* (Ref 25) and by Baik and White (Ref 26). Mukhopadahyay *et al.* observed Mg and Ca segregation to the surface of Mg-implanted sapphire that was strongly dependent on the O_2 activity in the atmosphere during annealing. This was attributed to the evaporation of MgO in low-O_2-activity atmospheres. Calcium segregation was observed with relative surface concentrations (Ca/Al) as high as 0.7, relative to Al, at a temperature of 1300 °C when Mg segregation did not occur. Magnesium concentrations (Mg/Al) did not exceed 0.05, relative to Al, at 1300 °C. Mukhopadhyay *et al.* also noted that the segregation was anisotropic, occurring predominantly on the $(10\bar{1}0)$ prismatic plane of sapphire, while previous observations of Mg segregation to the basal plane had been made (Ref 27, 28). Baik and White (Ref 26) noted that Ca did not segregate to the basal plane of sapphire.

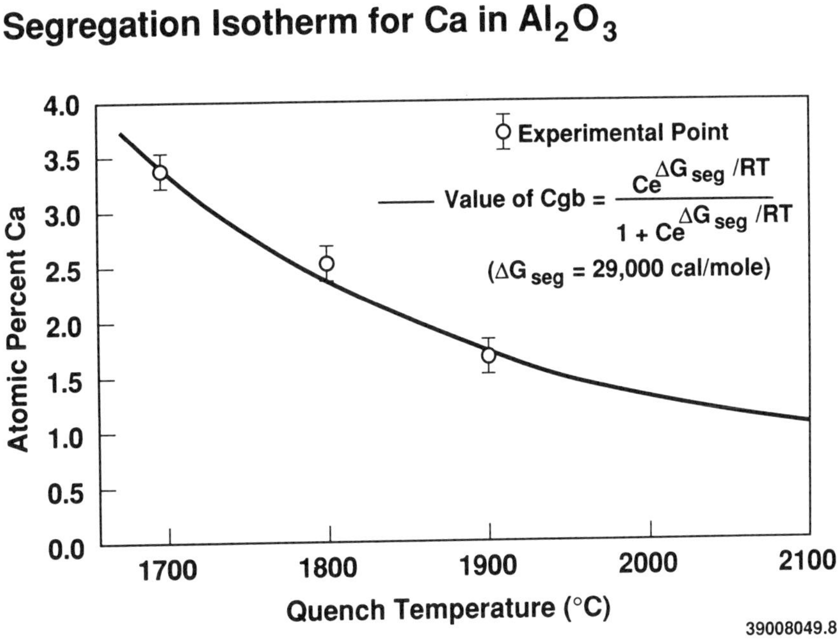

Fig. 2 Application of adsorption isotherm to Ca segregation (Ref 30)

Of greater interest to the study of interfaces is the lack of segregation of MgO to grain boundaries of Al_2O_3 (Ref 29-31). Marcus and Fine (Ref 29) analyzed grain boundaries of Al_2O_3 exposed by fracturing within the Auger electron spectrometer (AES) and found 2 to 5 at.% Ca segregation and no Mg segregation. Johnson and Stein (Ref 30) made a similar observation using AES and concluded that the Ca segregation obeyed equilibrium (Gibbsian) segregation theory. A binding energy of 29 kcal/mol was determined for Ca segregation to the interface of Al_2O_3 (Fig. 2). A maximum concentration of 3.5% Ca was obtained for an annealing temperature of 1700 °C; the grain-boundary concentration decreased with increasing annealing temperature. Sputter depths for the Ca from the intergranular fracture surface were observed to be six atomic layers by Johnson and Stein (Ref 30) (Fig. 3) and 12 nm by Marcus and Fine (Ref 29). These depths are greater than those seen in metals but may be the product of electrostatic and dipole interaction distributions at the interface. Some Mg was observed, but it was related to nonequilibrium spinel partitioning. Franken and Gehring (Ref 31) used x-ray photoemission

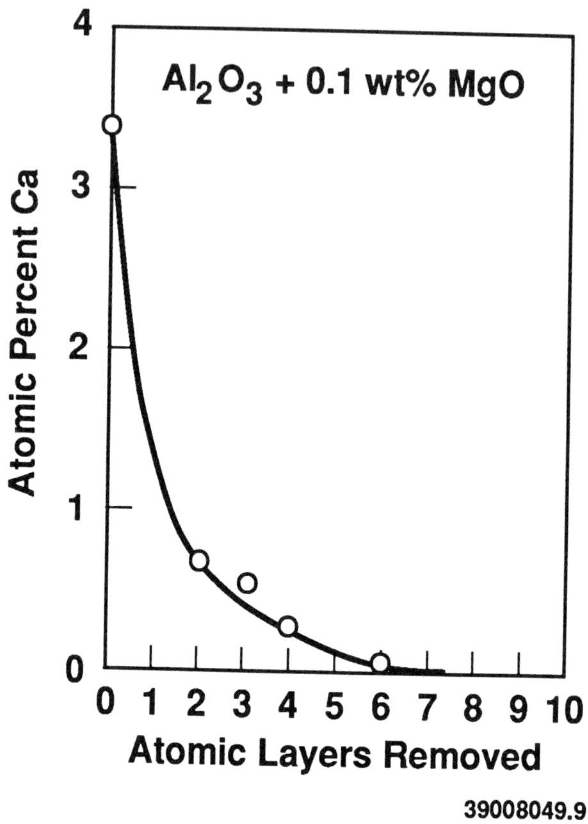

Fig. 3 Inert ion sputter etch profile from MgO-doped Al_2O_3 (Ref 30)

spectroscopy (XPS) to measure grain boundaries of Al_2O_3 and found significant Ca segregation, as had been observed in other investigations, but also found MgO segregation. They studied material that was doped with MgO and concluded that Mg is enriched at grain boundaries below the solubility limit of MgO in Al_2O_3, whereas above the solubility limit $MgAl_2O_4$ segregates or precipitates. Although Ca and in some cases Mg were found at the grain boundaries of Al_2O_3, these impurities are thought primarily to affect sintering and grain growth.

Chiang and Kingery (Ref 32) noted Ca and Si enrichment at the grain boundaries of manganese-zinc ferrites by using scanning transmission electron microscopy (STEM). An example of their results is given in Fig. 4. The Si enrichment was small and exhibited little variation between samples; the Ca

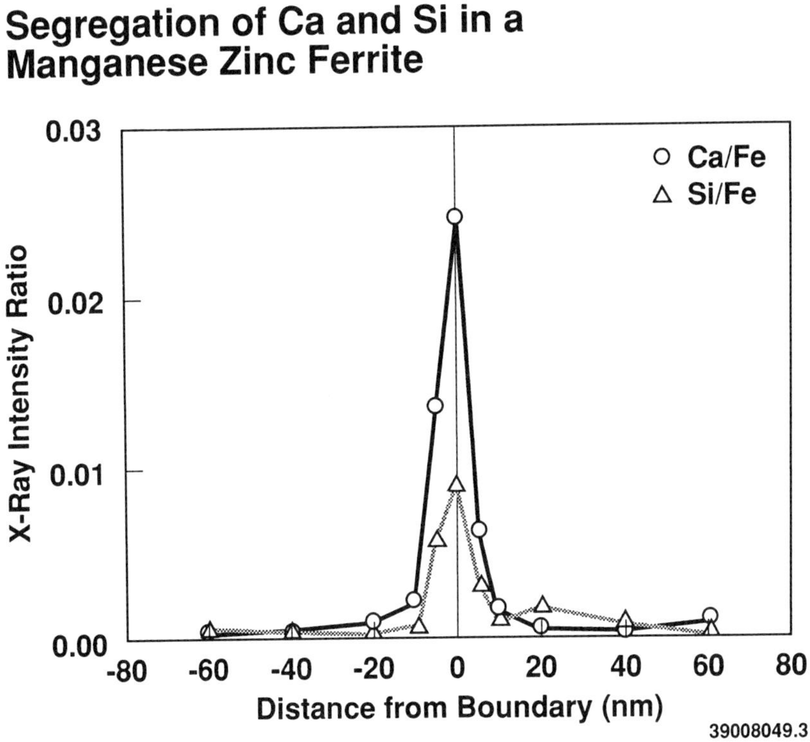

Fig. 4 Segregation of Ca and Si (500 ppm Ca sample) in a manganese-zinc ferrite, plotted as integrated x-ray intensity ratios Ca/Fe and Si/Fe against distance from the boundaries (Ref 32)

enrichment was greater in the interior of the sample than at the surface and increased with increasing bulk Ca content. Clear evidence for second-phase precipitation was given for samples with 2180 ppm of Ca, although narrow STEM profiles for material with 500 ppm Ca (about 20 nm) suggest segregation without precipitation. Franken and Stacy (Ref 33) reported a Ca profile of about 2 nm in a Mn-Zn ferrite using AES. The narrower profile observed using AES results from the beam broadening in the STEM analysis. A profile of 2 to 4 nm is consistent with monolayer-level segregation.

Grain-boundary enrichment of S, Fe, and O has been observed in TiB_2 by Tennery *et al.* (Ref 34). Enrichments as high as 1.4, 0.87, and 1.73 relative to Ti were observed for S, Fe, and O, respectively. The addition of Ni as a sintering

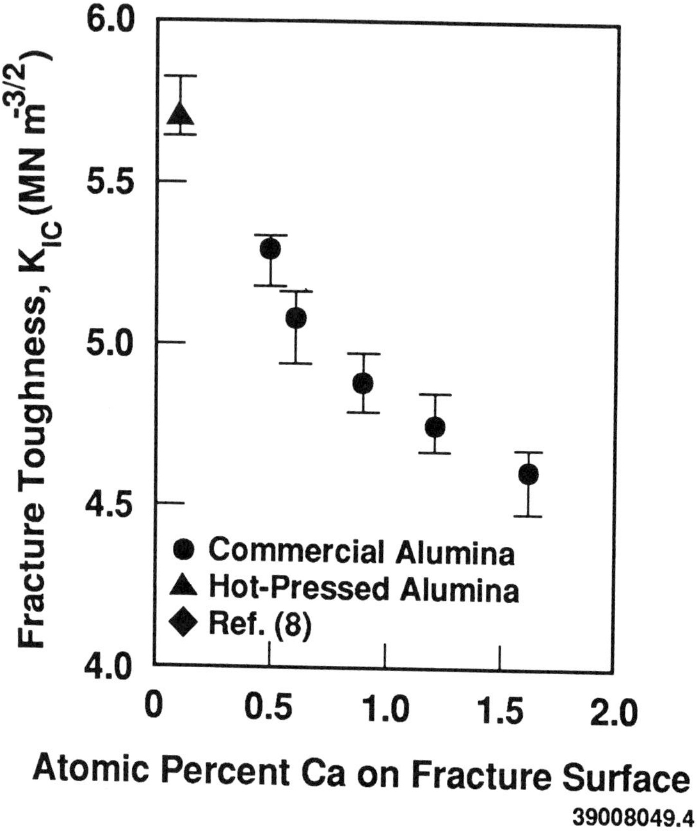

Fig. 5 Dependence of room-temperature fracture toughness (SENB tests) on grain-boundary Ca concentration from commercial and hot-pressed aluminas (Ref 36)

aid eliminated the Fe segregation and reduced the S and O segregation. The authors concluded that the O was present as a discrete second phase. No conclusion regarding the Fe and S was provided. Ion sputtering was used to demonstrate that the Fe and S were concentrated within the first 50 to 100 atomic layers around the grain boundaries. This broad profile does not clearly suggest atomic segregation; however, the distributions of the electrostatic and dipole interactions from the interface into the crystal establish this profile and could be of this magnitude.

Table 1 Room-temperature mechanical properties of Avco SiC-fiber-reinforced Si_3N_4

Sample	Maximum flexure strength, MPa	Strain to failure, %	Work of Fracture $\times10^6$ erg/cm^2
RBSN	325	0.18	4.62
RBSN + SiC fiber with 1 μm C coat	479	2.2	42.9
RBSN + SiC fiber with 3 μm C coat	755	3.0	94.5

Source: Ref 42

Few direct correlations between grain-boundary chemistry and the electrical, magnetic, and mechanical properties of ceramic materials have been reported. Two studies showed a direct correlation between the grain-boundary Ca concentration and the fracture strengths of Al_2O_3. Funkenbush and Smith (Ref 35) reported a decrease in K_{Ic} from 5.3 to 4.3 MPa$\sqrt{m}$ with a change in the grain-boundary Ca concentration of 0.5 to about 3 at.%. Jupp, Stein, and Smith (Ref 36) noted a decrease from 5.7 to 4.5 MPa$\sqrt{m}$ with a change in the grain-boundary Ca concentration of about 0.1 to 1.5 at.% (Fig. 5). Although these two studies covered different ranges of grain-boundary Ca concentrations, the results are consistent. The authors attempted to match the variation of K_{Ic} with grain-boundary Ca concentration using the interfacial strength model of Seah (Ref 37), but they found a very poor match between theory and experiment. Since the Seah model considers only atomic size effects on interatomic spacing at the interface, the electronic contribution to the interfacial strength is ignored. This contribution is clearly significant in ionic solids such as Al_2O_3.

Interfacial Chemistry and Structure of Fiber-Matrix Interfaces

Experimental research has been performed on a variety of ceramic-based composites to enhance their fracture toughness by modifying the filament-matrix interfaces produced during processing. Examples include ZrO_2 matrix composites reinforced with BN-coated SiC fibers (Ref 38), glass matrix composites reinforced with SiC fibers having C- and NbC-based layers at fiber-matrix interfaces (Ref 39-41), amorphous matrices reinforced with SiC-based fibers (Ref 42-48), alumina matrix composites reinforced with SiC filaments (Ref 49, 50), and several Si_3N_4-SiC composites.

Silicon Nitride Matrix Composites: SiC-Fiber-Reinforced Si_3N_4. The correlation between mechanical properties and the fiber-matrix interfacial phases in Nicalon-fiber-reinforced Si_3N_4 has been reported by Jones *et al.* (Ref 51) and Schoenlein *et al.* (Ref 17, 52). (Nicalon fibers are manufactured by Nippon

a

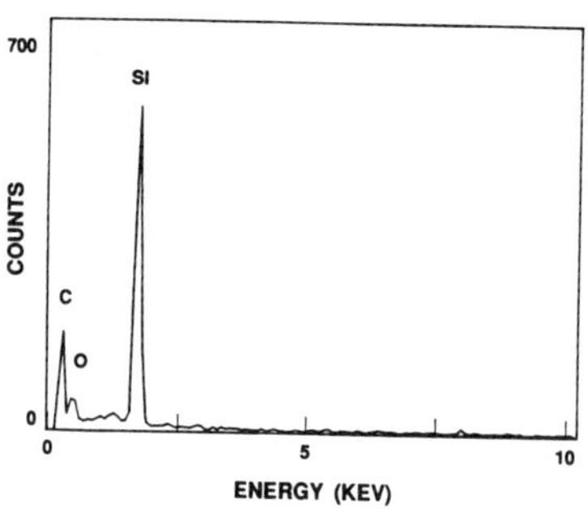

b

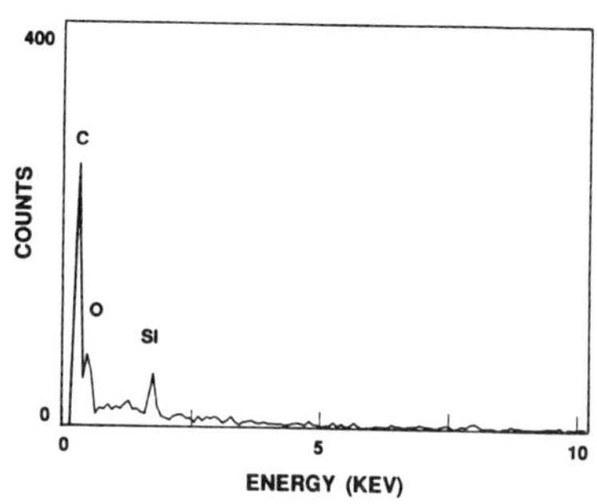

c

Fig. 6 Interface in hot-pressed Si_3N_4/SiC composite. (a) Transmission electron micrograph of interface between SiC fiber and Si_3N_4 matrix showing amorphous (A) and fibrous (F) carbon layers. Associated EDS spectrum of (b) amorphous and (c) fibrous C layers

Carbon Company, Ltd. and distributed by Dow Corning Corporation.) Their TEM observations of this material (Fig. 6) revealed that amorphous and graphitic C layers were present at the fiber-matrix interface, with the former adjacent to the fiber and the latter adjacent to the matrix. These C layers allowed some fiber pullout during fracture, which should have increased the toughness of the composite, but degradation of the fiber occurred during hot pressing and caused the fracture properties of the composite to be inferior to those of unreinforced Si_3N_4. The fiber degradation was attributed to SiC grain growth and the development of an amorphous intergranular phase in the fiber.

Microstructure/property relationships for SiC-fiber-reinforced reaction-bonded Si_3N_4 (RBSN) have been reported by Corbin *et al.* (Ref 53). Significant changes in the room-temperature strength, strain to failure, and work of fracture of Avco-fiber-reinforced RBSN were attributed to the thickness of a C layer present on the fibers before nitridation (Table 1). (Avco fibers are manufactured by Avco Specialty Materials, Textron, Inc.) Thick coatings (3 μm on Avco type SCS-6 fibers) enhanced toughening by weak interfacial bonding, fiber pullout, and crack deflection along the middle of the C layer at the fiber-matrix interface. This behavior was less pronounced in samples with thin C coatings (1 μm on Avco type SCS-2 fibers) because the thin coatings degraded during nitriding to form a strong fiber-matrix bond. In the case of the thin coatings, cracks propagated through the fibers, resulting in lower flexural strength and lower strain to failure. Fracture properties were not measured above room temperature, and it is expected that elevated temperatures may result in C-layer oxidation accompanied by the degradation of mechanical properties.

Corbin *et al.* (Ref 53) also reported the presence of a "two-layer structure" at fiber-matrix interfaces in nitrided samples containing the thick-coated fibers. This observation agrees with the surface structures observed on Avco type SCS-2 fibers by Marzik (Ref 54), Wawner *et al.* (Ref 55, 56), and Nutt and Wawner (Ref 57); the inner layer consisted of amorphous C and was approximately 0.5 μm thick, and the outer layer contained β SiC crystallites dispersed in C. The thick coatings on the SCS-6 fibers were reported by Marzik (Ref 54) to have a similar morphology. Scanning Auger microanalysis of fracture surfaces by Corbin *et al.* (Ref 53) showed that the thick-coated samples exhibited C layers bonded to the fibers and matrix troughs. In contrast, the thin-coated samples did not exhibit a C layer on fiber surfaces or in matrix troughs. It was believed that the C was consumed by the Si + C $\rightarrow$ SiC reaction during nitriding at 1380 °C for 48 h. Matrix troughs in the thin-coated samples did exhibit "small black deposits" believed to be SiC crystallites from the SiC/amorphous C surfaces of Avco fibers. However, the composition of these deposits was not reported.

Shetty *et al.* (Ref 58) evaluated room-temperature mechanical properties of Si_3N_4 reinforced with Avco type SCS-2 fibers. The samples were prepared by three methods: slurry coating the fibers with a mixture of Si_3N_4, Y_2O_3, and Al_2O_3 powders; cold pressing preforms of laminated sheets of unidirectionally aligned fibers; and binder removal followed by hot pressing at 1750 °C. Nonlinear stress-strain behavior was exhibited in the fiber-reinforced, as opposed to unreinforced, Si_3N_4. The authors reported that "the SiC filaments were compatible with the Si_3N_4 matrix, and there was no evidence of excessive fiber-matrix interaction based on SEM fractography and optical ceramography." It was also reported that the average crack growth resistance for the fiber-reinforced composites (7300 to 8150 J/m^2) was almost 100 times greater than the critical strain energy release rate for unreinforced Si_3N_4 (82.5 J/m^2). This suggests that large contributions from inelastic energy dissipation processes, such as fiber pullout and debonding, were operative. It is possible that toughening behavior was affected by the C layer, known to be present on the fiber surfaces in the as-received condition. Based on the reported data, it is uncertain to what extent the C layers underwent microchemical changes during the processing of the composites.

Bhatt (Ref 59-61) also reported enhanced toughening in reaction-bonded Si_3N_4 reinforced with Avco type SCS-2 fibers. He noted that the C coatings on the fibers resulted in weak bonding between the fibers and matrices, causing nonlinear stress-strain behavior and crack deflection along fiber-matrix interfaces. Microstructural examination indicated little evidence of chemical reaction between fiber and matrix. Bhatt mentioned that the use of these fibers is attractive because of the ability to specify various coatings by chemical vapor deposition. Interface microstructure and phase equilibriums were not analyzed in detail.

Corbin *et al.* reported interface-controlled toughening in reaction-bonded Si_3N_4 reinforced with four different types of SiC-based fibers: MPDZ and HPZ (experimental fibers produced by Dow Corning Corporation), Tyranno (Ube Industries, Ltd.), and Nicalon (Ref 62). The authors reported that "nitridation parameters can be controlled so that a useful interface will form in situ during matrix processing. Composites which do not fail in a brittle manner and can carry loads after matrix microcracking have been demonstrated. Due to the different composition and thermal stability of the fibers evaluated, each requires a different nitridation scheme to produce composites with desirable properties." Experiments primarily evaluated the way in which room-temperature fracture behavior was affected by the time-dependent formation of a reaction zone at fiber-matrix interfaces during nitriding. Reaction zones were produced in samples using 8 or 48 h of nitridation at 1350 °C.

The Nicalon-reinforced samples that were nitrided for 48 h exhibited nonlinear stress-strain behavior and noncatastrophic failure. In comparison, when the same material was nitrided for 8 h, it exhibited linear stress-strain behavior and catastrophic, brittle failure, with cracks propagating through the fibers. The 48-h nitridation caused the Nicalon fibers to become noncircular in cross section; the interface reaction zones grew to approximately 2 μm. The interface reaction zone that developed in the Nicalon fibers during the 48-h nitridation provided a crack deflection/debonding zone, causing cracks to propagate along the fiber length rather than penetrate the fibers. The 48-h nitridation of the Nicalon-containing samples also resulted in a trace amount of crystallized β-SiC, as determined by x-ray diffraction analysis. In contrast, β-SiC was not detected when the same material was nitrided for only 8 h. In addition, scanning Auger microanalysis of fractured samples indicated that after 48-h nitridation, the fiber surfaces contained less O_2 and exhibited a 1:1 atomic ratio between Si and C, suggesting stoichiometric SiC. Also, the surfaces of the matrix troughs in the same samples indicated C and low-O_2 concentrations. Apparently, two grades of Nicalon fiber were used, ceramic and nonceramic. The authors reported the formation of a smaller interface reaction zone during nitridation with the ceramic-grade Nicalon fiber.

Based on the data presented in this report, it is uncertain which chemical equilibriums and chemical transport processes were involved in the formation of interfacial reaction zones during nitridation of Nicalon-reinforced Si_3N_4. Possible reactions are likely to involve elemental Si in the matrix, N_2 gas in the nitriding furnace, and the Si-C-O-N glass known to coexist with β-SiC crystallites in the as-received Nicalon fibers (Ref 63, 64). Lucek *et al.* (Ref 65) evaluated the stability of Nicalon fibers under various RBSN processing conditions. Additionally, results have shown that thermal degradation of the Nicalon fibers may occur, resulting in the evolution of SiO and CO gases (Ref 62). It is plausible that these gases may react with elemental Si or N gas during the nitriding process.

In a companion paper, Corbin *et al.* (Ref 66) evaluated room-temperature mechanical properties of reaction-bonded Si_3N_4-matrix composites that were reinforced with Nicalon fibers having various chemical vapor deposited coatings. Coatings of SiC, Si_3N_4, and Al_2O_3 did not enhance toughness or strength because these coatings were strongly bonded to the Si_3N_4 matrix and caused brittle behavior. In contrast, toughening behavior was exhibited in samples containing C-coated fiber; cracks propagated along the coating-matrix interface, and the coatings adhered to the fibers after fracture. With the exception of the C and Al_2O_3 coatings, the coatings yielded composite tensile strengths that were lower than that of unreinforced Si_3N_4 prepared under similar conditions. In addition, all of the coatings resulted in slightly decreased tensile strengths of the fibers; tensile strengths were approximately 2.3 GPa for uncoated and 2.0 GPa for coated fibers. Chemical equilibriums at the interface were not reported in detail.

SiC-Whisker-Reinforced Si_3N_4. Room-temperature mechanical properties and the microstructure of whisker-matrix interfaces were correlated by Jones *et al.* (Ref 51) and Schoenlein *et al.* (Ref 17, 52) for RBSN reinforced with very large-scale (VLS) SiC whiskers. The samples were prepared at Los Alamos National Laboratory, and various coatings (including SiO_2, BN, and C) were applied to the whiskers. Analytical electron microscopy revealed significant changes to the microstructure of whisker-matrix interfaces. Composites that were reaction bonded with an Fe_2O_3 additive exhibited pitted or faceted whisker surfaces as a result of a reactive iron silicide phase, despite the C, BN, or SiO_2 coatings applied to the whiskers. When no sintering aid was used, the uncoated whiskers were not degraded and appeared to be strongly bonded to the matrix. Unfortunately, the samples containing no sintering aid possessed large areas of unreacted Si, which most likely acted as strength- and toughness-limiting defects. All whisker-reinforced samples exhibited reductions in flexure strength and toughness relative to unreinforced RBSN, which had been prepared under similar conditions. Reduced flexure strengths and toughness values of the whisker-reinforced composites containing the Fe_2O_3 additive were attributed to whisker degradation. The composites reinforced with SiO_2-coated whiskers exhibited the highest fracture strength and toughness, whereas the composites with BN-coated whiskers exhibited the lowest strength and toughness.

Matrix-whisker interfaces in hot-pressed Si_3N_4 reinforced with 10 vol% SiC (Tateho, manufactured by Tateho Chemical Industries, Ltd.) whiskers were also evaluated by Schoenlein *et al.* (Ref 17). These composites were prepared similarly to the Nicalon fiber-reinforced materials, but the interface contained only a thin 1- to 2-nm SiO_2 layer (Fig. 1). The whisker had transformed from predominantly β phase prior to consolidation to predominantly α SiC. Microdiffraction of the interface did not reveal any reflections other than Si_3N_4 and SiC, which suggests that the glass layer was amorphous. No faceting of the (111) SiC planes was observed.

Room-temperature mechanical properties of VLS whisker-reinforced Si_3N_4 prepared by hot pressing with MgO sintering aid were also evaluated by Shalek *et al.* (Ref 67). Slightly nonlinear stress-strain behavior was observed in whisker-reinforced samples, with little evidence of pullout. Nonetheless, a maximum toughness of 12.5 MPa$\sqrt{m}$ was observed in one sample containing 10 vol% whiskers. The authors reported that whisker-matrix interfaces were strongly bonded and contained a Mg-based glass phase. Large processing-related flaws became fracture origins and may have overshadowed the effects of interface-controlled toughening.

High-resolution TEM atomistic analyses of whisker-matrix interfaces in these samples were reported by Nutt and Phillips (Ref 68) and Davis *et al.* (Ref 69). These investigators reported three significant features of whisker-matrix interfaces in VLS-whisker-reinforced Si_3N_4: the VLS whiskers exhibited rounded-triangular cross sections with (112) faces adjacent to the Si_3N_4 matrix; whisker-interface planes were composed of (111) microfacets arranged in step-ledge structures that spiraled around the whisker circumference; and whisker-matrix interfaces contained an amorphous phase, which sometimes devitrified to form a Mg-Sialon phase that was epitaxial on the SiC whisker. Another significant observation was that whisker degradation occurred after hot pressing at 1850 °C, but not at lower temperatures. Nutt and Phillips noted that the effects of interface microstructure on mechanical properties were not clear and that higher toughness, the result of a weaker whisker-matrix bond, might be possible by increasing the thickness of the glass layer.

Wilkens and Corbin (Ref 70) prepared Tateho-whisker-reinforced RBSN and then used glass encapsulation and hot isostatic pressing. Various coatings were applied to the whiskers by chemical vapor deposition, and the effects on fracture toughness at room temperature were compared using indentation experiments. The coatings included a BN-based coating and four types of dual-layer coatings designed to have a debonding/parting layer (BN- or C-based coatings) protected by an overcoating layer (SiC- or Si_3N_4-based coatings). The dual-layer coatings were nonstoichiometric and included SiC/BN, Si_3N_4/BN, Si_3N_4/C, and SiC/C. The highest room-temperature toughness, 4.4 MPa$\sqrt{m}$, was exhibited by the sample containing whiskers coated with a single layer of BN. An analysis of this sample by TEM indicated that the BN had dissolved into the matrix. The SiC/BN and Si_3N_4/BN samples exhibited somewhat lower toughness values, of approximately 3.9 and 3.2 MPa$\sqrt{m}$, respectively. Samples containing uncoated whiskers and no whiskers exhibited slightly lower toughness values, of approximately 3.0 and 3.1 MPa$\sqrt{m}$, respectively. The lowest toughness values, of approximately 2.9 and 2.7 MPa$\sqrt{m}$, were exhibited by samples with Si_3N_4/C and SiC coatings, respectively.

Mechanical properties of AMC-whisker-reinforced Si_3N_4 hot-pressed with Y_2O_3 and Al_2O_3 sintering aids were evaluated by Buljian and coworkers (Ref 71-76) (AMC whiskers are manufactured by Advanced Materials Corporation). Increases in fracture toughness were seen to result from increases in the volume percent of SiC whiskers and from a prolonged time at maximum temperature during hot pressing to enlarge Si_3N_4 grain sizes and thereby increase crack deflection in the matrix. Fractography indicated limited amounts of whisker pullout, which suggests that the whiskers were strongly bonded to the matrices. Analyses by TEM revealed no evidence of whisker degradation or reaction phases at whisker-matrix interfaces. Fracture toughness anisotropy of SiC-whisker-reinforced Si_3N_4 was also reported as a result of preferred orientation of whiskers during hot pressing (Ref 73, 74). Preferred orientation of SiC whiskers was also reported in injection-molded samples of whisker-reinforced Si_3N_4. Changes in the geometry of the injection mold apparatus resulted in changes in preferred orientation of whiskers, which ultimately affected shrinkage anisotropy during densification by hot isostatic pressing (HIP).

The high-temperature deformation of SiC-whisker-reinforced Si_3N_4 was evaluated by Nixon *et al.* (Ref 77). Samples were prepared with Y_2O_3-Al_2O_3 sintering aids according to procedures reported by Buljian *et al.* (Ref 71-76). Samples were evaluated by steady-state creep tests, and results were correlated with TEM studies to evaluate mechanisms responsible for deformation. Nixon *et al.* reported that temperature and the crystallization of the amorphous grain-boundary phase in the Si_3N_4 matrix were the principal factors controlling high-temperature deformation behavior. Creep rates in the whisker-containing samples exceeded those of the unreinforced samples. These results were attributed to increased amounts of devitrified intergranular phase in the whisker-reinforced composites as revealed by TEM analysis. The authors suggested that the diffusion of glass modifiers (Ca, Mn, Fe, and Mg) from the whiskers into the matrix may have reduced the tendency of the intergranular glass phase to crystallize, thereby promoting creep. TEM analyses of creep-tested samples indicated that the glass continued to devitrify with time.

Haynes *et al.* (Ref 78) reported processing and room-temperature mechanical properties of SiC-whisker-reinforced Si_3N_4. Extensive degradation of Tateho whiskers but not of Arco whiskers occurred in composites as a result of reactions with N gas during a presintering process (Arco whiskers were manufactured by Atlantic Richfield Corporation. The manufacturing facility is currently operated by Advanced Composite Materials Corporation). This finding was verified in separate experiments. Tateho whiskers were heated in N to 1500 °C for 2 h; subsequent x-ray diffraction indicated the formation of extensive amounts of Si_3N_4. When the same experiment was performed with Arco whiskers, negligible amounts of Si_3N_4 were found. Analysis by STEM was performed on the composite containing Arco whiskers, and an amorphous film was detected at whisker-matrix interfaces. This film probably resulted in strong whisker-matrix bonds since pullout was not observed on fracture surfaces of the Arco whisker-reinforced Si_3N_4.

Extensive evaluation of the mechanical properties of Tokai-whisker-reinforced Si_3N_4 at temperatures between room temperature and 1300 °C was reported by Hayami *et al.* (Ref 79) and Ueno and Toibana (Ref 80) (Tokai whiskers are manufactured by Tokai Carbon Company). Samples were prepared by hot pressing using two types of sintering aids, Y_2O_3-Al_2O_3 and Y_2O_3-La_2O_3. Based on the analysis of whisker-matrix interfaces by SEM, the Al_2O_3-based sintering aid was found to degrade the whiskers, whereas the La_2O_3-based sintering aid prevented whisker attack. Fracture energies of whisker-reinforced samples were reported to exceed those of unreinforced Si_3N_4 by approximately 30%. Whisker-reinforced samples exhibited impressive tensile strengths of 600 MPa at room temperature and 650 MPa at 1300 °C. The authors also presented experiments indicating that both room-temperature tensile strength and Weibull modulus increased when large SiC particles and agglomerates were removed from the as-received whiskers by a filtration process.

Room-temperature mechanical properties were correlated with whisker-matrix interface structures by Kandori *et al.* (Ref 81). Samples of Tateho-whisker-reinforced Si_3N_4 containing Y_2O_3 and $MgAl_2O_4$ sintering aids were prepared by injection molding followed by HIP, and the fracture behavior of reinforced and unreinforced samples was correlated with analyses of whisker-matrix interfaces by fractography and TEM. Fracture surfaces did not show evidence of whisker pullout, indicating that the whiskers and matrices were strongly bonded. Before HIP, voids in samples were preferentially segregated at whisker-matrix interfaces based on TEM analyses reported earlier by Kobayashi *et al.* (Ref 82). After the HIP treatment, TEM indicated that the voids were eliminated and that a 1-nm-thick amorphous layer existed at whisker-matrix interfaces, with no evidence of a reaction layer. The authors reported that the good bonding provided by the amorphous layer at whisker-matrix interfaces enhanced load transfer but retarded pullout upon fracture. Whisker-reinforced samples exhibited a relatively high toughness of 9 to 10

MPa$\sqrt{m}$, despite the strong whisker-matrix bonding and the lack of evidence of pullout. The fracture surfaces of whisker-reinforced composites were rougher than were those of unreinforced samples, suggesting that crack deflection and/or crack branching existed because of the presence of the whiskers. The reinforced samples also possessed relatively high flexure strengths of 800 to 900 MPa. In addition, improvements in fracture toughness, fracture energy, and flexural strength were exhibited by reinforced, as opposed to unreinforced, samples.

Glass Matrix Composites. Glass matrix composites are being studied both for their commercial value and for their use as a model system for evaluating interfacial effects on ceramic composites. This common interest in glass matrix composite materials stems primarily from the low temperatures at which they can be fabricated, thereby minimizing damage to fibers and whiskers. A wide range in properties of the glass matrix phase is possible through significant compositional variations of the glasses. For commercial applications this results in the ability to tailor the matrix and hence the composite properties to specific applications for use as model composites, resulting in the ability to alter residual stresses on the fiber, interface character, etc. The combination of a low-modulus glass matrix reinforced by high-strength, high-stiffness fibers has resulted in composites with excellent properties. An example of a commercial composite material is lithium-aluminosilicate (LAS) glass reinforced with SiC yarn (produced by United Technologies). Brennan and Prewo (Ref 83) have reported a fracture strength of 700 MPa and a fracture toughness of 17 MPa$\sqrt{m}$ over the temperature range of 26 to 1000 °C. Other examples include matrices composed of calcium-aluminosilicate (CAS), barium-magnesium-aluminosilicate, and Corning Code 1723 glass reinforced with Nicalon fibers, as reported by Stewart *et al.* (Ref 84).

Chaim and Heuer (Ref 41) used analytical electron microscopy to evaluate the interface of a LAS-Nicalon fiber-reinforced composite (prepared by United Technologies). They found amorphous C adjacent to the fibers and NbC microcrystals between the carbon and the LAS matrix. Niobium was added to the glass to aid crystallization and presumably reacted with the C or SiC to precipitate NbC microcrystals. The amorphous C layer was 100 nm thick, and cracks in the thin TEM foils propagated between the two layers. Benson *et al.* (Ref 85) and Mah *et al.* (Ref 86) concluded that the development of a carbon layer on SiC embedded in glass matrices could be explained thermodynamically. Mah *et al.* suggested the following reaction for Nicalon in a Na silicate glass:

$$\text{Nicalon} + x\,O_2 \rightarrow a\,SiC + b\,SiO_2 + c\,CO + d\,C$$

The excellent fracture resistance of these glass matrix composites has been related to fiber pullout, which occurs because of the low interfacial shear strength of the C layers. Deshmukh and Coule (Ref 87) used single-fiber pullout tests to determine the interfacial shear strength of monofilament SiC in a borosilicate glass matrix. They found values ranging from 6 to 14 MPa for Avco SCS-6 fibers with a variety of coatings. These interfacial shear strength values are consistent with high toughness values from fiber pullout as predicted from fracture toughness models based on interfacial strength.

Oxide Matrix Composites. Aluminum oxide matrix composites reinforced with SiC and other fibers are also receiving considerable attention because of their commercial and model system value. Zirconia and mullite matrix composites are being evaluated for similar reasons, although the extent of research on these systems is more limited. Interface reactions between mullite, a mixture of Al_2O_3, B_2O_3 and SiO_2, and SiC or other reinforcements could be far more complex than the systems for which they are serving as a model. As in glass matrix composites, low sintering temperatures for Al_2O_3 and mullite matrix composites are key reasons for their use as model material. The evaluation of composite properties in a material nearly fully dense is clearly needed to prevent the domination of the composite properties by large processing flaws.

Ruhle *et al.* (Ref 88) evaluated a SiC-whisker-reinforced Al_2O_3 composite to determine the contribution of crack-bridging whiskers to the fracture toughness of the composite. With TEM analysis, a thin amorphous phase was revealed in the interface of the SiC whisker and Al_2O_3 matrix. *In situ* SEM fracture studies revealed crack bridging for a distance of about two whisker spacings but little evidence of whisker pullout. The amorphous phase (probably a glass phase) in the whisker-matrix interface probably produced such a strong interface that little whisker pullout occurred. A composite fracture toughness about 200% greater than that of the matrix was observed. Studies by Nair *et al.* (Ref 89) of Al_2O_3 reinforced with SiC fibers coated with borosilicate glass showed that fiber-matrix shear at temperatures of 800 to 1100 °C controlled the bridging zone behind the crack tip and the crack growth properties. Therefore, the presence of a glass phase in the interface between reinforcement and the matrix is a key factor in both the low- and high-temperature properties of ceramic matrix composites.

Vaughn *et al.* (Ref 90) examined the interface of SiC-whisker-reinforced Al_2O_3 composites by TEM and found no evidence of an interfacial phase.

Fig. 7 Transmitted-light optical micrograph of a longitudinal cross section of a SiC/SiC composite produced by CVI

Composite materials with both Silar SC-9 and Tateho SCW-1-S whiskers were examined. An increase in fracture strength and a 50% increase in fracture toughness of the Silar SC-9 SiC-whisker-reinforced Al_2O_3 composite relative to the matrix were correlated with whisker crack bridging and pullout. While SEM evidence of crack-microstructure interactions clearly showed these effects, the small increase in fracture toughness relative to the matrix leaves some question about the magnitude of this contribution. Crack deflection effects could also account for the increased fracture toughness. In addition, the nature of the interface between the whisker and matrix in the specimen exhibiting whisker pullout is uncertain because the TEM analysis was performed on samples hot pressed at 1900 °C and the whisker pullout occurred in a sample hot pressed at 1500 °C. Whisker pullout would not be expected for a well-bonded interface, as shown for the material hot pressed at 1900 °C.

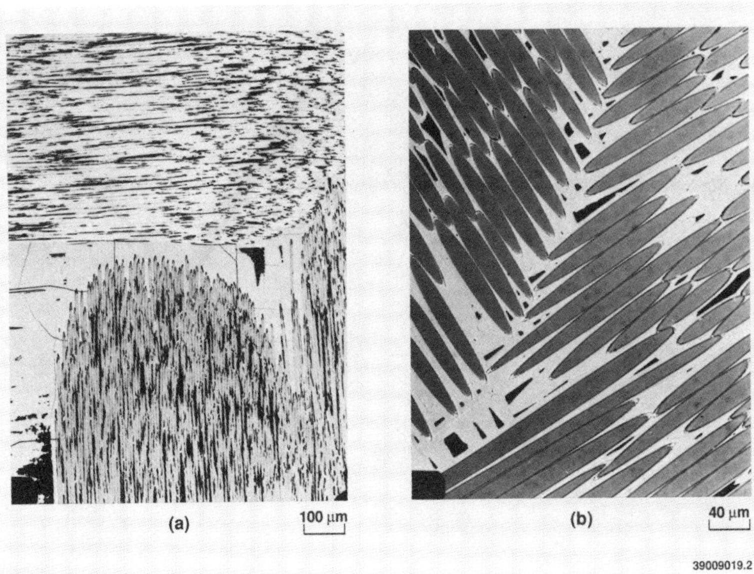

Fig. 8 Reflected-light optical micrographs showing bundle-bundle interface in a three-dimensional woven SiC/SiC composite

SiC-fiber-reinforced $ZrO_2 + SiO_2$ and $ZrO_2 + TiO_2$ matrix composites have been evaluated by Bender *et al.* (Ref 91). Composites containing uncoated or BN-coated fibers were evaluated, and the best fracture strengths and toughness values were obtained with the BN-coated fiber composites. The interface comprised a silica layer between the BN coating and the matrix, and fracture occurred readily along the BN-silica interface. The authors concluded that the BN coating eliminated strong bonding between the silica layer and the fiber and protected the fiber from mechanical damage and perhaps volatilization during processing.

Fiber-matrix interfacial effects in C (Magnamite As-4), SiC (SCS-6), and mullite (Nextel 440) fiber-reinforced mullite were evaluated by Singh and Brun (Ref 92). The fiber-matrix reaction and fracture behavior for uncoated and BN-coated fibers were examined. A BN coating prevented reaction between the mullite fibers and matrix during hot pressing at 1410 °C; however, fracture data on this composite were not reported. Some evidence of pullout during

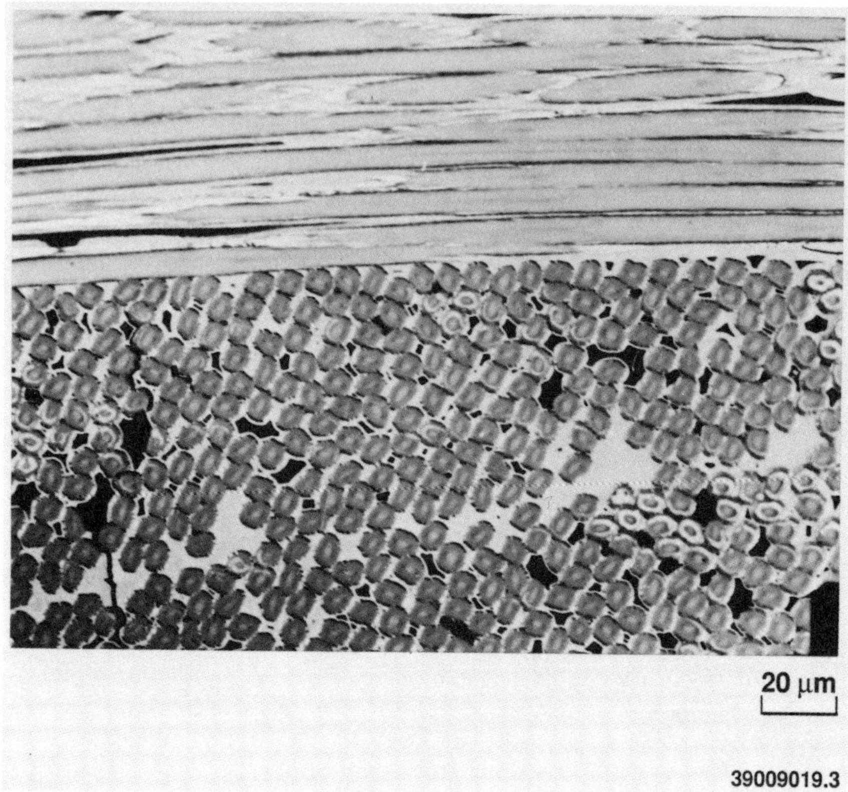

Fig. 9 Reflected-light optical micrograph showing bundle-bundle interface in a 0/90 SiC/SiC composite

fracture of the composite was observed for the uncoated and BN-coated SiC-fiber-reinforced material. A C coating, present on the uncoated fibers, is thought to have contributed to fiber pullout during fracture.

SiC Matrix Composites. Nicalon fiber-SiC matrix composites are currently being investigated at Oak Ridge National Laboratory. In work reported by Lowden and Stinton (Ref 93), the properties of composite material prepared by chemical vapor infiltration of a fiber preform have been evaluated. Fibers with a variety of coatings, including C, Si, B, SiC, and BN applied in thicknesses ranging from 30 to 500 nm, and fibers with no coating have been examined. A detailed analysis of the interface structure has not been reported; however, it was noted that the largest strains to failure correlated with the composites made with fibers having the thickest coatings. Interfacial shear strength measurements of 50, 12, and 0.5 MPa were reported (Ref 94) for the

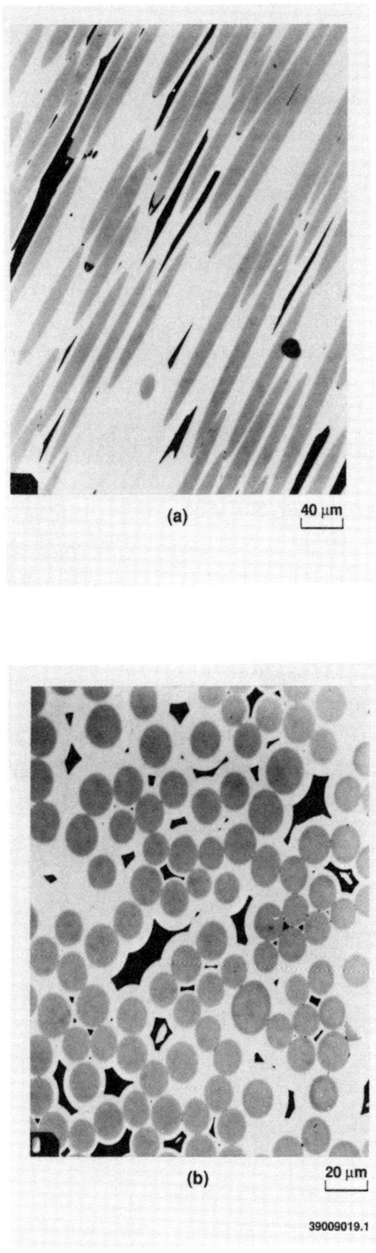

Fig. 10 Reflected-light optical micrographs showing details of the growth interface in a SiC/SiC composite produced by CVI

C-coated fibers with coating thicknesses of 30, 70, and 100 nm, respectively; the strain to failure increased with increasing C-layer thickness or with decreasing interfacial shear strength. An example of the structure of the SiC-fiber-reinforced CVI SiC material made with uncoated Nicalon fibers is shown in Fig. 7. Fiber-matrix and SiC-SiC growth interfaces are shown in this transmission optical micrograph.

Several types of interfaces are created in the process of manufacturing ceramic matrix composites. For composite fabrication processes such as hot pressing and resin infiltration and pyrolysis, the matrix-matrix interfaces are essentially grain boundaries, and the characteristics should be similar to those described in the section on grain boundaries. However, the production of composites from bundles of fibers creates bundle-bundle interfaces, while the fabrication of composites made by CVI techniques produces nonequilibrium matrix-matrix interfaces. Examples of bundle-bundle interfaces are shown in Fig. 8 and 9. The bundle-bundle interface in a three-dimensional woven SiC/SiC composite is shown in Fig. 8; that of a 0/90 composite is shown in Fig. 9. It is apparent from these micrographs that the nature of the bundle-bundle interface differs with the bundle lay-up method. There is a greater degree of porosity in the three-dimensional woven composite and less evidence of fiber-fiber contact than in the 0/90 material.

The preparation of ceramic composites by the infiltration of fiber bundles or an array of individual fibers creates nonequilibrium matrix-matrix interfaces. Examples of interfaces produced in SiC/SiC composites by CVI techniques are shown in Fig. 7 through 10. The longitudinal and transverse micrographs shown in Fig. 10 illustrate the shape of the porosity resulting from the CVI growth interface. Crack- and pore-shaped porosity results from the CVI growth process. Since these interfaces are created by the radial growth of SiC from the fiber that constrains the infiltration process, they have little opportunity to form an equilibrium interface. There is clear evidence in Fig. 7 that many of these interfaces are not intact but are making point contact only. Also, any impurities rejected during the growth of the SiC or trapped from the infiltrating gas will be enriched at these interfaces.

Summary

Ceramic composites contain a wide range of interfaces, each of which plays a role in the composite's properties. Grain boundaries, reinforcement-matrix interfaces, and a variety of growth interfaces (such as bundle-bundle and matrix-matrix) that can result from composite fabrication techniques are all present in CMCs. The chemistry and structure of these interfaces are varied; grain boundaries in some CMCs are similar to those in monolithic ceramics, while reinforcement-matrix interfaces engineered for controlled fiber pullout, as well as composite production interfaces ranging from equilibrium fiber-matrix interfaces to weakly bonded nonequilibrium interfaces with trapped impurities, are also present in many CMCs.

The chemistry and structure of grain boundaries in monolithic ceramics have been studied in some detail. The presence of glassy phases or other compounds in the grain boundaries of ceramics has been studied extensively and is shown to be a significant factor in high-temperature properties, while grain-boundary phases have less effect on low-temperature fracture properties (where cleavage fracture tends to dominate).

Chemical changes in grain boundaries resulting from equilibrium segregation have been reported for several ceramics. Calcium segregation in Al_2O_3 has been most studied; results show that up to 3.5 at.% Ca segregates to the grain boundaries of Al_2O_3 after an anneal at 1700 °C. A small but measurable decrease in the fracture toughness of Al_2O_3 was related to Ca segregation. Calcium and Si enrichment in the grain boundaries of Mn-Zn ferrites has been reported, while S, Fe, and O grain-boundary segregation has been reported in TiB_2.

The chemistry and structure of fiber-matrix interfaces in CMCs range from close similarity to the grain boundaries in monolithic ceramics to complex interfaces engineered to control fiber pullout properties. Examples of the phases used in the engineered interfaces include C, Si, BN, B_4C, SiO_2, and multiple layers of these compounds overlayered with SiC or Si_3N_4.

Composite production interfaces such as bundle-bundle or matrix growth interfaces range from fiber-fiber or matrix-matrix interfaces that are comparable to grain boundaries to nonequilibrium matrix-matrix interfaces produced by infiltration processes. Porosity in CMCs is distributed at these interfaces, with the size and shape of the porosity determined by the nature of the interface, the fiber weave, and the volume fraction. Matrix-matrix interfaces produced by infiltration contain microporosity, which may control the properties of CMCs. These interfaces are also likely to have trapped impurities because of the method by which they were created, and these impurities may affect the properties of CMCs. However, little is known about the chemistry and structure of these growth interfaces or their effect on the properties of CMCs. Knowledge about the effect of pore geometry on properties and about the relationship between fiber lay-up and pore geometry is also limited.

References

1. A.G. Evans and K.T. Faber, *J. Am. Ceram. Soc.*, Vol 67 (No. 4), 1984, p 255-260
2. K.T. Faber and A.G. Evans, *Acta Metall.*, Vol 31 (No. 4), 1983, p 565-576
3. D.B. Marshall, B.N. Cox, and A.G. Evans, *Acta Metall.*, Vol 33 (No. 11), 1985, p 2013-2021
4. A.G. Evans and R.M. McMeeking, *Acta Metall.*, Vol 34 (No 12), 1986, p 2435-2441
5. D.R. Clarke and G. Thomas, *J. Am. Ceram. Soc.*, Vol 60, 1977, p 491-495
6. L.K. Lou, T.E. Mitchell, and A. Heuer, *J. Am. Ceram. Soc.*, Vol 61, 1978, p 392-396
7. D.R. Clarke, *Ultramicroscopy*, Vol 4, 1979, p 33-44
8. P.E.D. Morgan, D.R. Clarke, C.M. Jantzen, and A.B. Harker, *J. Am. Ceram. Soc.*, Vol 64, 1981, p 249-258
9. M. Ruhle, N. Claussen, and A. Heuer, Ed., *Science and Technology of Zirconia*, American Ceramic Society, 1984
10. D.R. Clarke, N.J. Zaluzec, and R.W. Carpenter, *J. Am. Ceram. Soc.*, Vol 64, 1981, p 601-607
11. P.D. Powell and P. Drew, *J. Mater. Sci.*, Vol 9, 1967, p 70
12. F.F. Lange, D.R. Clarke, and B.I. Davis, *J. Mater. Sci.*, Vol 15, 1980, p 601-610
13. F.F. Lange, *J. Am. Ceram. Soc.*, Vol 62, 1979, p 222-223
14. B.S.B. Karunaratne and M.H. Lewis, *J. Mater. Sci.*, Vol 15, 1980, p 1781-1789
15. D.R. Clarke, *J. Am. Ceram. Soc.*, Vol 66, 1983, p 156-158
16. S.R. Nutt and D.S. Phillips, Interface Structure in Silicon Carbide-Silicon Nitride Composites, in *Interfaces in Metal-Matrix Composites*, A.K. Dhingra and S.G. Fishman, Ed., American Institute of Mining, Metallurgical, and Petroleum Engineers, 1986, p 111-120
17. L.H. Schoenlein, R.H. Jones, C.H. Henager, Jr., C.H. Schilling, and F.C. Gac, in *Proceedings of the 1988 Spring Materials Research Society Meeting*, April 1988
18. D.R. Clarke, *Annu. Rev. Mater. Sci.*, Vol 17, 1987, p 57-74
19. D.R. Clarke, *J. Am. Ceram. Soc.*, Vol 70, 1987, p 15-22
20. S.A. Bradley and K.R. Karasek, *J. Mater. Sci. Lett.*, Vol 6, 1987, p 791-794
21. F.F. Lange, B.I. Davis, and D.R. Clarke, *J. Mater. Sci.*, Vol 15, 1980, p 601-618
22. R.K. Govila, P. Beardmore, and K.R. Kinsman, Strength Characterization and Nature of Crack Propagation in Ceramic Materials, in *Fractography and Materials Science*, STP 733, L.N. Gilbertson and R.S. Zipp, Ed., American Society for Testing and Materials, 1981, p 225-245
23. R.K. Govila, A. Ezis, and E.C. Beckwith, in *Proceedings of Failure Mechanisms in High Performance Materials*, J.G. Early, T.R. Shives, and J.H. Smith, Ed., Cambridge University Press, 1984, p 22
24. M.F. Yan, R.M. Cannon, and H.K. Bowen, Solute Segregation at Ceramic Interfaces, in *Character of Grain Boundaries*, Vol 6, *Advances in Ceramics*, M.F. Yan and A.H. Heuer, Ed., The American Ceramic Society, 1983, p 255
25. S.M. Mukhopadhyay, A.P. Jardine, J.M. Blakely, and S. Baik, *J. Am. Ceram. Soc.*, Vol 71, 1988, p 358
26. S. Baik and C.L. White, *J. Am. Ceram. Soc.*, Vol 70, 1987, p 682

27. S. Baik, D.E. Fowler, J.M. Blakely, and R. Raj, Segregation of Mg to the (0001) Surface of Doped Sapphire, *J. Am. Ceram. Soc.*, Vol 68 (No. 5), 1985, p 281-286
28. S. Baik, Segregation of Mg to the (0001) Surface of Single-Crystal Alumina: Quantification of AES Results, *J. Am. Ceram. Soc.*, Vol 69 (No. 5), 1986, p C-101 to C-103
29. H.L. Marcus and M.E. Fine, Grain-Boundary Segregation in MgO-Doped Al_2O_3, *J. Am. Ceram. Soc.*, Vol 55 (No. 11), 1972, p 568-570
30. W.C. Johnson and D.F. Stein, Additive and Impurity Distributions at Grain Boundaries in Sintered Alumina, *J. Am. Ceram. Soc.*, Vol 58 (No. 11-12), 1975, p 485-488
31. P.E.C. Franken and A.P. Gehring, Grain Boundary Analysis of MgO-Doped Al_2O_3, *J. Mater. Sci.*, Vol 16, 1981, p 384-388
32. Y. Chiang and W.D. Kingery, Grain Boundary Composition in Manganese Zinc Ferrites, in *Character of Grain Boundaries*, Vol 6, *Advances in Ceramics*, M.F. Yan and A.H. Heuer, Ed., The American Ceramic Society, 1983, p 300
33. P.E.C. Franken and W.T. Stacy, Examination of Grain Boundaries of Mn-Zn Ferrites by AES and TEM, *J. Am. Ceram. Soc.*, Vol 63 (No. 5-6), 1980, p 315-319
34. V.J. Tennery, C.L. White, R.A. Padgett, and C.B. Finch, Auger Electron Spectroscopy Analyses of Fracture Surfaces in TiB2 Ceramics, *Character of Grain Boundaries*, Vol 6, *Advances in Ceramics*, M.F. Yan and A.H. Heuer, Ed., The American Ceramic Society, 1983, p 312
35. A.W. Funkenbush and D.W. Smith, Influence of Calcium on the Fracture Strength of Polycrystalline Alumina, *Metall. Trans. A*, Vol 6A, 1975, p 2299
36. R.S. Jupp, D.F. Stein, and D.W. Smith, Observations on the Effect of Calcium Segregation on the Fracture Behavior of Polycrystalline Alumina, *J. Mater. Sci.*, Vol 15, 1980, p 96-102
37. M.P. Seah, *Proc. R. Soc. (London) A*349, 1976, p 535
38. B. Bender, D. Shadwell, C. Bulik, L. Incorvati, and D. Lewis III, Effect of Fiber Coatings and Composite Processing on Properties of Zirconia-Based Matrix SiC Fiber Composites, *Am. Ceram. Soc. Bull.*, Vol 65 (No. 2), 1986, p 363-369
39. J.J. Brennan, Interfacial Chemistry and Bonding in Fiber Reinforced Glass and Glass-Ceramic Matrix Composites, in *Proceedings of the 22nd University Conference on Ceramics, and Ceramic Microstructures '86: Role of Interfaces*, Part a, LBL-21719, Lawrence Berkeley Laboratory, University of California, Berkeley, 1986, p 35
40. J.J. Brennan, Interfacial Characterization of Glass and Glass-Ceramic Matrix/Nicalon SiC Fiber Composites, in *Tailoring Multiphase and Composite Ceramics*, Part b, Vol 20, *Materials Science Research*, R.E. Tressler, G.L. Messing, C.G. Pantano, and R.E. Newnham, Ed., Plenum Press, 1986, p 549-560
41. R. Chaim and A.H. Heuer, The Interface between (Nicalon) SiC Fibers and a Glass-Ceramic Matrix, *Adv. Ceram. Mater.*, Vol 2 (No. 2), 1987, p 154-158
42. H.T. Godard, "Characterization of BN Coated Fiber Interfaces in Nicalon/Glass-Ceramic Matrix Composites," Paper presented at the 12th Annual Conference on Composites and Advanced Ceramics, Cocoa Beach, FL, American Ceramic Society, Jan 1988

43. R. Chaim, D.G. Brandon, and L. Baum, "Interface Interactions and Mechanical Properties of Nicalon-Reinforced Cordieritic Glass," Paper presented at the 12th Annual Conference on Composites and Advanced Ceramics, Cocoa Beach, FL, American Ceramic Society, Jan 1988
44. R. Singh, Role of Fiber-Matrix Interfacial Shear Stress on the Toughness of Reinforced Oxide Matrix Composites, in *Proceedings of Research Society Spring Meeting*, Materials Research Society, 1988, p G4-4
45. R.W. Goettler and K.T. Faber, "Interfacial Shear Stresses in Al_2O_3 and SiC Reinforced Glasses," Paper presented at the 12th Annual Conference on Composites and Advanced Ceramics, Cocoa Beach, FL, American Ceramic Society, Jan 1988
46. T. Mah and P.R. Subramanian, "Carbon-Layer Formation at the Fiber/Matrix Interface in Ceramic Matrix Composites," Paper presented at the 12th Annual Conference on Composites and Advanced Ceramics, Cocoa Beach, FL, American Ceramic Society, Jan 1988
47. R.F. Cooper and K. Chyung, "Structure and Chemistry of Fiber-Matrix Interfaces in Nicalon Fiber-Reinforced/Glass-Ceramic Composites: An Electron Microscopy Study," Paper presented at the 88th Annual Meeting, Chicago, IL, American Ceramic Society, April 1986
48. T.W. Coyle, E.R. Fuller, Jr., P. Swanson, and T. Palmides, "Fracture Mechanics Characterization of Crack/Fiber Interactions in Ceramic Matrix Composites," Paper presented at the 11th Annual Conference on Composites and Advanced Ceramics, Cocoa Beach, FL, American Ceramic Society, Jan 1987
49. W.L. Vaughn, J. Homeny, and M.K. Ferber, "Interfacial Effects on the Mechanical Properties of SiC Whisker/Al_2O_3 Composite," Paper presented at the 11th Annual Conference on Composites and Advanced Ceramics, Cocoa Beach, FL, American Ceramic Society, Jan 1987
50. W.L. Vaughn, J. Homeny, and M.K. Ferber, Processing and Mechanical Properties of SiC Whisker-Al_2O_3 Matrix Composites, *Am. Ceram. Soc. Bull.*, Vol 67 (No. 2), p 333-338
51. R.H. Jones, C.H. Henager, Jr., C.H. Schilling, L.H. Schoenlein, W.J. Weber, and F.C. Gac, Interfacial Chemistry-Structure and Fracture of Ceramic Composites, in *Proceedings of the 12th Annual Conference on Composites and Advanced Ceramics*, American Ceramic Society; also, *Ceram. Eng. Sci. Proc.*, in press
52. L.H. Schoenlein, R.H. Jones, C.H. Henager, Jr., C.H. Schilling, and F.C. Gac, Interfacial Chemistry-Structure and Fracture of Ceramic Composites, in *Proceedings of the Materials Research Society Spring Meeting*, Materials Research Society, 1988, p G5-2
53. N.D. Corbin, G.A. Rossetti, Jr., and S.D. Hartline, Microstructure/Property Relationships for SiC Filament-Reinforced RBSN, *Ceram. Eng. Soc. Proc.*, Vol 7 (No. 7-8), 1986, p 958-968
54. J.V. Marzik, CVD Fibers, in *Metal and Ceramic Matrix Composite Processing Conference Proceedings II*, Metals and Ceramics Information Center, Battelle Columbus Laboratories, 1984, p 39-65

55. F.E. Wawner, A.Y. Teng, and S.R. Nutt, Microstructural Characterization of SiC (SCS) Filaments, Part a, *SAMPE Q.*, April 1983, p 39-45
56. F.E. Wawner, A.Y. Teng, and S.R. Nutt, "Microstructural Characterization of SiC (SCS) Filaments, Part b," Paper presented at the Seventh Annual Conference on Composites and Advanced Ceramics, Cocoa Beach, FL, American Ceramic Society, Jan 1983
57. S.R. Nutt and F.E. Wawner, Characterization of the Surface Layer on SiC Filaments, Metal Matrix Composites II, J.D. Buckley, Ed., NASA Conference Publication 2252, National Aeronautics and Space Administration, 1982, p 27-42
58. D.K. Shetty, M.R. Pascucci, B.D. Mutsuddy, and R.R. Wills, SiC Monofilament-Reinforced Si_3N_4 Matrix Composites, *Ceram. Eng. Sci. Proc.*, Vol 6 (No. 7-8), 1985, p 632-645
59. R.T. Bhatt, "Mechanical Properties of SiC Fiber-Reinforced Reaction-Bonded Si_3N_4 Composites," NASA Technical Report 85-C-14, National Aeronautics and Space Administration, 1985
60. R.T. Bhatt, "Mechanical Properties of SiC Fiber-Reinforced Reaction-Bonded Si_3N_4 Composites," in *Tailoring Multiphase and Composite Ceramics*, Vol 20, *Materials Science Research*, R.E. Tressler, G.L. Messing, C.G. Pantano, and R.E. Newnham, Ed., Plenum Press, 1986, p 675-686
61. R.T. Bhatt, "Mechanical Properties of SiC/RBSN Composite Laminates," Paper presented at the 11th Annual Conference on Composites and Advanced Ceramic Materials, Cocoa Beach, FL, American Ceramic Society, Jan 1987
62. N.D. Corbin, C.A. Wilkens, and S.D. Hartline, RBSN Matrix Composites Reinforced with Polymer Derived Fibers, Metal Matrix, Carbon and Ceramic Matrix Composites 1987, in *Proceedings of the Joint NASA/DOD Conference*, 1988, NASA Conference Publication 2482, J.D. Buckley, Ed., National Aeronautics and Space Administration, 1987, p 351-364
63. R. Chaim, A.H. Heuer, and R.T. Chen, Microstructural and Microchemical Characterization of SiC-Based Fibers Produced from Polymer Precursors, *Adv. Ceram. Mater.*, in press
64. C.H. Anderson and R. Warren, Silicon Carbide Fibers and Their Potential for Use in Composite Materials, Part I, *Composites*, Vol 15 (No. 1), 1984, p 16-24
65. J.W. Lucek, G.A. Rossetti, and S.D. Hartline, Jr., Stability of Continuous Si-C (-O) Reinforcing Elements in Reaction Bonded Silicon Nitride Process Environments," NASA Publication 2406, National Aeronautics and Space Administration, 1985, p 27-38
66. N.D. Corbin, C.A. Wilkens, and S.D. Hartline, The Influence of Interfacial Modifiers on RBSN Matrix Composite Properties, Metal Matrix, Carbon and Ceramic Matrix Composites 1987, in *Proceedings of the Joint NASA/DOD Conference 1986*, NASA Conference Publication 2482, J.D. Buckley, Ed., National Aeronautics and Space Administration, 1987, p 351-379
67. P.D. Shalek, G.F. Hurley, D.E. Christiansen, W.J. Parkinson, and J.D. Katz, "Effect of Process Parameters on the Growth of VLS SiC Whiskers," Metal Matrix, Carbon and Ceramic Matrix Composites 1987, NASA Conference Publication 2482, National Aeronautics and Space Administration, 1987, p 127-144

68. S.R. Nutt and D.S. Phillips, "Interface Structure in Silicon Carbide-Silicon Nitride Composites," in *Interfaces in Metal-Matrix Composites*, A.K. Dhingra and S.G. Fishman, Ed., American Institute of Mining, Metallurgical, and Petroleum Engineers, 1986, p 111-120
69. R.J. Davis, C.H. Carter, Jr., S.R. Nutt, K.L. More, and S. Chevacharoenkul, "Structure and Chemistry of Interfaces in Silicon Carbide-Containing Materials," in *Ceramic Microstructures '86: Role of Interfaces*, J.A. Pask and A.G. Evans, Ed., Plenum Press, 1987, p 897-910
70. C.A. Wilkens and N.D. Corbin, Development of Toughened Si_3N_4 Composites by Glass Encapsulated Hot Isostatic Pressure, in Proceedings of the 25th Automotive Tech. Dev. Contractors Coord. Meeting, Society of Automotive Engineers, Oct 1987
71. S.T. Buljian, "Silicon Nitride-Metal Carbide Composites," Ceramic Technology for Advanced Heat Engines Project Semiannual Progress Report for April through September 1986, Report ORNL/TM-10308, U.S. DOE Cont. DE-AC05-840R21400, Oak Ridge National Laboratory, 1987, p 44-53
72. S.T. Buljian and J.G. Baldoni, Silicon Nitride-Based Composites, in *Preparation and Properties of Silicon Nitride*, D.A. Bonnell and T.Y. Tien, Ed., in press
73. S.T. Buljian, J.G. Baldoni, and M.L. Huckabee, Si_3N_4-SiC Composites, *Am. Ceram. Soc. Bull.*, Vol 66 (No. 2), 1987, p 347-352
74. S.T. Buljian, J.G. Baldoni, M.L. Huckabee, J.T. Neil, and G. Zilberstein, SiC Whisker-Reinforced Si_3N_4, in Proceedings of the 25th Automotive Tech. Dev. Contractors Coord. Meeting, Society of Automotive Engineers, 1987
75. S.T. Buljian, J.G. Baldoni, M.L. Huckabee, and G. Zilberstein, Dispersion Toughened Si_3N_4, in *Proceedings of the 24th Automotive Engineers, Inc.*, Society of Automotive Engineers, 1987, p 319-325
76. S.T. Buljian and G. Zilberstein, Microstructure Development in SiC Whisker-Reinforced Si_3N_4 Composites, in *Materials Research Society Symposium Proceedings 78,* Materials Research Society, 1987, p 273-281
77. R.D. Nixon, S. Chevacharoenkul, M.L. Huckabee, S.T. Buljian, and R.F. Davis, Deformation Behavior of SiC Whisker Reinforced Si_3N_4, in *Materials Research Society Symposium Proceedings 78*, Materials Research Society, 1987, p 295-302
78. K. Haynes, M. Martin, and H. Yeh, "SiC-Whisker-Toughened Silicon Nitride," Ceramic Technology for Advanced Heat Engines Project Semiannual Progress Report for April through September 1986, Report ORNL/TM-10308, U.S. DOE Cont. DE-AC05-840R21400, Oak Ridge National Laboratory, 1987, p 54-73
79. R. Hayami, K. Ueno, I. Kondou, N. Tamari, and Y. Toibaba, Si_3N_4-SiC Whisker Composite Material, in *Tailoring Multiphase and Composite Ceramics*, Vol 20, *Materials Science Research*, R.E. Tressler, G.L. Messing, C.G. Pantano, and R.E. Newnham, Ed., Plenum Press, 1986, p 663-674
80. K. Ueno and Y. Toibana, Mechanical Properties of Silicon Nitride Ceramic Composite Reinforced with Silicon Carbide Whiskers, *Yogyo-Kyokai-Shi* (*J. Ceram. Soc. Jpn.*), Vol 91 (No. 11), 1983, p 491-497
81. T. Kandori, S. Kobayashi, S. Wada, and O. Kamigaito, SiC Whisker Reinforced Si_3N_4 Composites, *J. Mater. Sci. Lett.*, Vol 6, 1987, p 1356-1358

82. S. Kobayashi, T. Kandori, and S. Wada, Microstructure of Si_3N_4 Composites Reinforced with SiC Whiskers, *J. Ceram. Soc.*, Vol 94, 1986, p 903-905
83. J.J. Brennan and K.M. Prewo, *J. Mater. Sci.*, Vol 17, 1982, p 2372-2383
84. R.L. Stewart, K. Chyung, M.P. Taylor, and R.F. Cooper, in *Fracture Mechanics of Ceramics*, R.C. Bradt, A.G. Evans, D.P.H. Hasselman, and F.F. Lange, Ed., Plenum Press, 1986, p 33
85. P.M. Benson, K.E. Spear, and C.G. Pantano, in *Proceedings of the 12th Annual Conference on Composites and Advanced Ceramics*, American Ceramic Society, 1988
86. T. Mah, P.R. Subramarian, and R.J. Kerans, in *Proceedings of the 12th Annual Conference on Composites and Advanced Ceramics*, American Ceramic Society, 1988
87. U.V. Deshmukh and T.W. Coyle, in *Proceedings of the 12th Annual Conference on Composites and Advanced Ceramics*, American Ceramic Society, 1988
88. M. Ruhle, B.J. Dalgleish, and A.G. Evans, *Scr. Metall.*, Vol 21, 1987, p 681-686
89. V. Nair, K. Johns, and C. Ostertag, in *Tailoring Multiphase and Composite Ceramics*, Vol 20, *Materials Science Research*, R.E. Tressler, G.L. Messing, C.G. Pantano, and R.E. Newnham, Ed., Plenum Press, 1986
90. W.L. Vaughn, J. Homeny, and M.K. Ferber, *Proc. Ceram. Eng. Soc.*, Vol 8, 1987, p 848-859
91. B. Bender, D. Shadwell, C. Bulik, L. Invorvat, and D. Lewis III, *Am. Ceram. Soc. Bull.*, Vol 65, 1986, p 363-369
92. R.N. Singh and M.K. Brun, *Adv. Ceram. Mater.*, Vol 3, 1988, p 235
93. R.A. Lowden and D.P. Stinton, "The Influence of the Fiber-Matrix Bond on the Mechanical Behavior of Nicalon/SiC Composites," Report ORNL/TM-10667, Oak Ridge National Laboratory, 1987
94. R.A. Lowden and D.P. Stinton, in *Tailoring Multiphase and Composite Ceramics*, Vol 20, *Materials Science Research*, R.E. Tressler, G.L. Messing, C.G. Pantano, and R.E. Newnham, Ed., Plenum Press, 1986

9

Structure-Property Relationships for $YBa_2Cu_3O_{7-x}$ Grain Boundaries: Comments on Possible Microstructural Origins for Their Critical Current Characteristics

S.E. Babcock
Department of Materials Science and Engineering and Applied Superconductivity Center
University of Wisconsin—Madison
Madison, WI 53706

In the wake of the exciting discovery of $YBa_2Cu_3O_{7-x}$, it has become disappointingly clear that the critical current characteristics of this high-temperature superconductor are controlled by the detrimental properties of high-angle grain boundaries. Despite the severity of this problem for future applications of the material, a comprehensive understanding of the sources for the low grain-boundary critical current density values and proof of their inevitability has yet to be established. This chapter summarizes key points of $YBa_2Cu_3O_{7-x}$

grain-boundary studies that have been conducted both by the author and her collaborators and by numerous other researchers. However, it is neither a complete nor an unbiased review of all the published literature on this subject.

The two aims of this chapter are to build a picture of what is known about the structure and properties of $YBa_2Cu_3O_{7-x}$ grain boundaries and to stimulate thought about what must be established more concretely. Because research in this area is very active, the answers to many basic issues are incomplete or speculative in nature, and more questions are posed than are resolved. Credit for a number of ideas presented must be given to the referenced authors.

The Nature of $YBa_2Cu_3O_{7-x}$ Grain-Boundary Studies

The $YBa_2Cu_3O_{7-x}$ Grain-Boundary Problem. Many futuristic applications of superconducting materials (for example, high-speed levitated trains and energy storage devices), as well as present technologies (magnetic resonance imaging body scanners and particle accelerators), require the use of kilometers of wire-form superconductor under conditions of high magnetic field and high current density. Consequently, the transport critical current density (J_{ct}) is the crucial design parameter for these technologies. Unlike the other critical parameters used to describe the superconducting properties of a material, the J_{ct} value is almost always controlled by microstructural features. Depending on their size and distribution, crystal defects can either increase the measured J_{ct} value, by acting as flux-pinning sites, or decrease the J_{ct} value by acting as weak links. Weak links are defects at which the J_{ct} is locally reduced or is equal to zero. They pose a serious problem for high current density applications when the weak-linking defects form a network throughout the material that diminishes the macroscopic critical current density either in a homogeneous way or by restricting the current to a few percolative paths.

If the inevitability of uniform weak-link behavior at high-angle $YBa_2Cu_3O_{7-x}$ grain boundaries were established beyond a doubt, detailed grain-boundary studies would largely be of academic interest only. However, a number of experiments have indicated that a variety of mechanisms can control the J_{ct} characteristics of high-angle grain boundaries. The resulting J_{ct} characteristics differ in very important ways. For example, thin film bicrystal studies (Ref 1) have revealed high-angle grain-boundary weak links that exhibit a Josephson junction characteristic. Josephson junction behavior suggests that the superconducting grains are separated by a very thin layer of material with normal metal properties. This microstructure results in a highly field-dependent J_{ct} in the weak-field regime and in zero-field J_{ct} values that are one to two orders of magnitude lower than those of the abutting grains. On the other hand, direct evidence that some high-angle grain boundaries have a flux-pinning controlled J_{ct} has been found in studies of bulk-scale bicrystals (Ref 2) and highly textured polycrystalline thin films (Ref 3). Flux-pinning limited J_{ct} characteristics are of particular interest for two reasons. First, they suggest the possibility of fabricating materials that sustain appreciable J_{ct} values in fields approaching the upper critical field. Second, since magnetic flux is generally pinned by microstructural defects, the possibility arises for improved engineered materials through microstructure design. In fact, microstructural control is responsible for a dramatic improvement of the properties of the low-temperature superconductors that are used in applications today (Ref 4). The observed variability of the grain-boundary J_{ct} characteristics makes the problem of particular interest to grain-boundary researchers. It also kindles some optimism regarding the fabrication of bulk-scale polycrystalline materials that are free from the weak-link character.

The ultimate goal of many $YBa_2Cu_3O_{7-x}$ grain-boundary studies is the improvement of the J_{ct} properties of polycrystalline materials through control of the microstructure. To accomplish this in a premeditated fashion requires very detailed descriptions of the mechanism or mechanisms that limit the grain-boundary J_{ct} or $J_{ct(GB)}$. In principle, this objective can be reached by first identifying the crucial features of the microstructure and then determining how to manipulate them for optimum J_{ct} performance. In practice, however, these steps are very complicated. As discussed below, a myriad of factors may contribute to the J_{ct} characteristics in complex, interconnected ways, and controlling the microstructure on the required spatial scale is extremely challenging.

Pertinent Microstructural Information. The types of microstructural features that can determine grain-boundary properties are many and varied.

Types of Features. Thin layers of wetting second-phase, impurity segregation, or nonstoichiometric material could act as continuous barriers to the supercurrent. The superconducting transition temperature (T_c) of $YBa_2Cu_3O_{7-x}$ is very sensitive to oxygen concentration in the range of $0.6 > x > 0$ (Ref 5) and is also reduced in the presence of substitutional impurities such as Fe (Ref 6). Furthermore, other superconducting phases with reduced T_cs (~80 K) are formed at the compositions $YBa_2Cu_4O_{8-x}$ and $Y_2Ba_4Cu_7O_{15-x}$ (Ref 7). Concomitant changes in the transport properties of these compositions are anticipated but not presently known in sufficient detail. The grain-boundary properties may depend on the macroscopic and microscopic crystallographic parameters of boundaries that are free from J_{ct} limiting second phases or nonstoichiometric layers. Because the transport properties of high-temperature superconductors can be influenced by defects with near atomic-scale dimensions, variations in the atomic arrangement in the various structural units that compose the boundary and their layout may determine the boundary properties. If different structural units have widely differing properties, the $J_{ct(GB)}$ characteristics will be sensitive to the crystal misorientation and the boundary plane inclination. A mapping of any crystallographic dependencies is of particular interest for $YBa_2Cu_3O_{7-x}$ because of both the known anisotropy of the physical properties and the strong tendency for plate-shaped grains. The electrical properties of superconductors are generally sensitive to lattice strain. Therefore, the distribution of structural and/or extrinsic grain-boundary dislocations may also strongly influence the $J_{ct(GB)}$ characteristics through the elastic strain profile of the boundary (Ref 8, 9). Any one of these features may control the electrical properties of the grain boundaries, or all may contribute in complicated, interconnected ways. The independent testing of the influence of each factor is a considerable experimental challenge in itself.

Spatial Scale Required for Microstructure Evaluation. The maximum dimensions of defects that can be tolerated by the supercurrent are on the order of the superconducting coherence length (ξ). Networked weakly superconducting or nonsuperconducting regions that are significantly larger than ξ suppress the J_{ct} by acting as barriers to electrical transport. On the other hand, isolated defects with near-ξ-scale dimensions can play the beneficial role of flux-pinning sites (Ref 4). The superconducting coherence length thus dictates the spatial resolution with which microstructural characterizations ultimately must be carried out. It is the distribution of weakly superconducting microstructural features that determines their consequences for the macroscopic J_{ct} properties that are measured.

The superconducting coherence length for $YBa_2Cu_3O_{7-x}$ is approximately 0.3 nm along the [001] direction and 1.5 nm within the (001) plane of its orthorhombic structure. These spatial dimensions are on the order of the structural width of the grain boundary. Thus, the intrinsic structure and composition of the boundary core are expected to control the $J_{ct(GB)}$ characteristics, provided that the boundary is free from extrinsic defects (for example, second phase, impurity contamination, or cracks). The anisotropic nature of ξ directs further attention to the crystallographic parameters of the grain boundary. In particular, the grain-boundary plane inclination at the atomic level may strongly influence the $J_{ct(GB)}$ value.

A battery of techniques has already been applied to the grain-boundary weak-link problem. Lower-spatial-resolution techniques such as light microscopy and scanning electron microscopy have provided important information about the general nature of the microstructure. Higher-resolution techniques such as scanning Auger microanalysis, microanalysis in the transmission electron microscope (TEM), and diffraction contrast TEM imaging, indicate that the $J_{ct(GB)}$ characteristics are, indeed, determined by the near-ξ-scale (micro)structure. However, even these techniques may not have sufficient resolution to provide a clear understanding of the microstructural origin of the weak-link character. The fine scale on which the $J_{ct(GB)}$ appears to be controlled has thus pushed researchers to use the highest-resolution investigative techniques available.

Present Knowledge about $YBa_2Cu_3O_{7-x}$ Grain Boundaries

Grain-Boundary Superconducting Properties. Grain-boundary J_{ct} properties have been deduced indirectly from measurements on polycrystalline materials that are contrasted with single-crystal measurements and directly from the characterization of thin film and bulk-scale bicrystals.

The highest J_{ct} values (~10^7 A · cm^{-2} at 77 K, 0 T) are obtained in oriented thin film single crystals (Ref 10). Their J_{ct} behavior is characteristic of a flux-pinning mechanism. In particular, J_{ct} is only moderately dependent on the applied magnetic field for fields sufficiently below the upper critical magnetic field (H_{c2}). The 77 K zero-field J_{ct} values observed in bulk-scale single crystals, significantly lower than those of thin films, range from ~ 10^3 A · cm^{-2} for flux-grown single crystals (Ref 2) to somewhat greater than 10^4 A · cm^{-2} for melt-processed quasi-single crystals (Ref 11). Nevertheless, the field dependencies of J_{ct} for these materials also show characteristics of a flux-pinning-limited J_{ct}. Values for J_{ct} as high as 10^3 A · cm^{-2} have been measured for sintering polycrystalline $YBa_2Cu_3O_{7-x}$ at 77 K and 0 T, but the field dependence is different from that observed in single crystals in very important ways (Ref 12). In these materials, J_{ct} falls rapidly to less than 50 A · cm^{-2} in very weak fields of a few mT, a characteristic that is typical of a Josephson junction weak-linked system. Experiments that probed the average grain-boundary properties for large sets of boundaries have therefore shown that the $J_{ct(GB)}$ on average is rather low in zero field and drops rapidly to very low values in small magnetic fields.

Subsequent studies that provided more detailed information about $J_{ct(GB)}$ have indicated that the superconducting properties actually vary substantially from boundary to boundary. For example, improved high-field J_{ct} characteristics were observed for thin, silver-clad tapes (Ref 13) (>10^3 A · cm^{-2} at 4.2 K and 15 T). Their higher J_{ct} values suggest that a significant fraction of the grain boundaries in this microstructure may have a flux-pinning-controlled J_{ct}

with more favorable J_{ct} characteristics than the average boundary has. Presumably, the grains of these tapes are largely aligned with their high-conductivity (001) planes parallel to the current direction (Ref 14). Bicrystal studies have shown explicitly that the mechanism controlling J_{ct} varies from boundary to boundary in ways that are not yet predictable (Ref 1-3, 15). The corresponding measured properties range from a flux-pinning-controlled J_{ct} to Josephson junction character to those of a resistive barrier. Complicating the picture even more are indications that the superconducting properties vary within individual grain boundaries in a scale that approaches ξ (Ref 3, 16). Experimental evidence in support of these conclusions is reviewed in the following paragraphs.

The first bicrystal studies were conducted with thin film grain boundaries. Systematic sets of samples with a wide range of misorientation relationships were produced by the epitaxial deposition of $YBa_2Cu_3O_{7-x}$ onto prefabricated $SrTiO_3$ bicrystal templates (Ref 1, 17). These bicrystals clearly showed the highly field-dependent J_{ct} that is characteristic of a Josephson junction-controlled weak link for all high-angle grain boundaries, even at 4.2 K. The transition from strong-coupling to weak-linking behavior occurred over the range of 5 to 10°. In practice, this experiment has defined the low-angle and high-angle regimes for all subsequent investigations.

The conclusions of these studies were particularly discouraging. First, they indicated that all high-angle boundaries were Josephson junction weak linked in a uniform manner that was largely independent of the detailed boundary of crystallography. Second, the crystal misorientation angle (θ) dependence of $J_{ct(GB)}$ suggested that the weak-link behavior was attributable to the intrinsic primary grain-boundary dislocations. Third, the low-angle regime, in which at least the zero-field $J_{ct(GB)}$ was comparable to the intragrain J_{ct} value, was very limited in extent. Finally, they demonstrated a weak-link character for the set of boundaries predicted least likely to be weak linked. On the basis of ξ anisotropy alone, the [001] tilt boundaries probed in the first study (Ref 17) are expected to show the most favorable J_{ct} characteristics because the higher-conductivity (001) planes remain parallel throughout the bicrystal and ξ is at or near a maximum in the current direction.

Subsequent characterization of grain boundaries in thin films (Ref 3) and bulk-scale bicrystals (Ref 2) has shown the grain-boundary weak-link problem to be more complicated than a uniform Josephson junction picture can explain. In the thin film experiments, the voltage current characteristics for 45° [001] tilt boundaries, most likely with a range of boundary plane inclinations, were analyzed in detail. Tilt boundaries, from a single boundary to a few, were

Table 1 Characteristic behavior observed in various flux-grown bicrystals

C-type, degree	Flux-pinning	Josephson junction	Resistive
3	×		
10		×	
10	×		
14	×		
22 - 37(a)	×		
22		×	
26		×	
28	×		
33		×	
34		×	
38		×	
40		×	
42			×
43		×	
43		×	
45			×
P-type			
3	×		
15	×		
15			×

(a) Flux-pinning current in this multicrystal must have passed along a path containing one of several high-angle boundaries: a C24; a C29; a C37 and C29; or a C16, C22, and C37. The exact path could not be determined.

isolated for characterization by patterning narrow tracks in highly textured thin films. The requisite texture was obtained by deposition onto [001]-oriented MgO under conditions for which nearly all of the $YBa_2Cu_3O_{7-x}$ grains grow with *c* parallel to the surface normal and *a* or *b* parallel to either [100] or [110] of the cubic substrate (Ref 3). Two important conclusions were drawn from the electrical behavior of the tracks. First, a substantial variety in properties was observed from boundary to boundary, even though all boundaries had nominally the same misorientation relationship. The weak-link J_{ct} varied from 10^4 to 10^6 A · cm^{-2} at 4.2 K and 0 *T*. Perhaps more importantly, the different tracks showed characteristics ranging from those of a resistively shunted Josephson junction at the low-J_{ct} end to those of one-dimensional flux creep along the weak link at the high-J_{ct} end. Second, the resistively shunted Josephson junction weak links showed field-dependent behavior that indicated a nonhomogeneity of the properties within the boundary as well.

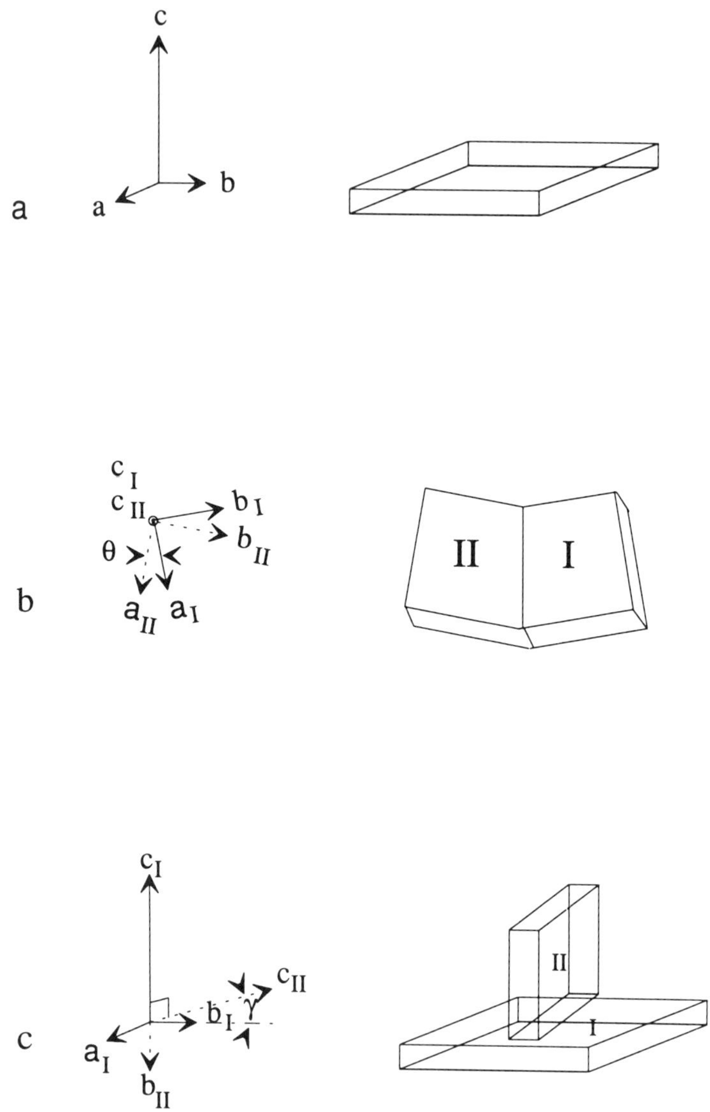

Fig. 1 C-type and P-type bicrystal geometries. (a) Common rectangular-plate shape of flux-grown crystals. (b) Misorientation relationships of C-type bicrystals are θ [001] (with possible small misalignments of the [001] axis). Nomenclature Cθ gives macroscopic θ. (c) P-type bicrystals, with two components of misorientation. Common component of misorientation is 90° rotation about *a*-axis, which corresponds to one of the adjacent crystals standing on large plane surface, the other on plate edge. Second component is rotation about $c_I = b_{II}$ axis by angle γ. Nomenclature Pγ describes misorientation angle of second component of misorientation

Evidence for boundary-dependent J_{ct} properties and for local J_{ct} variation within individual boundaries also has been obtained in bulk studies (Ref 2, 15, 16). Studies of the bulk-scale bicrystals by the authors and coworkers have focused on distinguishing boundaries that show flux-pinning, Josephson junction, and resistive properties. A strongly field-dependent J_{ct} in weak fields is interpreted as the signature of Josephson junction character. Boundaries described as flux pinning have J_{ct} characteristics that largely mimic those of the component single crystals. Their properties include a field-independent J_{ct} in weak fields (up to a few tens of mT) and weakly field-dependent J_{ct} in fields nearly up to the irreversibility field. Resistive boundaries show resistive behavior down to 4.2 K. As seen in Table 1, all three types of behavior have been observed in the high-angle regime (Ref 15). Since boundary-plane inclination varies for all of the bicrystals measured, only the macroscopic misorientation relationship is given. The C-type and P-type bicrystal geometries are described in Fig. 1.

Perhaps the most exciting result presented in Table 1 is the flux-pinning character observed for bicrystals P3 (see Fig. 2) and P15. For both of these bicrystals, the high-conductivity (001) planes of one crystal are oriented 90° away from those in the other crystal. Thus, from the viewpoint of the anisotropy in ξ, these boundaries must be identified as very high-angle ones whose J_{ct} properties might be expected to be the weakest. Furthermore, the field-orientation dependence of the high-field (0 to 7 T) properties of P3 suggested, but, unfortunately, did not prove, that the macroscopically (001)-faced facet in the bicrystal (Fig. 2) had essentially the same flux-pinning character as the (010) twist facet (Ref 2, 15). The discovery of any high-angle grain boundaries with flux-pinning-controlled J_{ct} properties that are equivalent to those of their single-crystal counterparts (Ref 2) is, in its own right, very stimulating because it breaks with the bleak conclusions of the original thin film bicrystal studies.

A second observation of interest is the two cases in which two bicrystals with nominally the same crystal misorientation relationship showed the characteristics of different controlling mechanisms. One flux-pinning and one Josephson junction-coupled bicrystal was found for both the C10 and the P15 geometries (Ref 15). It is noted here that the θ values are measured on a macroscopic scale with a confidence of $\pm$ 1° and that small (a few degrees) misalignments of the c-axes for the C-type geometry could go undetected. Furthermore, the boundary plane inclinations are different for the two bicrystals in each pair. Subtle differences in the crystal misorientation relationship or differences in the boundary plane, therefore, could account for the critically

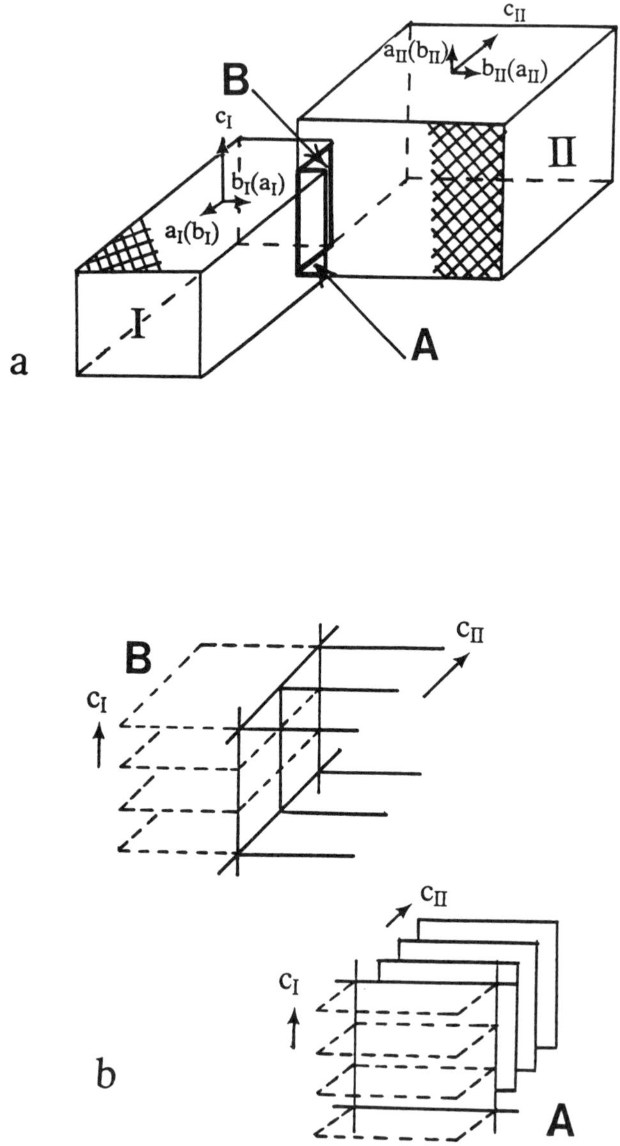

Fig. 2 Grain-boundary geometry for P3 flux-pinning boundary. (a) A bicrystal. (b) Relative orientation of higher-conductivity (001) planes at macroscopic facets A and B. Source: Ref 2

diverse J_{ct} properties. However, the observation is of note because of the analogous behavior observed at 45° [001] tilt boundaries in thin films (Ref 3). From the standpoint of applications, evidence that the $J_{ct(GB)}$-controlling mechanism can be manipulated is very encouraging because it opens up the possibility of the further development of polycrystalline materials.

Grain Morphology, Grain Alignment, and Important Grain Boundaries that Result. This section describes the features of typical $YBa_2Cu_3O_{7-x}$ grain microstructures that direct attention to the study of particular families of grain boundaries.

In almost all cases, the grains in polycrystalline $YBa_2Cu_3O_{7-x}$ are far from equiaxed. On the contrary, they generally are plate shaped with large (001) surface facets. In sintered material, the aspect ratio of the linear dimensions in the (001) basal plane to the thickness of the [001] direction is typically six to one. As a result, the distribution of grain-boundary types is skewed toward a predominance of grains that have (001) boundary planes in one or both crystals. In fact, the areal percentage of basal-plane-faced boundaries in randomly aligned sintered material has been estimated to be as high as 70% (Ref 18). Because ξ is minimum in the *c*-direction, much stricter limits are placed on the dimensions of a boundary region that can still support a flux-pinning-controlled supercurrent for these boundaries than is the case for others. Both their prevalence and their potential for controlling the bulk J_{ct} properties have focused considerable attention on the characteristic of this class of boundaries.

In addition to the limits that the intrinsic electrical properties of the grains impose on transport across (001)-faced boundaries, deleterious microstructural defects have been associated with them. Early work identified the presence of cracks (Ref 18), highly defective regions (Ref 18), wetting amorphous second phase (Ref 19), carbon contamination (Ref 18), and evidence for thin layers of related superconducting compounds at these boundaries (Ref 19-22). As the processing conditions were refined, however, the fraction of boundaries containing some of these relatively large-scale features was reduced to the point that they no longer could clearly account for the poor J_{ct} properties of the material. For example, material that was free from observable grain-boundary-segregated carbon (Ref 23) and amorphous grain-boundary phases still showed weak-link character.

Some disagreement exists about the crystallographic parameters of grain boundaries that tend to be wet with layers of $Y_2Ba_4Cu_7O_{15-x}$ and $YBa_2Cu_4O_{8-x}$. Some authors have argued that they form preferentially at (001)-faced grain boundaries (Ref 21, 22), whereas others suggest that they grow only at grain boundaries that are inclined to the (001) basal plane (Ref 20). Observations made at the University of Wisconsin have suggested that they can form all types of boundaries (Ref 24). Unfortunately, the resolution of this conflict is somewhat complicated because these phases are detected only by TEM in lattice fringe and structure images. These imaging techniques generally require that a sample be oriented with the electron beam parallel to the grain boundary and the (001) planes of the grain under study. The types of boundary that can be investigated are thus limited by experimental concerns. In addition, the possible effects of these superconductor phases are, at present, difficult to predict. With the exception of their T_c values (~80 K) (Ref 7), little is known about the superconducting properties of these phases. It is not at all clear that they differ sufficiently from those $YBa_2Cu_3O_{7-x}$ to account for its weak-link character, especially at lower temperatures.

The somewhat higher J_{ct} values obtained in fine-grained sintered material in which the average aspect ratio of the grains is reduced to approximately 2 suggests that (001)-faced boundaries may dominate the weak-link behavior (Ref 25). However, it is not known whether the distribution of grain-boundary types is responsible for the improved J_{ct} characteristics of the finer-grained material or whether another factor, such as reduced microcracking, controls the properties. Nevertheless, understanding the properties of basal-plane-faced boundaries remains an important goal, especially if the properties are unique, because they may well determine the J_{ct} characteristics of sintered $YBa_2Cu_3O_{7-x}$.

A platelike morphology of the grains and texturing leads to the predominance of a different type of grain boundary in thin films and in silver-clad tapes. The grains often have their *c*-axes aligned parallel to the thin dimension of the sample and perpendicular to the current direction. Boundaries that have large [001] tilt components are thus expected to dominate the properties of these materials. For polycrystalline thin films on cubic substrates, in particular, MgO and ZrO, the [001] axis alignment is nearly perfect. Furthermore, the misorientation angle for most of the grain boundaries belongs to a limited set of values. These preferred θs are correlated to the formation of near-coincidence site lattices (NCSL) of the [001]-oriented cubic substrate and $YBa_2Cu_3O_{7-x}$ overlayer. The resulting $YBa_2Cu_3O_{7-x}$ grain boundaries are largely either a low-angle or near-low-Σ[001] tilt in character (Ref 26-28). Thus, in principle, boundaries with only a limited number of misorientations must be characterized in detail in order to understand the properties of polycrystalline thin films. The [001] axis alignment is less complete in the case of silver-clad wires, and the distribution of misorientation angles is not well known. Never-

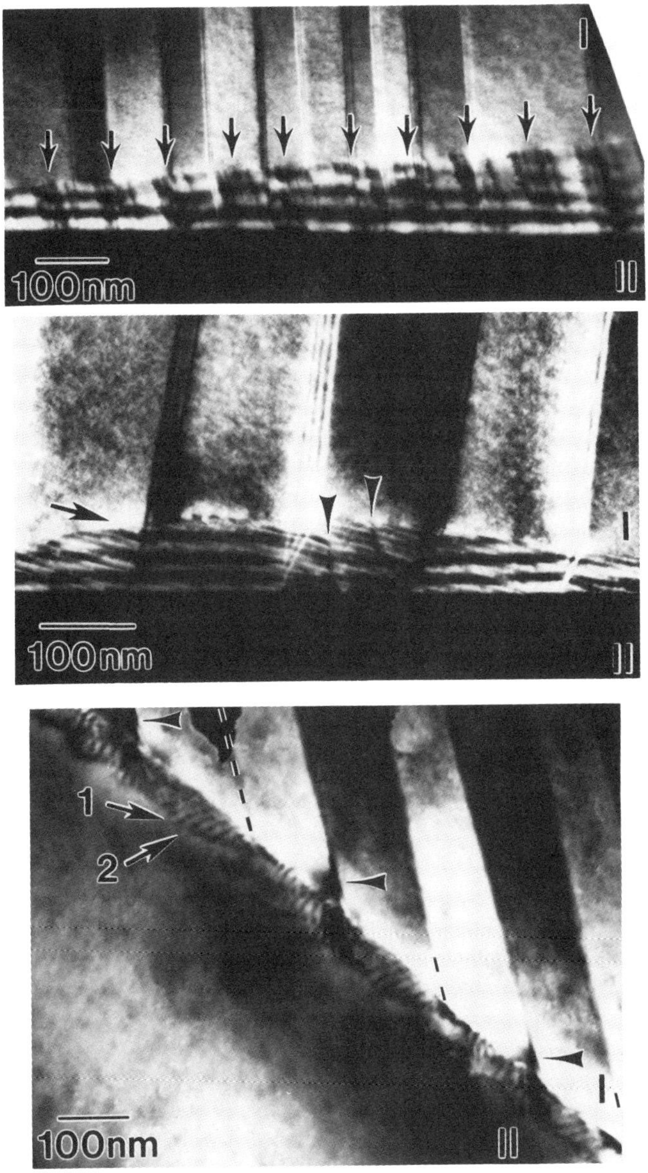

a

b

c

Fig. 3 Examples of SGBD networks in a variety of grain-boundary types. (a) 45° [001] twist boundary with small (0.7°) superimposed tilt component. Clusters of SGBDs indicated by arrows appear to accommodate tilt misorientation. (b) SGBD network in near-Σ51 (001)-faced boundary. (c) SGBD network in 147° [111] near-Σ9 boundary. Boundary is not (001) faced. Source: Ref 8 theless, it is tempting to speculate that similar types of boundaries control the properties of high-J_{ct} tapes and polycrystalline thin films. Correlations between their superior properties relative to sintered compacts and the types of grain boundaries they contain remain an intriguing possibility.

Boundaries with a misorientation relationship of 90° about the *a*- (or *b*-) axis also occur rather frequently in epitaxial thin film samples deposited under certain conditions. Because the orthorhombic $YBa_2Cu_3O_{7-x}$ structure is derived from three cubic perovskite subcells stacked along the *c*-axis, it can be well matched to a cubic substrate in three different orientations. The boundaries that result would be designated P0 in the nomenclature of Table 1. These boundaries have been described as particularly special because they preserve a Σ1 misorientation relationship of the cubic subcell (Ref 29). There is some evidence from thin film studies that P0-type boundaries have special $J_{ct(GB)}$ properties (Ref 29). Clear evidence for a flux-pinning-controlled J_{ct} behavior has been found in a P3-type bulk bicrystal (Ref 2) (see Table 1 and the section "Grain-Boundary Superconducting Properties").

Grain-Boundary Energy. A variety of evidence has indicated that grain-boundary energy varies with boundary crystallography.

The Role of Crystal Misorientation. Low energy appears to be correlated with low-Σ NCSL misorientations of the orthorhombic lattice. Macroscopic and microscopic faceting indicate that certain boundary planes have singularly lower energy as well, but the crystallographic variable that correlates with the special boundary planes is less clear.

Indications for the relevance of the NCSL construction to the high-angle $YBa_2Cu_3O_{7-x}$ grain boundaries are of the types often cited as evidence for Wulff plot minima. For example, Smith, Chisholm, and Clabes (Ref 30) investigated the distribution of misorientation angles about [001] in flux-grown bicrystals. Because the crystals grew in relatively low densities in a melt, it was assumed that they could rotate freely into energy-minimizing misorientations upon impingement. A histogram of the number of observations versus θ in 1° increments suggests that the energy-minimizing misorientations are associated with low-Σ coincident site lattices (CSLs) of the tetragonal parent phase. Values of Σ up to at least 29 and possibly 41 appear to correlate with low energy. Evidence for preferred misorientation angles and misorientation axes has also been found in melt-processed textured samples (Ref 31).

The existence of special NCSL misorientations of the orthorhombic unit cell is further supported by observations of regular secondary grain-boundary dislocation (SGBD) networks in high-angle grain boundaries. SGBD networks have been observed in boundaries that could be described as NCSL-related

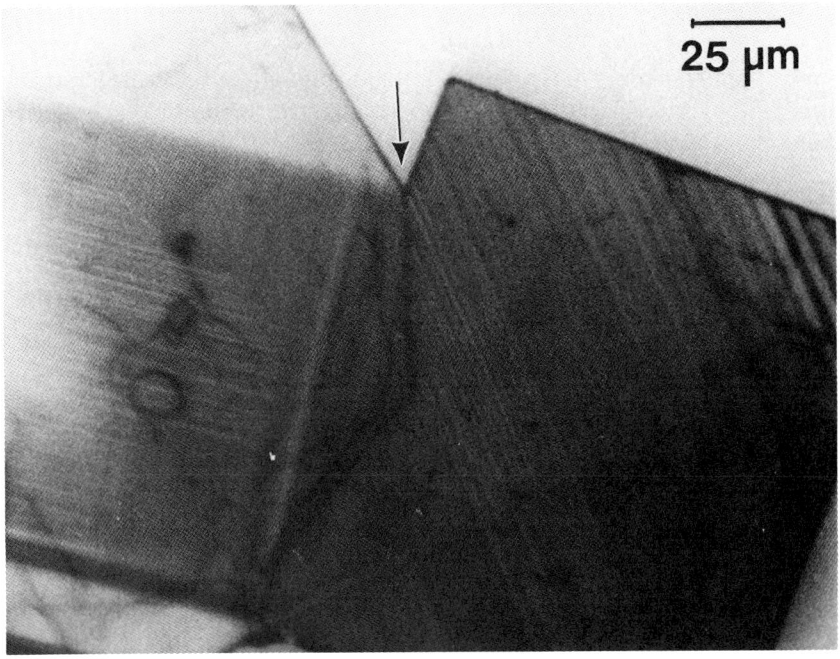

Fig. 4 Light micrograph showing macroscopic faceting in flux-grown bicrystal. Grain boundary indicated by arrows and twin microstructure

with Σ values up to 51 (Ref 8). When compared with systematic searches for SGBD networks in cubic metals and oxides, these Σs are clearly in the "low" range. For example, SGBD networks have been associated with $\Sigma \leq 29$ and $\Sigma \leq 53$ misorientations for Au (Ref 32) and MgO (Ref 33), respectively. Evidence for low-energy NCSL misorientations for rotations other than [001] was also found in the $YBa_2Cu_3O_{7-x}$ study (Fig. 3). Such observations are important because the (001) basal planes of the layered orthorhombic structure are no longer maintained parallel in these cases. Furthermore, the formation of exact CSLs like those that exist for [001] rotations of the tetragonal precursor phase are rare, if they exist at all, for rotations about other axes.

The Role of Grain-Boundary Plane Inclination. Macroscopic and microscopic faceting observations provide evidence for low-energy boundary plane inclinations. On the macroscopic scale, the grain boundaries present in flux-grown bicrystals tend not to lie on the boundary-area-minimizing boundary plane. On the contrary, they facet to form relatively large-scale straight segments (linear dimensions of tens of microns) separated by macroscopically curved segments (Fig. 4). A rather thorough study of boundary trace orientations in flux-grown bicrystals suggested that the energy-minimizing boundary planes are associated with preferred crystal planes {(100), (110), (130), (150)} rather than preferred planes of the CSL (Ref 34). However, the frequency (8 to 14%) for some of the apparently preferred planes was not largely different from that expected for a random distribution of grain-boundary planes (5.6%). A lack of correlation between the macroscopic crystal misorientation and boundary plane inclination was also noted (Ref 34).

Microscopic faceting (facet dimensions apparently on the order of tens of unit cells) has been observed in high-resolution TEM images of low-angle [001] tilt boundaries in thin films that have been deposited on Mg (Ref 35)and $SrTiO_3$ (Ref 36). In this case, the favored boundary planes are (100), (010), (110), and (110). High-resolution TEM observations have indicated that high-angle grain boundaries are also faceted onto low-index crystal planes on the microscopic scale. For example, two approximately 40° [001] tilt grain boundaries were faceted on (100), (110), (210), and (310) at various positions along the boundary. All these observations are fully consistent with the macroscopic evidence for favored boundary plane inclinations at grain boundaries for which the rotation axis is [001]. A microfaceted grain-boundary morphology was also observed in a 90° [100] (P0-type) grain boundary that resulted from the triple positioning of the orthorhombic unit cell during the epitaxial growth of thin films on cubic substrates (Ref 37). This high-angle grain boundary was sharply faceted onto two types of low-index planes in the abutting crystals, the (001) basal plane, and the (013)(013) symmetrical-tilt plane.

The observed propensity for faceting on the macroscopic and microscopic scale suggests a considerable anisotropy in grain-boundary energy with respect to boundary plane inclination. Anisotropy of boundary energy is not surprising because of the layered nature of the crystal structure. From the vantage point of understanding grain-boundary structure-properties relationships, a strong tendency for microfaceting on a few low-index planes might greatly reduce the number of boundaries that must be investigated in detail in order to develop a clear picture of how the grain-boundary J_{ct} is determined.

Twin Matching. Twin domain matching is a relatively frequently observed phenomenon in which the twin boundaries of two abutting crystals intersect the grain boundary at the same place. An example of a twin-matched boundary is shown in Fig. 5. Matched twin patterns have been observed at the same location in a flux-grown bicrystal by light microscopy and TEM imaging. Twin matching has also been detected, but not studied in detail, at high-angle grain boundaries in sintered materials (Ref 38).

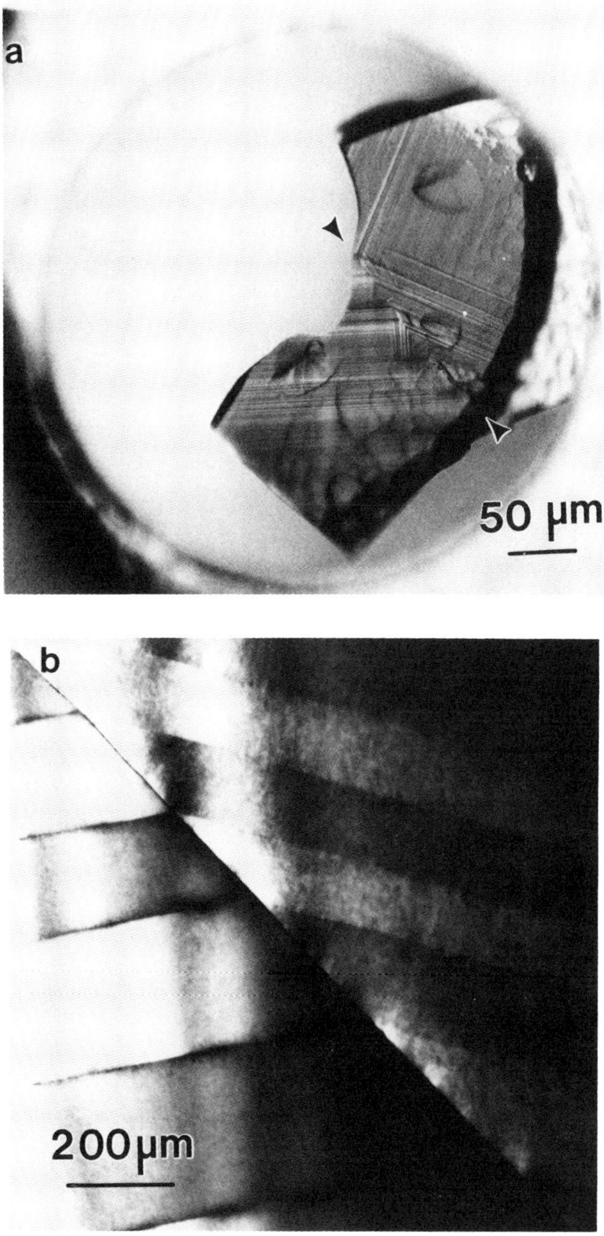

Fig. 5 Twin-matched grain boundary in flux-grown bicrystal (C26 of Table 1). Boundary shows Josephson junction character. (a) Light micrograph of transmission electron microscopy (TEM) sample. Arrows locate grain boundary. (b) Diffraction contrast TEM image of boundary

Twin matching at [001] tilt boundaries has been examined extensively by light microscopy of flux-grown bicrystals (Ref 34). For these boundaries, twin matching occurs most frequently when the angle formed by the boundary and the twin planes is not widely different in the two grains. Under these conditions, the matched twin domains can retain approximately the same spacing in both crystals. Furthermore, for both crystals, the twin boundary traces at the grain-boundary plane of a [001] tilt boundary are parallel to [001] and therefore to each other. This condition may enhance the likelihood of twin domain matching at [001] tilt boundaries.

The twin matching observed in flux-grown bicrystals has been attributed to the elastic strain profile that arises at the grain boundary during the tetragonal-to-orthorhombic phase transition (Ref 34). Gayle and Kaiser's strain-energy model, indeed, predicts the observed details of twin matching. For example, the predicted driving force for twin matching at [001] tilt boundaries is maximum for the (100) and (010) boundary plane and is zero for (110)-faced and (110)-faced boundaries. Twin matching was observed frequently as symmetrical-tilt boundaries at (100)-faced and (010)-faced boundaries. It was observed infrequently at nearly (110)-faced or (110)-faced boundaries. Consistent with observations of twin matching throughout the range of possible misorientation angles, the model also predicts no dependence of twin matching on θ. It also seems possible that the twin plane matching is driven by differences in the crystal misorientation that occur because of the intersection of slightly differently orientated twin domains with the boundary plane. At a twin-matched boundary, each of the twin domains of one crystal abuts only one of the two possible twin orientations of the other crystal. If the boundary energy is strongly dependent on crystal misorientation, the twin-matched structure could correspond to a minimum in the boundary surface energy. Twin matching could also be favored at symmetrical tilt boundaries over other boundaries because it corresponds to the case of equal twin spacings in both crystals.

The observations outlined in the section "Grain-Boundary Energy" provide evidence for the existence of special low-energy misorientation relationships and low-energy planes for $YBa_2Cu_3O_{7-x}$ grain boundaries. Perhaps more important, they also demonstrate consistency between observed characteristics of $YBa_2Cu_3O_{7-x}$ grain boundaries and the basic principles of the grain-boundary dislocation/structural unit (GBD/SU) model for grain-boundary structure. A central theme of this model is that certain low-energy grain-boundary structures act as reference structures from which the structures of more general boundaries are derived. This would suggest that a detailed description of the structure and properties of low-energy grain boundaries could be invaluable for understanding the $J_{ct(GB)}$ characteristics of a broad range of grain boundaries.

Grain-Boundary Dislocations. This section discusses low-angle and high-angle grain boundaries with respect to grain-boundary dislocations.

Low-Angle Grain Boundaries and Primary Grain-Boundary Dislocations. The classic picture of a wall of dislocations that produce a macroscopic misorientation and separate regions of perfect crystal describes well the observed structure of low-angle grain boundaries in $YBa_2Cu_3O_{7-x}$. High-resolution TEM (HREM) images indicate that the disorder is concentrated at primary grain-boundary dislocation (PGBD) cores and that structural integrity is retained in the single-crystal regions between them. As in simpler model oxides such as NiO (Ref 39), the continuity of lattice planes across the boundary is observed between the PGBD cores (Ref 35, 36).

The most complete studies of low-angle ($\theta \leq 7.5°$ in this case) grain-boundary core structures have been conducted on thin films deposited on MgO by a metallo-organic chemical vapor deposition (MOCVD) process (Ref 35, 37, 40). These films are, on average, slightly rich in copper relative to the stoichiometric cation composition. Details of the structural features of low-angle boundaries found in stoichiometric sintered materials have also been investigated directly by HREM (Ref 36).

The boundaries in the MOCVD films contain PGBDs with core structures and compositions that differ greatly from the single-crystal regions between them. The best-matching image simulations model the core as a localized cylinder of highly distorted CuO that is about 2 nm in diameter. In the vicinity of the PGBD core, the lattice appears only slightly strained. The deduced core composition is, at minimum, in qualitative agreement with the copper richness that is detected at the grain boundary in nominally stoichiometric sintered material (Ref 41-43) and, to a greater extent, in CuO-doped sintered material (Ref 44). The PGBD core composition, perhaps including the spatial extent, may be the mechanism by which the excess copper is incorporated into the grain-boundary core. The cumulative evidence suggests, but does not yet prove, that the PGBD cores may be intrinsically CuO rich.

It seems appropriate to describe the PGBD cores as localized, although their sizes somewhat exceed the approximately one-lattice-parameter dimension that is usually associated with a dislocation core. Localized cores with the observed sizes, however, could have a profound effect on grain-boundary superconducting properties because their dimensions exceed ξ. Depending on the details of their distribution, such cores could have either a beneficial or a detrimental effect on the $J_{ct(GB)}$ (Ref 1, 36, 37, 40). A continuous layer of disordered material could produce a barrier to the supercurrent and concomitant weak-link characteristics. Alternatively, a regular network of electrically separated PGBD cores could act as an effective array of flux-pinning sites. "Core overlap" of 2-nm-diam PGBD cores is predicted to occur at about 11° for [001] symmetrical-tilt boundaries with primitive PGBDs (Ref 40). This value is in very good agreement with the J_{ct} plateau and Josephson junction character that were observed in the thin film bicrystal studies for θ greater than about 10°. However, observations of flux-pinning characteristics for grain boundaries with substantially higher θ values (Table 1) indicate that this explanation for grain-boundary weak-link character does not apply in a completely general way.

HREM images of microfaceted low-angle grain boundaries have suggested that the facet junction dislocation cores are generally less localized than are the PGBD cores (Ref 35). The lattice strains in the vicinity of junction dislocations also appear stronger, on average, than those near PGBDs. Their magnitude, however, varies from boundary to boundary.

Evidence that PGBD cores may become amorphous when the misorientation angle exceeds 5 to 7.5° has been found in studies of grain boundaries in thin films and sintered materials (Ref 36). However, the observations described above and in the section below show that not all PGBD cores are intrinsically amorphous. The reason that grain-boundary cores are amorphous is not known, but impurity atoms at trace levels are one possible cause (Ref 19, 45).

Observations of low-angle [100] tilt boundaries have shown PGBDs with Burgers vectors equal to [001] often dissociate into three partial dislocations with $b \cong 1/3[001]$, leaving ribbons of stacking fault in the grain boundary (Ref 36). Crystalline order is retained in the regions between the dissociated cores. The effect of stacking faults on superconducting properties is not known, but it is reasonable to suspect that it may be small because of the extreme localization of the defect. Nevertheless, dislocation dissociation may effectively spread the dislocation core region.

Secondary Grain-Boundary Dislocation Networks in High-Angle Grain Boundaries. Regular networks of secondary grain-boundary dislocations (SGBDs) have been observed by diffraction contrast TEM in a variety of high-angle grain boundaries (Fig. 3) (Ref 8). The range of boundaries in which they are found spans each of the major grain-boundary families that are suggested by crystal structure and anisotropy. Examples of SGBD networks have been found in, first, a boundary for which the rotation axis is largely [001] and boundaries that have other rotation axes and, second, boundaries that are faced by the (001) plane of one or both crystals and boundaries that are not. These observations provide useful information about the boundary structure and microstructure, in addition to insight into their role as evidence for low-energy preferred structures (see the section "The Role of Crystal Misorientation").

The strain contrast produced by SGBDs implies that deviations from low-Σ misorientations are accommodated by physically significant dislocations that concentrate the misfit at their cores and separate areas in which low-energy (low-Σ) atomic configurations remain largely intact. Since this relaxation is a focal point of the GBD/SU model for grain-boundary structure, the detection of localized SGBDs is important evidence in support of its use for describing $YBa_2Cu_3O_{7-x}$ grain boundaries. The resulting microstructure draws attention to two major features of the boundary structure and their possible influences on the $J_{ct(GB)}$. The first of these is the structural units that exist in the regions between the dislocations and at the dislocation cores, and the second is the medium-range strain field that is produced by the SGBD networks. At present, it is not known whether the core structure or the lattice strain has the greater influence on the grain-boundary superconducting properties.

Diffraction contrast images have suggested that SGBD strain fields have physical significance for spacings at least as small as 10 nm. No evidence for the minimum spacing at which physically distinguishable SGBDs exist in $YBa_2Cu_3O_{7-x}$ is presently available, but spacings down to 3.3 nm have been observed in Σ5 GBs in Au (Ref 46) and MgO (Ref 33). Structure images have shown that good crystalline order is maintained right up to the boundary core (Ref 20, 35, 37, 40, 45) and that the boundary core is localized and dense (Ref 35). In fact, images of two approximately 40° tilt boundaries have indicated that the core is thinner than the 2 nm diameter observed for the PGBDs in low-angle grain boundaries (Ref 35). The formation of SGBD networks also suggests that the grain-boundary core is ordered (Ref 8). Atomically well-matched structures have been observed for a number of high-angle grain boundaries, including a faceted 90° [100] triple-positioning boundary in a thin film (Ref 40).

Properties and Application of CSL Constructions. As described previously, several observations have indicated that the NCSL construction is useful for understanding at least some aspects of the structure and properties of $YBa_2Cu_3O_{7-x}$ grain boundaries. In view of the success in using (N)CSL constructions to describe grain boundaries in a wide range of cubic materials and in hexagonal Zn, this outcome is not surprising. Nevertheless, to the author's knowledge, $YBa_2Cu_3O_{7-x}$ is the most complicated crystal structure for which so much clear evidence for the applicability of the NCSL construction has been obtained. The uninterrupted network of SGBDs found on a smoothly curved boundary provided evidence that the three-dimensional (N)CSL has relevance for $YBa_2Cu_3O_{7-x}$. This observation is of interest because of the highly anisotropic nature of the bonding and properties in the layered $YBa_2Cu_3O_{7-x}$ structure.

To date, it has generally been assumed that the pertinent NCSLs are associated with the exact CSLs of a tetragonal unit cell with $c = 3a$. Since the mismatch of the a and b lattice parameters of the orthorhombic structure is about 1.8% and since c is within 0.9% of $3a$, this association seems reasonable (Ref 47). Furthermore, the micron-to-submicron scale of the twin microstructure implies that the a and b directions cannot be differentiated in most physical property measurements. The suggestion that CSLs based on the parent cubic perovskite unit cell are suitable for describing NCSL misorientation relationships in $YBa_2Cu_3O_{7-x}$ also has been proposed (Ref 29). Arguments in favor of this approach are based on the observation that both structures are derived from the same oxygen sublattice. The argument is, in some respects, a matter of semantics, since a CSL of the tetragonal unit cell is always found for each congruent misorientation relationship of the cubic unit cell. The Σ values for the two CSLs are either the same or differ by a factor of three (Ref 48). However, some considerations indicate that the study of tetragonal CSLs may provide more physical insight into the structure and properties of $YBa_2Cu_3O_{7-x}$ grain boundaries. First, the tetragonal unit cell better reflects the layered nature of the crystal structure and the anisotropy of the material's bonding and physical properties. In particular, it seems more appropriate to describe the 90° about [100]-type triple-positioning boundary found in thin films as related to a $\Sigma 3$ CSL of the tetragonal unit cell rather than to a $\Sigma 1$ CSL of the cubic perovskite subcell because this misorientation relationship brings the extrema of the properties' anisotropy into registry. Second, Singh, Chandrasekhar, and King made the important point that the translation vectors of the cubic-based displacement shift complete lattice (DSCL) would not conserve the grain-boundary structure (unless they were coincidentally also translation vectors of the tetragonal-based DSCL) (Ref 47).

Singh *et al.* (Ref 47) have computed all of the CSLs with $\Sigma \leq 50$ for the $c = 3a$ tetragonal unit cell. These CSLs are distributed throughout the high-angle regime such that very broad, angular ranges that are free from CSL misorientations are not found. There is some evidence for the clustering of CSLs. However, the clustering is greatly reduced from that observed for the hexagonal unit cell (Ref 47).

An NCSL construction has also been applied successfully to explain the limited set of misorientation relationships that are found for grain boundaries in polycrystalline $YBa_2Cu_3O_{7-x}$ thin films on single-crystal MgO substrates (Ref 26, 27). The NCSL misorientation angles for [001] MgO//[001] $YBa_2Cu_3O_{7-x}$ were computed and compared with the explicitly determined local crystallographic relationships between the substrate and the thin film. Essentially all of the $YBa_2Cu_3O_{7-x}$ grains formed a low-Σ NCSL with the substrate. Furthermore, and conceivably more important from the J_{ct} standpoint, a significant fraction of the high-angle grain boundaries had low-Σ NCSL misorientation relationships. (In this case, the tetragonal unit cell was assumed for computing the NCSL misorientations.)

In sum, it appears that the CSL and NCSL concepts are, indeed, useful for understanding certain features of grain-boundary morphology and grain-boundary structure in both bulk and thin film $YBa_2Cu_3O_{7-x}$.

Grain-Boundary Composition. The details of the rather complex grain-boundary composition question have yet to be understood, but some general trends have been identified. Scanning Auger microanalysis (SAM) of fracture surfaces and energy-dispersive x-ray analysis in the scanning transmission electron microscope (STEM-EDS) have shown that the average grain-boundary core is rich in copper relative to grain interiors. STEM-EDS studies have shown that the copper excess is present even at high-angle grain boundaries that are clearly free from wetting second-phase layers and other microstructural defects that can be detected in a diffraction contrast image. This result is, at minimum, qualitatively consistent with the observation of CuO-rich PGBD cores in thin film grain boundaries. Furthermore, the apparent dimensions of the CuO core are in reasonable agreement with the dimensions of the copper-rich zone, as deduced from SAM and STEM-EDS experiments (<5 nm) (Ref 41, 43). However, both the SAM (Ref 41-43) and STEM-EDS studies (Ref 41, 49) have suggested that there is considerable compositional variety from boundary to boundary. In fact, within the sensitivity of the available techniques, the compositions of some grain boundaries and subsections of grain boundaries are the same as those of the grain interiors (Ref 41-43). High-resolution Z-contrast images have suggested that low-angle ($\theta \leq 13^\circ$) tilt boundaries in polycrystalline thin films are structurally and compositionally homogeneous right to the boundary. They show no evidence for compositional changes at the grain boundary and GBD cores (Ref 9). STEM-EDS studies of low-angle grain boundaries in sintered materials (Ref 45) and, apparently, low-Σ grain boundaries in thin films (Ref 49) also have indicated that these boundaries have the same composition as the grain interiors. These observations are easily

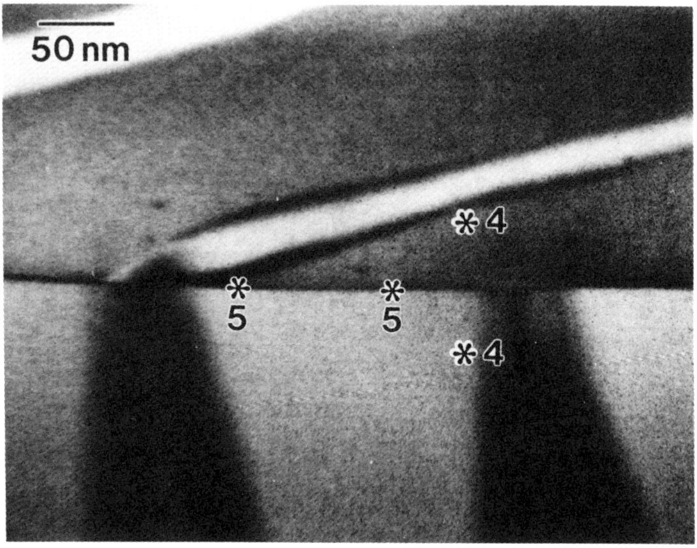

a

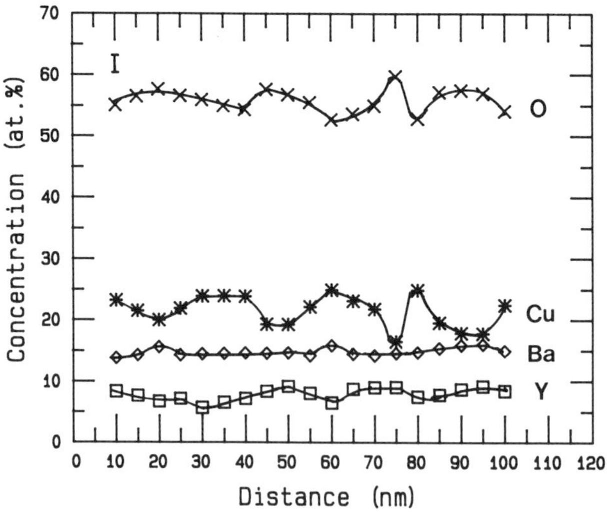

b

Fig. 6 Boundary showing regular composition modulations. (a) Diffraction contrast scanning transmission electron microscope image of grain boundary. (b) Composition profile obtained between asterisks marked 5 in (a). (Notice lack of contrast producing features at profile position in image.) Source: Ref 41 reconciled with the evidence for compositional variety among boundaries. The reports of conflicting compositions for low-angle grain boundaries and PGBD cores could have their origin in either the processing of the materials or the sensitivities of the experimental methods used.

To further complicate the grain-boundary compositional picture, evidence that individual grain boundaries themselves are compositionally nonhomogeneous has been found in STEM-EDS (Ref 41) and SAM (Ref 43) experiments. Rather short-period oscillatory cation composition variations along the boundary plane have been observed in one boundary (Fig. 6). A nonhomogeneous distribution of the $YBa_2Cu_4O_{8-x}$ or $Y_2Ba_4Cu_7O_{15-x}$ phase could be responsible for the observed compositional nonhomogeneity, but this has not been shown directly. Alternatively, observations of oscillatory compositional variations have been interpreted as possible evidence of intrinsic compositional changes associated with secondary grain-boundary dislocations. Although regular compositional variations are consistent with observations of CuO PGBD cores, the observed effect is too large to explain in terms of the intrinsic uneven distribution of PGBDs that occurs at most high-angle grain boundaries.

Trace-level impurity atoms such as Mo, Zr, and Sr have been shown to segregate preferentially to low-symmetry boundaries. They are not detected at high-symmetry high-angle grain boundaries, twin boundaries, or low-angle grain boundaries (Ref 45). Wetting layers of a second phase, often amorphous, are frequently associated with impurity-contaminated grain boundaries. In particular, amorphous layers 1 to 5 nm thick containing Si and excess Y have been attributed to the Si impurity (Ref 19). Second-phase formation has also tentatively been attributed to carbon segregation (Ref 18, 21).

The $YBa_2Cu_3O_{7-x}$ grain-boundary stoichiometry problem is clearly complicated and is incompletely understood. However, at least two important features of grain-boundary composition have been uncovered. First, compositional deviations and variety are associated with high-angle grain boundaries in this material. Second, the compositionally modified grain-boundary region is very narrow. Since its dimensions approach ξ, it can strongly influence grain-boundary properties. Further investigation challenges the techniques presently available.

Summary. The key points of the above discussion are:

- $YBa_2Cu_3O_{7-x}$ grain boundaries are "conventional." They contain structural features and have dimensions that are characteristic of grain boundaries in more completely studied materials with cubic crystal structures.

They demonstrate key principles of the models that have been developed to describe grain boundaries in crystalline materials

- The J_{ct} properties of grain boundaries and the mechanisms by which they are determined vary from boundary to boundary within the high-angle regime. Some evidence indicates that properties vary markedly on the nanometer scale within individual grain boundaries
- The structure and composition of grain boundaries is heterogeneous on the scale of superconducting coherence length. Dislocation core structure and composition, strain fields associated with intrinsic grain-boundary dislocation networks and facet junction dislocations, and microfacet morphology are among the ways in which heterogeneity is introduced into the grain-boundary region. Any or all of these features could control the J_{ct} characteristic of the grain boundary

Possible (Micro)structure-J_{ct} Property Relationships

This section describes possible correlations between observed (micro)structural features and $J_{ct(GB)}$ properties. The suggested mechanisms for the control of $J_{ct(GB)}$ are generally still speculative in nature. Many have their basis in the rather strong supporting evidence for the applicability of the GBD/SU model. However, definitive evidence for the singular importance of any of the noted features has not been produced. On the contrary, counterexamples of each of the proposed mechanisms can be found in the available data, and the fundamental origins for grain-boundary properties remain concealed

The PGBD versus the SGBD Content. A (micro)structure-property relationship that has been proposed frequently is that $J_{ct(GB)}$ is controlled by the PGBD content of the boundary (for example, Ref 1, 9, 40). This dependence would result in an inverse relationship between the misorientation angle and the $J_{ct(GB)}$ value in the low-angle regime, where isolated PGBDs can be identified, and a uniformly diminished $J_{ct(GB)}$ in the high-angle regime, where the PGBDs effectively overlap.

An early proposal speculated that the PGBD cores were not superconducting because of large, localized distortions of the crystal lattice (Ref 1). The superconducting area of the grain boundary would then be reduced by an amount directly proportional to the PGBD content of the boundary, and a concomitant decrease in $J_{ct(GB)}$ would result. Once the PGBD cores overlapped, a thin but continuous layer of normal material would act as a barrier to the supercurrent, causing Josephson junction-type coupling across the grain boundary. Both $J_{ct(GB)}$ characteristics and microstructural observations that support this mechanism have been found. The strong correlation between θ and the normalized $J_{ct(GB)}$ observed in the low-angle regime for thin film bicrystals prompted the nonsuperconducting PGBD core hypothesis as a possible explanation for observed behavior (Ref 1). Josephson junction characteristics have also been observed for a large fraction of the high-angle grain boundaries characterized in bicrystal measurements (Ref 1, 2, 15). TEM investigations of the thin film bicrystals showed a continuity of lattice fringes right up to the boundary plane for all boundaries investigated. They thereby ruled out the presence of a wetting layer of second phase as the J_{ct} reducing defect and pointed to the intrinsic features of the boundary as the culprits. More recent HREM studies showing that large atomic displacements and compositional changes occur within an approximately 2 nm PGBD core (see the section "Low-Angle Grain Boundaries and Primary Grain-Boundary Dislocations") explicitly support the PGBD-core, limited-grain-boundary-area mechanism (Ref 40). Dimos *et al.* (Ref 1) were, however, careful to point out that this was not the only mechanism to show at least qualitative consistency with their data. For example, PGBD-content-dependent easy flux flow along the grain-boundary plane could also explain the observed J_{ct} behavior.

In parallel with the PGBD core model, it is also possible that local variations in the electrical properties of high-angle grain boundaries are determined by the types and densities of the various structural units that make up the boundary core. An electrically heterogeneous microstructure could result if certain structural units that occur in high-angle grain boundaries were superconducting units with good J_{ct} properties whereas others were not. The fraction of superconducting units with arrangement in the boundary would then determine the nature of the electromagnetic properties on the macroscopic (measurable) scale. Although this model remains speculative, it is an attractive one because it describes both the variety of properties that is observed from boundary to boundary in the high-angle regime and the variation of properties within the boundary core that is indicated by some electromagnetic measurements (Ref 3, 16).

Alternatively, the elastic strain field associated with the grain-boundary dislocation network, rather than the severe distortion at the core alone, could be responsible for the grain-boundary superconducting properties. A simple model that assumes that the material behaves like a normal metal if the local strain exceeds 0.1% has shown rather good agreement with the thin film bicrystal data for the low-angle regime (Ref 9). It also could conceivably explain the observed flux-pinning characteristics of certain high-angle grain boundaries in which the PGBDs were uniformly spaced. It had been proposed previously that the elastic strain field associated with localized SGBD networks might produce the electrically heterogeneous "microbridge" microstructure in certain high-angle grain boundaries (Ref 8). This hypothesis was based on two complementary observations on sintered materials. First, the measured temperature dependence of J_{ct} could be modeled best if the grain boundary were described as a largely isolating barrier that is interrupted by superconducting microbridges. Second, observations of localized SGBDs and their associated periodic strain fields provided a possible mechanism for dramatic local variations of superconducting properties within the boundary core (Ref 8).

Observations of high-angle grain boundaries that show flux-pinning characteristics and J_{ct} values that are comparable to those of their single-crystal counterparts rule out the general validity of simple PGBD models. Such characteristics have been observed for 14° [001] and P15 bicrystals and for a crystal cluster for which the current must pass through one or more boundaries with misorientation angles between 22 and 37°. As a result, it must be concluded that although the PGBD content may be important, it is not the only factor that contributes to the grain-boundary J_{ct} value.

"Special" Properties. Grain boundaries that have low-Σ misorientations and lie on low index planes of the crystal lattice or the CSL are generally believed to be special in terms of their structures and properties. As already described, indicators of the boundary energy suggest that such boundaries in $YBa_2Cu_3O_{7-x}$ do have particularly low-energy structures that act as reference structures for other related boundaries. Such observations led several researchers to propose, without proof, that these microstructurally special boundaries have high J_{ct} values and/or flux-pinning character (Ref 8, 26, 27, 29). In particular, the Bellcore group has pointed out that such boundaries should, at minimum, be free from second-phase and trace-level impurity atoms (Ref 26, 27, 29). The present author submits that the controlling mechanism may be more fundamental. Structural units into which it is energetically unfavorable to include impurity atoms may also have unique superconducting properties that are not necessarily controlled by the presence of the impurity atoms.

The strongest supporting evidence for special properties at low-Σ boundaries comes from textured polycrystalline thin films deposited on MgO. As described in the section "Grain Morphology, Grain Alignment, and Important Grain Boundaries that Result," the favored misorientation relationships between the MgO substrate and the $YBa_2Cu_3O_{7-x}$ grains leads to a preponderance of low-angle, low-Σ, and near-low-Σ grain boundaries in these relatively fine-grained (~1 μm) thin films. The 0 T, 77 K J_{ct} values (5×10^5 to

1×10^6 A · cm^{-2}) (Ref 27, 49, 50) measured for films that are patterned such that the current almost certainly crosses at least one high-angle grain boundary are as much as two orders of magnitude higher than those obtained for individual grain boundaries in thin film bicrystals (10^3 to 5×10^4 A · cm^{-2}). More specifically, some of the ubiquitous 40° [001] tilt boundaries in thin films grown on MgO under certain conditions appear to have the favored flux-pinning-controlled J_{ct} characteristics, but their J_{ct} values are limited by flux creep (Ref 3). The J_{ct} values for these boundaries are on the order of 10^6 A · cm^{-2} at 4.2 K and 0 *T*. Although this is not a low-Σ boundary, it is a high-symmetry misorientation that causes low index planes of one crystal to be parallel to low index planes of the other (Ref 51).

Evidence that grain boundaries with the Σ3 90° [100] misorientation relationship have special properties has also been found in both thin films and flux-grown bicrystals. Chan *et al.* (Ref 29) measured the J_{ct} value of boundaries with this misorientation relationship by patterning a thin film grown on the polished (014) surface of $SrTiO_3$. Films grown on this surface have a much higher fraction of [100]-oriented $YBa_2Cu_3O_{7-x}$ grains among the [001]-oriented grains. The resulting microstructure contains a relatively high density of 90° [100] grain boundaries. Values for J_{ct} as high as 10^6 A · cm^{-2} were obtained in zero field at 77 K, which indicates that these boundaries may have favorable properties. However, isolated zero field measurements unfortunately supply little information about the mechanism controlling the J_{ct} behavior. Work done by the author and coexperimenters on bulk-scale bicrystals has also indicated that boundaries with this misorientation relationship may be special (Ref 2). The low-field and high-field characteristics of J_{ct} for a bicrystal with this geometry supplied explicit evidence that its J_{ct} properties were flux-pinning controlled, like those of its single-crystal counterparts, not Josephson junction weak linked.

Evidence for special properties — in particular flux-pinning-controlled J_{ct} and a lack of weak-link character — at low-Σ boundaries other than the 90° [100] Σ3 misorientation is inconclusive. Boundaries with macroscopic misorientation relationships within a few degrees of the tetragonal-based low-Σ misorientations have shown Josephson junction weak-link characteristics in both thin films and bulk bicrystals. For example, Dimos *et al.* (Ref 1) have characterized a near-Σ13 grain boundary ($\theta = 22°, \phi = 0°, \gamma = 2°$), a near-Σ5 (36, 4, 0) grain boundary, and a near-Σ41 (13, 2, 0) grain boundary in their thin film studies and found reduced J_{ct} and Josephson junction characteristics in each one (θ, ϕ, and γ are the [001] tilt, [100] tilt, and [100] twist components of the misorientation, respectively). Likewise, the author and cohorts have detected Josephson junction characteristics in their own bulk bicrystal studies for bicrystals that are nominally near-Σ5 (37° [001]) and near-Σ13 (22° [001]). Evidence for low boundary energy is found for all these misorientations (Ref 30). The given angles are believed to be accurate to ± 0.5° in the thin film case and ± 1° in the bulk-scale case. The misorientation angles about the other principal axes were not measured explicitly in the bulk case, but they are believed not to exceed a few degrees. The values Σ5, Σ13, and Σ41 occur in any simple tetragonal unit cell for 36.87°, 22.62°, and 12.47° rotations about [001], respectively. In contrast to the above results, flux-pinning-controlled J_{ct} characteristics were obtained for a 14° [001] and for P3- and P15-type bulk-scale bicrystals. Applying the same sort of limitations of Δθ as above, these boundaries could be described as near-Σ41 (12.47° [001]), near-Σ3 (90° [001]), and, possibly, near-Σ123 (at P12.47°, which is Σ41 for the cubic parent cell) grain boundaries, respectively.

The bicrystal experiments provide no compelling evidence that the Σ value alone determines the electromagnetic character of the boundary. However, neither do they deny the hypothesis. For example, it remains possible that the special properties occur over such a small range of Δθ that the two types of properties observed above for near-low-Σ boundaries merely reflect the inability of the author and her colleagues to measure the misorientation angle with sufficient accuracy to test the supposition. This explanation for the conflicting data is not unreasonable if the strain fields of localized SGBDs are responsible for the Josephson junction character of near-low-Σ boundaries. Localized SGBDs with spacings as small as 10 nm have been observed in $YBa_2Cu_3O_{7-x}$. This spacing corresponds to deviations of about 0.7°, 0.4°, and 0.2° for dislocations with primitive DSC lattice Burgers vectors in symmetrical Σ5, Σ13, and Σ41 [001] tilt boundaries, respectively. The minimum spacings observed for SGBDs in the other oxide materials is about 3 nm. This spacing would correspond to deviations of approximately 2.3°, 1.4°, and 0.8° for the above boundaries. These representative dimensions suggest that the superconducting properties might be rather sensitive to the exact misorientation angle, with respect to the accuracy of misorientation measurements in the preceding experiments.

Finally, it remains possible that the Σ value controls the grain-boundary properties indirectly. For example, high-surface-energy (and, presumably, non-NCSL-related) boundaries could have weak-link characteristics because of an impurity segregation mechanism (Ref 45). An intriguing alternative NCSL-related mechanism has been proposed by Zhu *et al.* (Ref 31). They suggest that some boundaries may accommodate the deviation between the macroscopic NCSL misorientation and a perfect CSL by adjusting the lattice parameters through local compositional variations instead of by introducing compensating grain-boundary dislocations (GBDs). In particular, the *c*-lattice parameter is sensitive to oxygen concentration. Intrinsic nonstoichiometry of this type could explain the variety of properties observed in the high-angle regime and the apparent lack of consistency in the properties observed for near-low-Σ boundaries.

Anisotropy. The bulk bicrystal studies have shown that boundaries for which the crystal misorientation is the maximum with respect to the superconducting properties, anisotropy can still have flux-pinning-controlled J_{ct} values that are comparable to those of their single-crystal counterparts. This result suggests that the anisotropy, or, more specifically, the relative orientation, of the high-conductivity planes, does not play as critical a role in limiting the $J_{ct(GB)}$ as has been proposed. The thin film bicrystal results also show J_{ct} properties that are largely independent of the *axis* of misorientation, suggesting that a misorientation *angle*-dependent structural feature, rather than the anisotropic nature of the crystal structure, controls the superconducting properties of the boundaries. In fact, thin-film [001] tilt boundaries, in which the (001) planes are parallel throughout the bicrystal and intersect the boundary with the same frequency in both crystals, clearly show the characteristically rapid reduction of J_{ct} in the $5^\circ < \theta < 10^\circ$ range. It is of note that the thin film [100] tilt boundaries with no other components of misorientation did not show strongly depressed $J_{ct(GB)}$ out to be the maximum angle measured (10°). In this family of misorientation relationships, the high-conductivity planes are rotated out of parallel by the angle θ. In this family of misorientation relationships, the high-conductivity planes are rotated out of parallel by the angle θ. The flux-pinning P3-type boundary of the flux-grown bicrystal studies is essentially the member of this family with the maximum possible θ value (90°). On the basis of anisotropy arguments alone, this type of boundary might be expected to exemplify the worst-case scenario, yet its measured properties were in fact more favorable.

Boundary Plane and Facet Structure. The arguments presented in the above section are by no means intended to imply that the boundary plane is not an important factor in $J_{ct(GB)}$ determination. Rather, they propose that the mechanism by which the boundary plane may influence $J_{ct(GB)}$ is not the geometric alignment of the (001) planes. Most likely, the detailed atomic arrangement at the boundary, which undoubtedly depends on the local boundary plane and the specific facet structure, determines the J_{ct} properties of the boundary. However, explicit correlations between boundary planes and superconducting properties are not yet evident. On the contrary, boundaries that at the outset might be predicted to have favorable or special properties because of their short-period nature appear not to have superior J_{ct} characteristics. Among these boundaries are the nominally symmetrical-tilt boundaries examined by Dimos *et al.* (Ref 1). The bulk bicrystal boundary shown in Fig. 5 is also a symmetrical-tilt boundary on the scale of a diffraction contrast TEM image and, in addition, is twin matched. Nevertheless, this boundary explicitly showed Josephson junction characteristics (Ref 15).

An alternative mechanism by which the macroscopic boundary plane may influence the grain-boundary properties is suggested by the detailed structural studies of low-angle grain boundaries (Ref 40). The observed propensity for microfaceting and the delocalization of the cores and elastic strain fields of the associated facet junction dislocations provide a second means by which the facet microstructure can determine the $J_{ct(GB)}$ characteristics. Because local superconducting properties can be sensitive to local lattice distortion, it is possible that facet junction morphology influences the superconducting properties of the grain boundary as much as or more than the atomic arrangement along the component facet planes.

Impurity Segregation. The role of grain-boundary nonstoichiometry and impurity segregation in determining grain-boundary J_{ct} characteristics must also be understood. As discussed previously, impurity segregation is correlated with boundary crystallography. However, the consequences of impurity segregation at boundaries where the formation of a wetting layer of second phase is not induced are not known. The commonly observed grain-boundary nonstoichiometry undoubtedly also influences grain-boundary properties. At present, the origin of the variable grain-boundary composition is not known; it may be determined by grain-boundary crystallography. It is of note, however, that high-resolution TEM images of a 90° [100] tilt boundary in a thin film showed no indication of Cu enrichment at grain boundaries (Ref 37). (The reader may recall that a bulk-scale bicrystal with nearly this misorientation relationship showed J_{ct} characteristics that mirrored its single-crystal counterpart) (Ref 2).

Heterogeneity at the Boundary Core. A most intriguing aspect of grain-boundary structure-properties relationships for $YBa_2Cu_3O_{7-x}$ grain boundaries is the indication of local $J_{ct(GB)}$ variation within the boundary plane. Evidence for the heterogeneity of both the electrical properties and the microstructure on the scale of the superconducting coherence length has been found out. Electromagnetic property characterization has suggested that at least some grain boundaries are composed of nonsuperconducting and superconducting patches with dimensions on the order of the ξ (Ref 3, 16). From a microstructural standpoint, the localized, oscillatory strain contrast produced by structural SGBD networks (Ref 8), the localized composition variation (Ref 41), the very localized nonstoichiometry and structural modification at PGBD cores (Ref 40), and the mixtures of localized and delocalized GBDs that result from facet junction morphology are all possible origins for the nonuniform electrical properties. Such microstructural variations undoubtedly produce local variations in the electrical properties of the boundary. The lingering question for the average grain boundary is whether these microstructural nonuniformities determine their J_{ct} characteristic or simply modify to a small degree properties that are limited by a different mechanism. It remains possible that the cited electrical character and microstructural features are pertinent to the properties of just a few special grain boundaries that have extraordinary properties and form a percolative path for the supercurrent.

Closing Comments

The $YBa_2Cu_3O_{7-x}$ grain-boundary problem is clearly very complex. Possible correlations between the properties and specific (micro)structural features have been proposed on the basis of the $J_{ct(GB)}$ behavior observed indirectly and in the limited number of experiments that explicitly probe structure-properties relationships for grain boundaries. Although some more straightforward associations appear to be nearly general (for example, the development of Josephson junction character as θ exceeds ~ 10°), clear exceptions to each of the proposed (micro)structural mechanisms for $J_{ct(GB)}$ control have been found. These special cases are of great interest because they suggest that the weak-link behavior is not inevitable. They also clearly illustrate that a comprehensive understanding of the microstructural features and mechanism (or mechanisms) that determine grain-boundary superconducting properties is yet to be realized.

Several aspects of the structure and the superconducting properties of $YBa_2Cu_3O_{7-x}$ grain boundaries complicate the identification of their structure-properties relationships. First, it is not clear that the true intrinsic properties of single $YBa_2Cu_3O_{7-x}$ grain boundaries have been probed. On the contrary, in all of the bicrystal techniques used to date, the boundary is either constrained in some way or exposed to soluble impurity elements during fabrication. Second, in well-prepared material, the $J_{ct(GB)}$ properties appear to be controlled on the spatial scale of ξ. Furthermore, considerable macroscopic evidence suggests that the superconducting properties of individual bicrystals vary within the grain-boundary region on a scale that approaches ξ. The very localized features that control the superconducting properties thus present formidable challenges to the available characterization techniques. Finally, a variety of defects, ranging from bonding to strain to oxygen concentration to impurity segregation, could influence grain-boundary properties. Most likely, the key to understanding grain-boundary properties lies in the interconnected nature of a number of these factors.

Acknowledgment

The critical roles of Xue Yu Cai and David C. Larbalestier of the University of Wisconsin and Debra L. Kaiser of the National Institute for Standards and Technology in the bulk bicrystal studies are gratefully acknowledged. David C. Larbalestier's many contributions to the studies of grain boundaries in sintered materials by the author and her colleagues are also acknowledged. The research conducted at the University of Wisconsin has been supported by the Electric Power Research Institute, the U.S. Department of Energy — Division of High-Energy Physics, and the National Science Foundation — Materials Research Group Program.

References

1. D. Dimos, P. Chaudhari, and J. Mannhart, *Phys. Rev. B*, Vol 41, 1990, p 4038
2. S.E. Babcock, X.-Y. Cai, D.L. Kaiser, and D.C. Larbalestier, *Nature*, Vol 347, 1990, p 167
3. S.E. Russek, D.K. Lathrop, B.H. Moeckly, R.A. Buhrman, D.H. Shin, and J. Sillcox, *Appl. Phys. Lett.*, Vol 57, 1990, p 1155
4. C. Meingast, P.J. Lee, and D.C. Larbalestier, *J. Appl. Phys*, Vol 66, 1989, p 5962; C. Meingast and D.C. Larbalestier, *J. Appl. Phys.*, Vol 66, 1989, p 5971
5. R.J. Cava, A.W. Hewat, E.A. Hewat, B. Batlogg, M. Marezio, K.M. Raabe, J.J. Krajewski, W.F. Peck, Jr., and L.W. Rupp, Jr., *Phys. C*, Vol 165, 1990, p 419
6. O. Bremert, C. Michaelson, and H.U. Krebs, *J. Appl. Phys.*, Vol 65, 1989, p 1018
7. J. Karpinski, E. Kaldis, E. Jilek, S. Rusiecki, and B. Bucher, *Nature*, Vol 336, 1988, p 660
8. S.E. Babcock and D.C. Larbalestier, *J. Mater. Res.*, Vol 5, 1990, p 919
9. M.F. Chisholm and S.J. Pennycook, *Nature*, to be published
10. T. Terashima, K. Iijima, K. Yamamoto, Y. Bando, and H. Mazaki, *Jpn. J. Appl. Phys.*, Vol 27, 1988, p L91
11. K. Selama, V. Selvamanickam, L. Gao, and K. Sun, *Appl. Phys. Lett.*, Vol 54, 1989, p 2352
12. R.L. Peterson and J.W. Ekin, *Phys. C*, Vol 157, 1989, p 325
13. K. Osamura, T. Takayama, and S. Ochiai, *Appl. Phys. Lett.*, Vol 55, 1989, p 396
14. K. Osamura, T. Takayama, and S. Ochiai, *Supercond. Sci. Technol.*, Vol 2, 1989, p 111

15. X.Y. Cai, Applied Superconductivity Center, University of Wisconsin—Madison, unpublished research
16. D.C. Larbalestier, S.E. Babcock, X. Cai, L.D. Cooley, M. Daeumling, D.P. Hampshire, J. McKinnell, and J.M. Seuntjens, Progress in High Temperature Superconductivity, in *Proc. Tokai Univ. Int. Workshop Sci. Supercond. New Mater.*, Vol 18; S. Nakajima, Ed., World Scientific, 1989, p 128
17. D. Dimos, P. Chaudhari, J. Mannhart, and F.K. LeGoues, *Phys. Rev. Lett.*, Vol 61, 1988, p 219
18. S. Nakahara, G.J. Fisanik, M.F. Yan, R.B. van Dover, and T. Boone, *Appl. Phys. Lett.*, Vol 53, 1988, p 2105
19. R.A. Camps, J.E. Evetts, B.A. Glowacki, S.B. Newcomb, and W.M. Stobbs, *J. Mater. Res.*, Vol 2, 1987, p 750
20. H. Zandbergen and G. Ven Tendeloo, *Proc. MRS Symp.*, Vol 156, 1989, p 209
21. J.E. Blendell, C.A. Handwerker, M.D. Vaudin, and E.R. Fuller, Jr., *J. Cryst. Growth*, Vol 89, 1988, p 93
22. T. Kogure, Y. Zhang, R. Levonmaa, R. Kontra, W.-X. Wang, D.A. Rudman, G.J. Yurek, and J.B. Vander Sande, *Phys. C*, Vol 156, 1988, p 707
23. S.E. Babcock, T.F. Kelly, P.J. Lee, J.M. Seuntjens, L.A. LaVanier, D.C. Larbalestier, *Phys. C*, Vol 152, 1988, p 25
24. J.L. Vargas, Applied Superconductivity Center, University of Wisconsin—Madison, unpublished research
25. J.M. Seuntjens and D.C. Larbalestier, *J. Appl. Phys.*, Vol 67, 1990, p 2007
26. D.M. Hwang, T.S. Ravi, R. Ramesh, S.-W. Chan, C.Y. Chen, and L. Nazar, *Appl. Phys. Lett.*, Vol 57, 1990, p 1690
27. T.S. Ravi, D.M. Hwang, R. Ramesh, S.-W. Chan, L. Nazar, C.Y. Chen, A. Inam, and T. Venkatesan, *Phys. Rev. B.*, Vol 42, 1990, p 10141
28. H. Moeckly, S.E. Russek, D.K. Lathrop, R.A. Buhrman, J. Li, and J.W. Mayer, *Appl. Phys. Lett.*, Vol 57, 1990, p 1687
29. S.-W. Chan, D.M. Hwang, R. Ramesh, S.M. Sampere, and L. Nazar, in *AIP Conference Proceedings 200, High T_c Superconducting Thin Films*, R. Stockbauer, Ed., American Institute of Physics, 1990, p 172
30. D.A. Smith, M.F. Chisholm, and J. Clabes, *Appl. Phys. Lett.*, Vol 53, 1988, p 2344
31. Y. Zhu, H. Zhang, H. Wang, and M. Suenaga, *J. Mater. Res.*, submitted for publication
32. S.E. Babcock and R.W. Balluffi, *Philos. Mag. A*, Vol 55, 1987, p 643
33. C.P. Sun and R.W. Balluffi, *Philos. Mag A*, Vol 46, 1982, p 49, 63
34. F.W. Gayle and D.L. Kaiser, *J. Mater. Res.*, Vol 6, 1991, p 908
35. Y. Gao, G. Bai, H.L.M. Chang, K.L. Merkle, and D.J. Lam, *Ultramicroscopy*, to be published
36. M.F. Chisholm and D.A. Smith, *Philos. Mag. A*, Vol 59, 1989, p 181
37. Y. Gao, G. Bai, D.J. Lam, and K.L. Merkle, *Phys. C*, Vol 173, 1991, p 487
38. S.E. Babcock, Department of Materials Science and Engineering and Applied Superconductivity Center, University of Wisconsin—Madison, unpublished research
39. K.L. Merkle and D.J. Smith, *Ultramicroscopy*, Vol 22, 1987, p 57

40. Y. Gao, K.L. Merkle, G. Bai, H.L.M. Chang, and D.J. Lam, *Phys. C*, Vol 174, 1991, p 1
41. S.E. Babcock and D.C. Larbalestier, *Appl. Phys. Lett.*, Vol 55, 1989, p 393
42. Y.M. Chiang, J.A.S. Ikeda, and A. Roshko, in *Ceramic Superconductors*, Y.F. Yan, Ed., The American Ceramic Society, 1988, p 607-618
43. D.M. Kroeger, A. Choudhury, J. Brynestad, R.K. Williams, R.A. Padgett, and W.A. Coghlan, *J. Appl. Phys.*, Vol 64, 1988, p 331
44. D.M. Kroeger, J. Brynestad, R.K. Williams, R.A. Padgett, and J.O. Scarbrough, *Proc. MRS Symp.*, Vol 122, 1988, p 521
45. L.T. Romano, P.R. Wilshaw, N.J. Long, and C.R.M. Grovenor, *Supercond. Sci. Technol.*, Vol 1, 1989, p 285
46. E.P. Kvam, Ph. D. thesis, Massachusetts Institute of Technology, 1985
47. A. Singh, N. Chandrasekhar, and A.H. King, *Acta Crystallogr. B*, Vol 46, 1990, p 117
48. M.H. Berger, C. Delamarre, and J.Y. Laval, *Rev. Phys. Appl.*, Vol 25, 1990, p 35
49. D.H. Shin, J. Silcox, S.E. Russek, D.K. Lathrop, B. Moeckly, and R.A. Buhrman, *Appl. Phys. Lett.*, Vol 57, 1990, p 508
50. Q. Li, O. Meyer, X.X. Xi, J. Geerk, and G. Linker, *Appl. Phys. Lett.*, Vol 55, 1989, p 310
51. D. Wolf, *J. Phys.*, Vol 46, 1985, p C4-197

10

Interfaces and Twin Boundaries in $BaTiO_3$- and Fe-Doped $Y_1Ba_2Cu_3O_{7-y}$

L.T. Romano, O.F. Schilling, and C.R.M. Grovenor
Department of Materials, University of Oxford, Parks Road, OX13PH, United Kingdom
M.G. Smith and R.D. Taylor
Physics Division, Los Alamos National Laboratory, Los Alamos, New Mexico 87545
H. Oesterricher
Department of Chemistry, University of California, San Diego, La Jolla, California 92093

Heavily strained $Y_1Ba_2Cu_3O_{7-y}$-Ba_2TiO_4 interfaces were found in $BaTiO_3$-doped samples by transmission electron microscopy (TEM). These interfaces were found to increase the dislocation density in the YBCO and to cause the formation of microcracks. In another study involving $Y_1Ba_2(Cu_{1-x}Fe_x)_3O_{7-y}$ alloys, a twin and tweed structure was found by TEM to coexist for $x < 0.1$. Before this study, only tweed was present for $x > 0.02$.

Doping in $Y_1Ba_2Cu_3O_{7-y}$ (YBCO) has been used to study the fundamental mechanism of high-temperature superconductivity (Ref 1, 2). Introducing second phases and defects is a way to improve the critical currents densities (J_c) of superconducting materials by generating strains in the lattice and by providing additional interfaces that interact with the flux lattice (Ref 3). This has been achieved in conventional superconductors by adding precipitates (Ref 4) and dislocations (Ref 5). In the high-T_c materials, dislocations (Ref 6), precipitates (Ref 7), and twin boundaries (Ref 8) have all been suggested as flux-pinning sites. This chapter characterizes interfaces in the YBCO that are strained because of the presence of $BaTiO_3$ doping, as well as twin-boundary interfaces whose character has been altered by the partial substitution of Cu by Fe.

$BaTiO_3$ had been chosen initially as an impurity phase because of its sequence of structural-phase transitions (Ref 9), which could generate strains in the YBCO matrix. Instead, however, strains were generated by a reaction with the YBCO to form a Ba_2TiO_4 phase. Magnetization measurements of the composite YBCO-$BaTiO_3$ samples showed that the introduction of 5% by weight of $BaTiO_3$ improved the field dependence of J_c by an order of magnitude, in contrast to the undoped material (Ref 10). TEM was used to study the interface of these two materials and relate it to the improved J_c characteristics.

The substitution of Cu by transition metals with magnetic moments has been found to decrease the transition temperature (T_c) (Ref 1), whereas substitutions of the other cations by rare earths with magnetic moments have little effect on the T_c (Ref 2). The structure and spacing of the twin boundaries in $Y_1Ba_2(Cu_{1-x}Fe_x)_3O_{7-y}$ alloys (Ref 11) has been found to be drastically altered for Fe compositions where $x > 0.015$, and therefore this system has been used as a model to determine the pinning effect of twin boundaries (Ref 12). The present authors found that the structure of the twin boundary varied according to the heat treatment of the sample and did not depend only on the Fe concentration. Mossbauer spectroscopy performed on these same samples showed that the sample treatment changes the occupancy of the Fe on the two different Cu sites available and also affects the T_c (Ref 13). A model is proposed for the twin boundaries using the difference in Fe site occupancies.

Experimental Details

Preparation of the $BaTiO_3$-doped samples has been thoroughly described elsewhere (Ref 10). The samples were grown by a melt-texturing method using a mixture of 5% by weight $BaTiO_3$ to YBCO. Grain sizes were up to 2 mm wide and 5 mm long and contained microcracks. The size of the $BaTiO_3$ inclusions were typically 1×10 μm, and the distribution was fairly uniform as determined by elemental maps using electron microprobe analysis (EPMA). EPMA also showed that the final composition of the Ba inclusions was close to

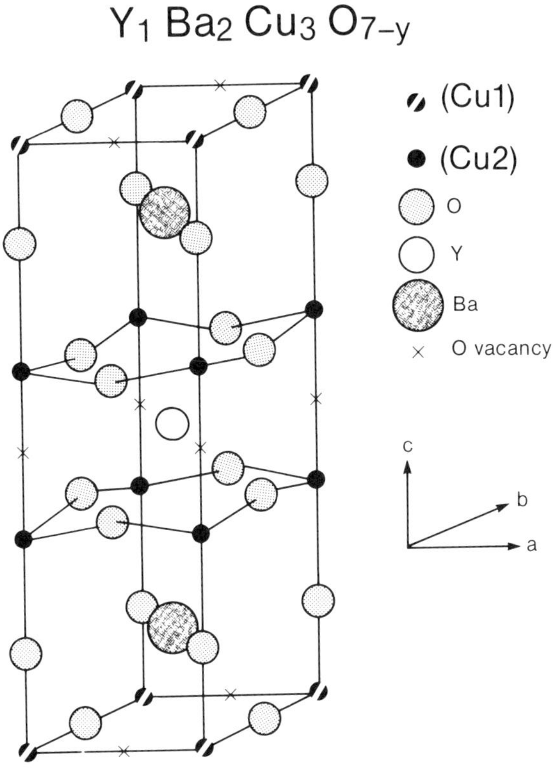

Fig. 1 Unit cell of $Y_1Ba_2Cu_3O_{7-y}$ designating the Cu1 and Cu2 positions

Ba_2TiO_4 with small concentrations of yttrium and copper. Field-dependent J_c measurements showed an order of magnitude improvement in the J_c of the doped samples, compared to melt-textured undoped samples at 4 T and 77 K (Ref 10).

Polycrystalline $Y_1Ba_2(Cu_{1-x}Fe_x)_3O_{7-y}$ samples were prepared under reducing conditions at high temperature and later oxygenated at low temperatures, as described in detail in Ref 13. Mossbauer measurements showed that the ratio of the Fe occupancy on the Cu2 to Cu1 sites (refer to Fig. 1 for Cu site positions) was greater in these samples than in samples prepared in the conventional high-temperature sintering process in air. Samples made under reducing conditions also exhibited higher values of T_c (Ref 13).

TEM samples of the YBCO-$BaTiO_3$ composites were prepared by mechanically grinding and dimpling to less than 30 μm and then ion milled for less than 10 h from both sides at 5 kV and 0.4 mA until electron transparency was reached. The Fe samples were prepared for TEM by crushing in acetone and floating onto carbon-coated copper grids. Samples were observed in Philips 120 kV and 200 kV microscopes. The higher kV was particularly essential for greater electron penetration of the $BaTiO_3$ composites. Energy-dispersive x-ray analysis in the TEM was used to verify the presence of the Ba particles and the relative concentrations of Fe in each of the samples.

Results and Discussion

$BaTiO_3$-YBCO Composites. The major difference between the melt-textured undoped and $BaTiO_3$-doped materials was the density of dislocations generated in the YBCO matrix. Undoped melt-textured YBCO typically contained (110) twins, $Y_2Ba_1Cu_1O_5$ (211) inclusions, and (001) basal-plane microcracks. (Microcracks are associated with anisotropic grain growth and are expected to occur for grain sizes greater than 1 μm, Ref 14.) The 211-YBCO interface may show strain and in some cases may generate a low density of dislocations. Figure 2(a) shows that strain around the 211 interface induced 90° twin domains in the vicinity of the precipitate and a few [100] dislocations. A similar effect occurred in the $BaTiO_3$-doped samples when the Ba_2TiO_4 inclusions were less than 2 to 3 μm in size, as shown in Fig. 2(b). The top end of the same precipitate is shown in Fig. 2(c) surrounded by a low density of [100] dislocations.

Figure 3 shows a distinct difference in the dislocation density near the interfaces of much larger 211 and Ba_2TiO_4 precipitates. Figure 3(a) shows that the 211-YBCO interface has relatively few dislocations, compared to the Ba_2TiO_4-YBCO interface In Fig. 3(b). The twins in the YBCO matrix near the Ba_2TiO_4 interface are also greatly deformed, and a small crack is observed as the YBCO tries to pull away from the Ba_2TiO_4. The Ba_2TiO_4 particles, however, show very little deformation.

Figure 4 shows a grain-boundary region between two YBCO grains. Contained at the grain boundary is Ba_2TiO_4 and CuO (Fig. 4a). Figures 4(b) and 4(c) are higher-magnification micrographs of the Ba-YBCO interface, in which the YBCO is shown to contain a large number of defects, unlike the relatively perfect Ba_2TiO_4 grain. A small amount of strain is detected in the Ba_2TiO_4 (see arrows in Fig. 4c).

Additional near-basal-plane microcracks were also generated by the presence of the Ba_2TiO_4 inclusions. Figure 5 shows a crack connecting two

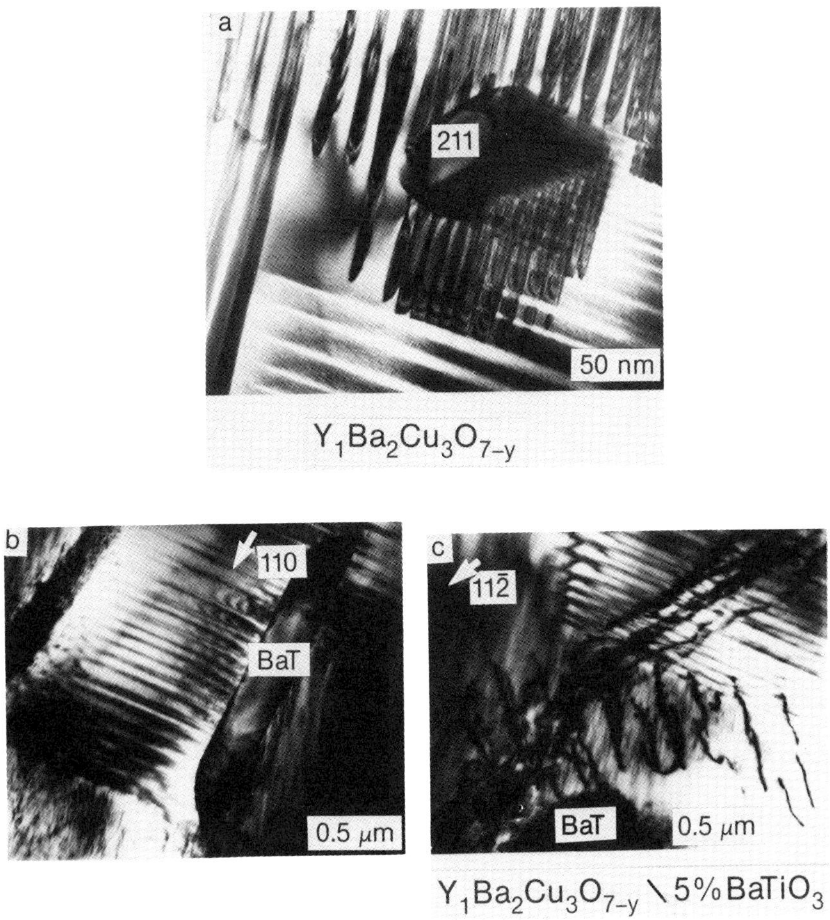

Fig. 2 TEM bright-field micrographs. (a) 211 precipitate in YBCO matrix of an undoped sample. (b) Ba_2TiO_4 (BaT) precipitate in YBCO matrix of a $BaTiO_3$-YBCO composite. (c) The top end of the same Ba_2TiO_4 (BaT) precipitate shown in (b) but at a tilt to show [100] dislocations

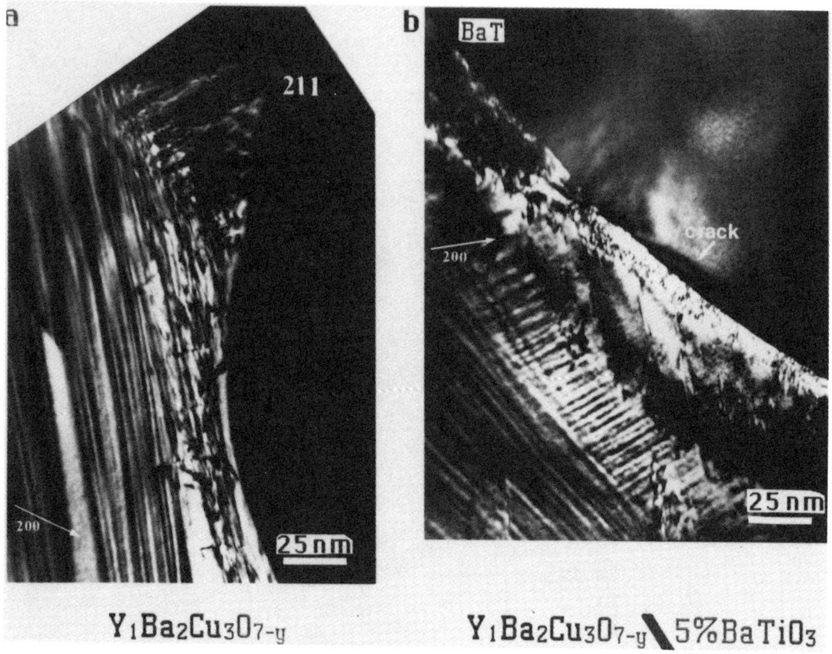

Fig. 3 TEM dark-field micrographs. (a) 211-YBCO interface from an undoped sample. (b) Ba_2TiO_4 (BaT)-YBCO interface from the doped sample

Ba_2TiO_4 grains in the same area. Figure 5(a) shows the crack surrounded by dislocations extending from one of the Ba_2TiO_4 grains into a deformed YBCO matrix. The micrograph in Fig. 5(c) shows the continuation of this crack taken at a different sample tilt to show the high density of dislocations surrounding the other Ba_2TiO_4 grain. The Fig. 5(b) micrograph was taken along the (210) pole to show that the "dislocated" band was indeed a crack of a width close to 30 nm.

The generation of defects in the YBCO matrix by the presence of the Ba_2TiO_4 particles is thought to be due to two major effects. First, the melting point of the Ba_2TiO_4 phase (1323 °C) (Ref 9) is much higher than the incongruent melting temperature of the YBCO (1005 °C) (Ref 15). It was found, from quench experiments, that the transformation to the Ba_2TiO_4 phase occurs in the first 10 min of a 48-h heat treatment and that the growth of the Ba_2TiO_4 grains nears completion after only 5 h. Therefore the YBCO is forced to grow from the melt around much harder ceramic particles. The growth temperature is also

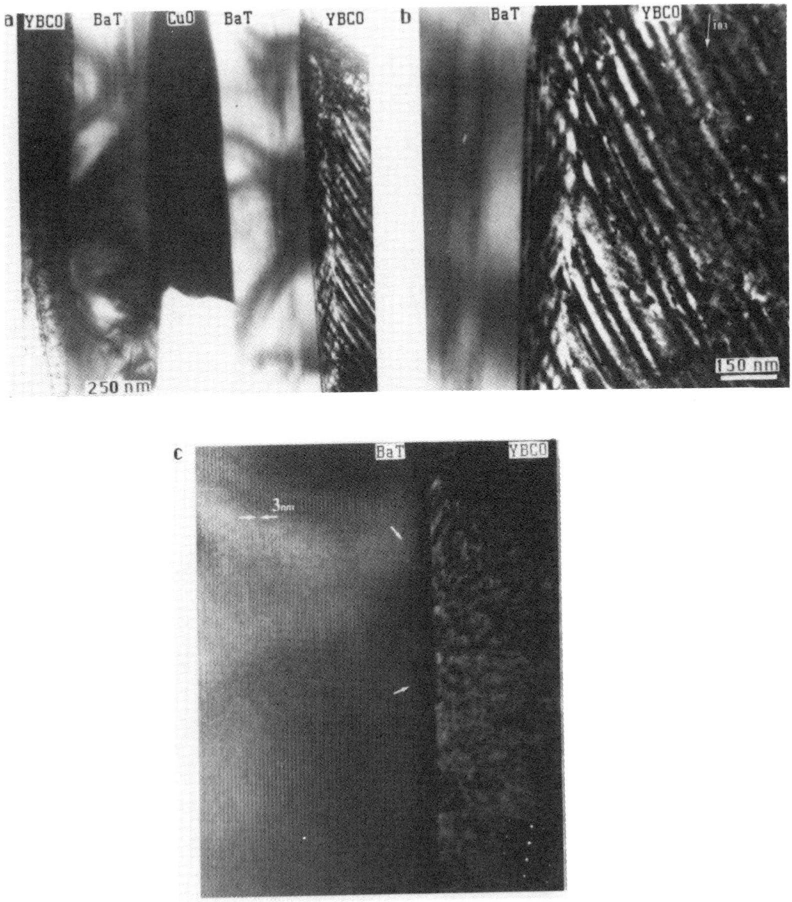

Fig. 4 TEM bright-field micrographs. (a) Grain-boundary phase between two YBCO grains in $BaTiO_3$-doped sample. (b) Higher magnification of the Ba_2TiO_4 (BaT)-YBCO interface from the right part of (a). (c) Lattice image of the same interface shown in (b). Arrows indicate strain in the BaT phase

held very close to the melt temperature of the YBCO, making plastic deformation in the YBCO more energetically favorable. Second, there is a difference in thermal expansion between the two phases. The values for YBCO can be estimated from recent data to be approximately 1 and 4×10^{-5} K^{-1} for the *a*-, (*b*-), and *c*-axes, respectively (Ref 16). Typical values for the barium titanates are about 0.85 to 1.0×10^{-5} K^{-1} (Ref 17). (A value for Ba_2TiO_4 was not found.) The large anisotropy in the thermal expansion of the YBCO induces basal-

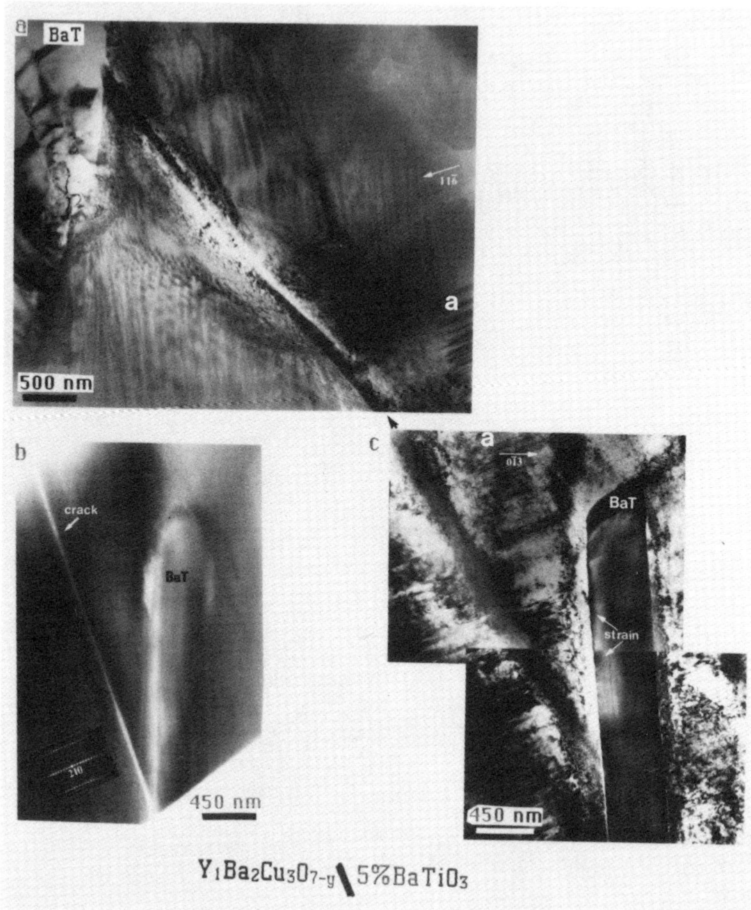

Fig. 5 TEM bright-field micrographs. (a) Crack surrounded by dislocations extending from a Ba_2TiO_4 (BaT) precipitate. (b) The same part of crack shown in (c) tilted along the (210) zone. (c) The same crack shown in (a) extended into another Ba_2TiO_4 (BaT) precipitate. The region labeled "a" is where pictures (a) and (c) overlap

plane cracking independent of doping. However, the Ba_2TiO_4 resists further contraction of the YBCO at the YBCO-Ba_2TiO_4 interface, inducing additional cracking and the generation of dislocations. The combination of these two effects causes a greater density of dislocations in the YBCO doped with $BaTiO_3$ than in the plain sample. These additional defects are thought to give rise to

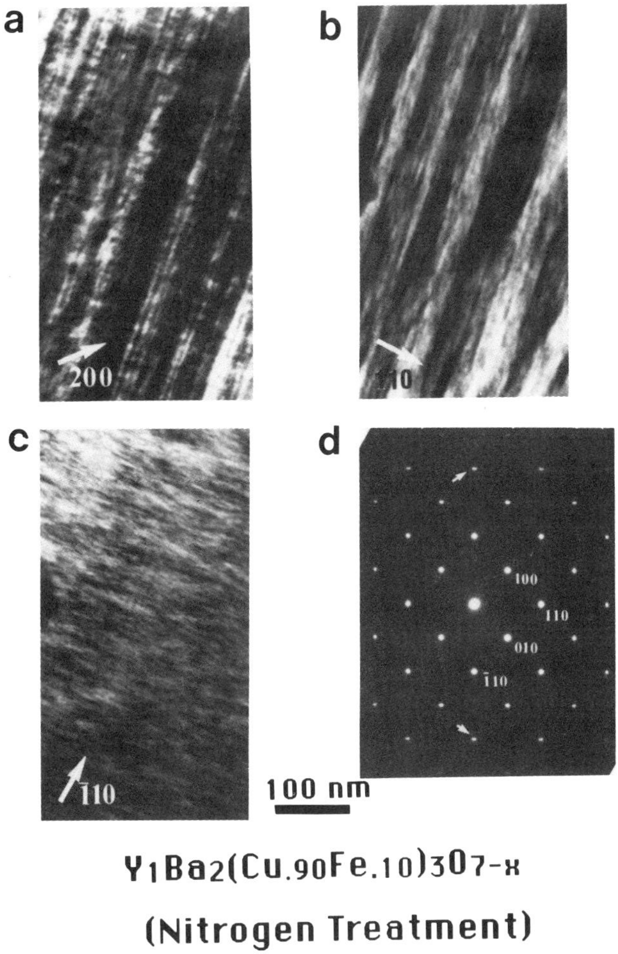

Fig. 6 Dark-field micrographs of a $Y_1Ba_2(Cu_{1-x}Fe_x)_3O_{7-y}$ sample. (a) Tilted to show twin and tweed. (b) Tilted to show twins and one direction of the tweed. (c) Tilted to show tweed in direction perpendicular to (b). (d) Selected-area diffraction pattern along [001] zone. Arrows indicate the twin spots

the enhanced flux pinning that is similar to cold working a metallic, conventional superconductor.

Fe-Doped YBCO. Undoped YBCO typically contains (110) twins that are approximately 50 nm wide, depending on the grain size and oxygenation temperature (Ref 14). An extensive TEM study of conventionally treated

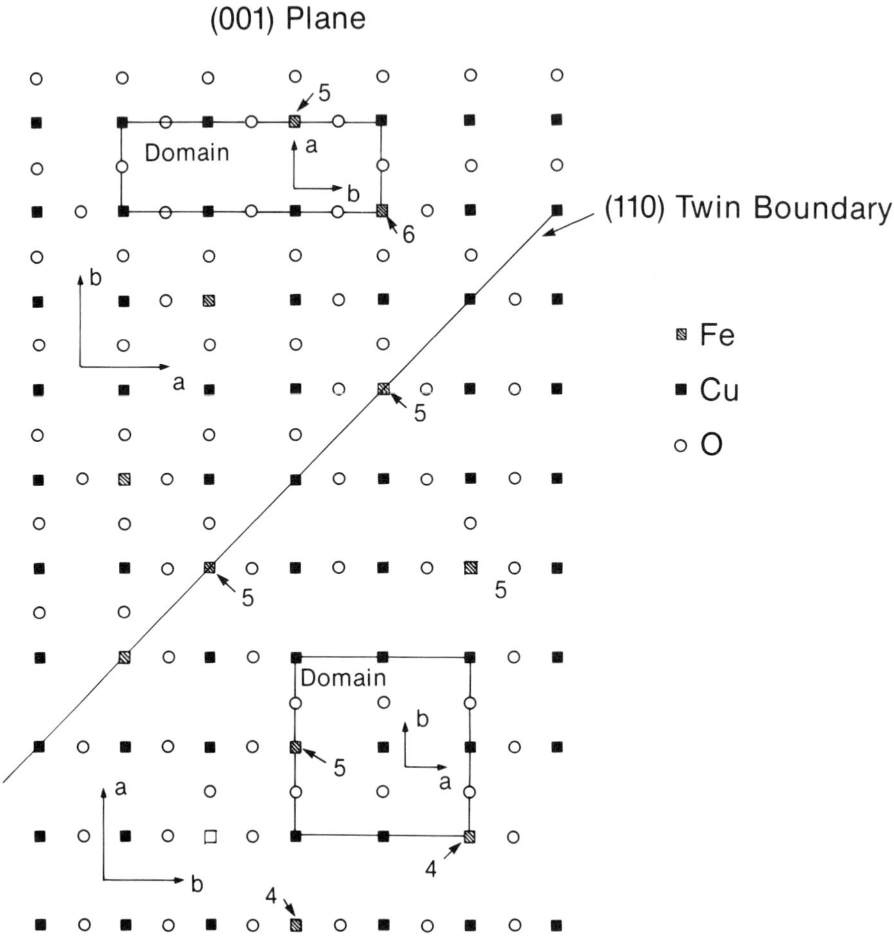

Fig. 7 Model for $Y_1Ba_2(Cu_{1-x}Fe_x)_3O_{7-y}$ showing Fe with fivefold coordination at a twin boundary in a matrix where the *a*- and *b*-axes can be rotated with respect to the average *a*- and *b*-axes of the region on either side of the boundary. Numbers 4, 5, and 6 represent the oxygen coordination of the Fe

$Y_1Ba_2(Cu_{1-x}Fe_x)_3O_{7-y}$ showed that the twin structure transforms to twin and tweed at $x = 0.015$ and finally to a complete tweed structure for $x > 0.02$ (Ref 11). Mossbauer spectroscopy (Ref 18) and neutron diffraction (Ref 19) have determined that the Fe substitutes preferentially on the Cu1 chain site. (Discrepancies in the amount of Fe on the Cu1 or Cu2 site have been associated with differences in sample preparation.) Hence the combination of these techniques led to a well-agreed domain model, explaining the tweed as an ordered array of twins on a fine scale (<10 nm). However, Zhu *et al.* (Ref 20) showed by computer-imaging techniques that the tweed is actually associated with shear displacements along the (110) direction because of the off-stoichiometric oxygen concentration.

Figure 6 shows the existence of twin and tweed at $x = 0.1$ for a sample prepared under reducing conditions. Contrast from only the twin bands is seen by using a (110) diffraction condition, as shown in Fig. 6(b). (Compare to Fig. 6c taken with the ($\bar{1}10$) reflection, which shows the other direction of only the tweed structure.) The twin bands are perpendicular to the direction of the splitting in the diffraction spots shown in the diffraction pattern in Fig. 6(d). Mossbauer spectroscopy measurements on the same samples found that the ratio of the percentage of Fe on the Cu1 to that on the Cu2 position decreased and that the Fe has more fivefold coordinated sites on the Cu1 than did conventionally treated material for the same composition (Ref 13). A model for the twin boundary in the (001) plane is shown in Fig. 7. If Fe is placed at the twin boundary, it can be fivefold with the surrounding oxygens. Fe is also substituted in the matrix either as four-, five-, or sixfold coordinated with the oxygen, in agreement with coordinations found by the Mossbauer experiments. For this material, the twin boundary divides two regions that on average are mirror related across the boundary. It is an average because the superimposed tweed implies a matrix of smaller twins (domain model) or periodic changes in the *a*- and *b*-axes of the crystal (Zhu model). The presence of twins at this higher-Fe composition shows that the YBCO microstructure can be controlled by influencing the Cu site on which the Fe substitutes. Enhanced flux pinning was not observed in the pure tweed structure (Ref 12), indicating either that these are not true twin boundaries or that twin boundaries do not contribute to pinning. Measurements of the flux-pinning parameters remain to be done on the reduced samples.

Conclusions

Interfaces in $BaTiO_3$-YBCO composites and twin boundaries in Fe-YBCO doped materials have been investigated by TEM. $BaTiO_3$ was originally used as a dopant to improve flux pinning in superconducting YBCO by introducing strain due to its known structural transitions. Instead, however, $BaTiO_3$ transformed to a Ba_2TiO_4. The Ba_2TiO_4-YBCO interface was found to be highly strained, and the differences in the thermal expansion coefficients of the two materials, as well as their plasticity at high temperatures, led to increased dislocation densities in the YBCO. The substitution of Fe in YBCO by a reduction method promoted the stability of the twin structure and the presence of tweed at higher concentrations than previously reported. Twin boundaries are important interfaces in the YBCO because of their possible role as a flux-pinning center. A model for the twin boundary is proposed that puts Fe at the twin boundary in a matrix of tweed structure.

Acknowledgment

The authors are grateful to Prof. Sir Peter Hirsch for the provision of laboratory facilities and acknowledge the support of the Science and Engineering Research Council and of the Los Alamos Laboratory under the UC-LANL INCOR Program and the U.S. Department of Energy.

References

1. J.M. Taracson, P. Barboux, P.F. Miceli, L.H. Greene, G.W. Hull, M. Eibschutz, and S.A. Sunshine, *Phys. Rev. B*, Vol 37, 1988, p 7458
2. M. Gasgnier and M.-O. Ruault, *J. Appl. Phys.*, Vol 62, 1987, p 4935
3. A.M. Campbell and J.E. Evetts, *Adv. Phys.*, Vol 21, 1972, p 199
4. R.I. Coot, J.E. Evetts, A.M. Campbell, *Can. J. Phys.*, Vol 50, 1972, p 421
5. D. Dew-Highes, *Philos. Mag. B*, Vol 55, 1987, p 459
6. W. Ruikan, R. Hongtao, X. Ling, H. Qing, W. Chaoqun, and Y. Dingan, *Supercond. Sci. Technol.*, Vol 3, 1990, p 344
7. M. Murakami, M. Morita, K. Doi, and K. Miyamoto, *Jpn. J. Appl. Phys.*, Vol 28, 1989, p 1189
8. M.M. Fang, V.G. Kogan, D.K. Finnemore, J.R. Clem, L.S. Chumbley, and D.E. Farrell, *Phys. Ref. B*, Vol 37, 1988, p 2234
9. D.E. Rase and R. Roy, *J. Am. Ceram. Soc.*, Vol 38, 1955, p 102
10. O.F. Schilling, Y. Yang, C.R.M. Grovenor, and C. Beduz, *Phys. C*, Vol 170, 1990, p 123
11. Y. Xu, M. Suenaga, J. Tafto, R.L. Sabatini, L.H. Greene, A.R. Moodenbaugh, and P. Zolliker, *Phys. Ref. B*, Vol 39, 1989, p 6667
12. R. Wordenweber, G.V.S. Sastry, K. Heinemann, and H.C. Freyhardt, *J Appl. Phys.*, Vol 65, 1989, p 1648
13. M.G. Smith, R.D. Taylor, and H. Oesterreicher, *Phys. Ref. B*, accepted for publication
14. T.M. Shaw, S.L. Shinde, D. Dimos, R.F. Cook, and P.R. Duncombe, *J. Mater. Res.*, Vol 4, 1989, p 248
15. B.-J. Lee and D.N. Lee, *J. Am. Ceram. Soc.*, Vol 72, 1989, p 314
16. E.D. Specht, C.J. Sparks, A.G. Dhere, J. Bynestead, O.B. Cavin, D.M. Kroger, and A. Oye, *Phys. Rev. B*, Vol 37, 1988, p 7426
17. J.A. Bland, *Can. J. Phys.*, Vol 37, 1959, p 417

18. O. Bremert, C. Michaelson, and H.U. Krebs, *J. Appl. Phys.*, Vol 65, 1988, p 1018
19. G. Roth, G. Heger, B. Renker, J. Pannetier, V. Caignaert, M. Hervieu, and B. Raveau, *Z Phys. B*, Vol 71, 1988, p 43
20. Y. Zhu, M. Suenaga, and A.R. Moodenbaugh, *Philos. Mag. Lett.*, Vol 62, 1990, p 41

Section IV:

Mechanical Properties of Interfaces

11

Grain Boundaries and Yielding Behavior of Disordered and Ordered fcc Alloys

R.A. Varin
Department of Mechanical Engineering
University of Waterloo, Waterloo, Ontario, Canada, N2L 3G1

This chapter presents recent developments in the understanding of the role of grain boundaries in the yielding behavior of ordered and disordered face-centered-cubic (fcc) polycrystalline alloys. A resurgence of interest in the study of ordered intermetallic compounds has taken place in the past several years because of their potential application as high-temperature structural materials. The alloys Zr_3Al, Ni_3Fe, and Ni_3Al are examples of ordered fcc ($L1_2$) materials. A highly successful attempt to ductilize Ni_3Al by boron doping (Ref 1, 2) triggered a considerable amount of research on this intermetallic. The role of grain boundaries in discontinuous yielding and the dependence of yield strength on grain size (Hall-Petch relation) are discussed for both ordered and disordered alloys. It is shown that the magnitude of the Lüders strain is usually greater in ordered alloys than in disordered fcc alloys and that it persists to larger grain sizes. The applicability of various n exponents in the Hall-Petch relation ($\sigma = \sigma_0 + k_y \ell^{-n}$) for ordered and disordered fcc alloys, as well as the

Table 1 The magnitude of Lüders strain, ε_L, for various ordered and disordered fcc alloys

Alloy	Grain size range, μm	Lüders strain, × 10^{-2}	Strain rate	Reference
Ordered				
Zr_3Al	1.6 - 65	3.8 - 0.55	2.7×10^{-4}	5
Ni_3Al	8 - 50	2.5 - 0.5	1×10^{-4}	7
Ni_3Al + 750 ppm B	2 - 80	5 - 0.7	1×10^{-4}	8
Ni_3Fe	17	~3	2.5×10^{-4}	4
Disordered				
Ni_3Fe	17	~1	2.5×10^{-4}	4
Al-0.8 wt% Mn - 2 wt% Fe	1.1-1.45	5-2.5	6.7×10^{-4}	10
Al-6 wt% Ni	1.2 - 2.1	2.6 - 1.3	2×10^{-3}	9
AISI 316	1.2 - 1.6	1.8 ± 0.2	3×10^{-5}	12
AISI 316	1.6	2.2 ± 0.6	3×10^{-5}	12
AISI 316	1.6	1.5 ± 0.2	3×10^{-5}	12
AISI 304	1.7	1.1 ± 0.2	3×10^{-3}	12
AISI 304L	0.9	2.4 ± 0.2	1×10^{-3}	13

Note: The Lüders strain in austenitic steels was measured as depicted in Fig. 1 (Ref 12)

segregation effects of boron, nitrogen, carbon, and other elements on the grain-boundary-strengthening parameter (k_y), are discussed.

The mechanical properties of ordered and disordered fcc alloys have been widely studied. However, a comparison of the yielding behavior in ordered and disordered fcc alloys, with a special emphasis on the role of grain boundaries, has never been attempted. The present work is the first attempt to provide a comprehensive overview of this subject. Disordered alloys will be mainly represented by the austenitic stainless steels that the author has thoroughly investigated in the last several years. Only the results obtained at ambient temperatures will be discussed. The first section presents several aspects of discontinuous yielding in both ordered and disordered fcc alloys; the second considers the effect of grain boundaries on the yield strength, expressed by the well-known Hall-Petch relationship.

Grain Boundaries and Discontinuous Yielding

A discontinuous yielding (Lüders type) is commonly observed in ordered fcc alloys (Ref 3-8), but so far it has been assumed to be rather uncommon in disordered fcc alloys. However, some earlier studies of aluminum alloys (Ref 9, 10) and subsequent works on ultrafine-grained austenitic stainless steels (Ref 11-13) have clearly shown that quite pronounced Lüders-type yielding can also occur in disordered fcc alloys.

Table 1 lists the magnitude of Lüders strain (ε_L) as measured from stress-strain curves for various ordered and disordered fcc alloys. A general conclusion is that ordered fcc alloys are much more prone to develop a substantial Lüders deformation than are disordered fcc alloys (which was intuitively expected). This is manifested in a twofold way. First, at comparable grain size, ordered alloys exhibit greater magnitudes of Lüders strain (ε_L). Second, in disordered alloys, perhaps with the exception of disordered Ni_3Fe (Ref 4), the Lüders strain disappears on the stress-strain curve if grain size increases to just above 2 to 4 μm (Ref 9-13). In ordered alloys it persists up to the medium range of grain sizes, usually exhibiting the dependence:

$$\varepsilon_L = \lambda \ell^{-m} \qquad \text{(Eq 1)}$$

where ℓ is the mean linear intercept of grain size, $m = 0.54$ for Zr_3Al (Ref 5, 6), and $m = 0.55$ for Ni_3Al with and without boron (Ref 7, 8). Schulson and Roy (Ref 5) argued that the material undergoes strain ε_L to recover, by strain hardening, the strength it lost through the $k_y \ell^{-1/2}$ term when it yielded. This leads directly to the equation:

$$\varepsilon_L = \frac{k_y \ell^{-1/2}}{\theta} \qquad \text{(Eq 2)}$$

where θ is the work-hardening rate at low strains and $\lambda = k_y / \theta$.

A common experimental observation for both ordered and disordered fcc alloys is a very low initial density of dislocations in the grain interiors (Ref 5, 11-13). This implies that yielding is initiated by dislocation generation from the sources rather than by a classical mechanism of dislocation unpinning from impurity atmosphere (Ref 14). Schulson and Roy (Ref 5) have postulated that dislocation pileups either penetrate the grain boundaries or activate the grain-boundary sources, providing mobile dislocations to propagate yield from grain to grain. Varin *et al.* (Ref 11-13) did not invoke the involvement of pileups during discontinuous yielding but proposed that the grain boundaries, as well as triple points, might act as inherent sources of dislocations, for example, by the mechanisms discussed in Ref 15.

A direct role of grain-boundary sources in controlling the magnitude of the Lüders strain (ε_L) during discontinuous yielding of AISI 316 and 304 ultrafine-grained austenitic stainless steels has recently received some experimental support. Figure 1 shows the magnitude of the Lüders strain as a function of the number of particles (carbides) per unit area of grain-boundary surface. It is seen that the Lüders strain (ε_L) increases with an increasing number of carbides precipitated at grain boundaries and triple points. This dependence may be

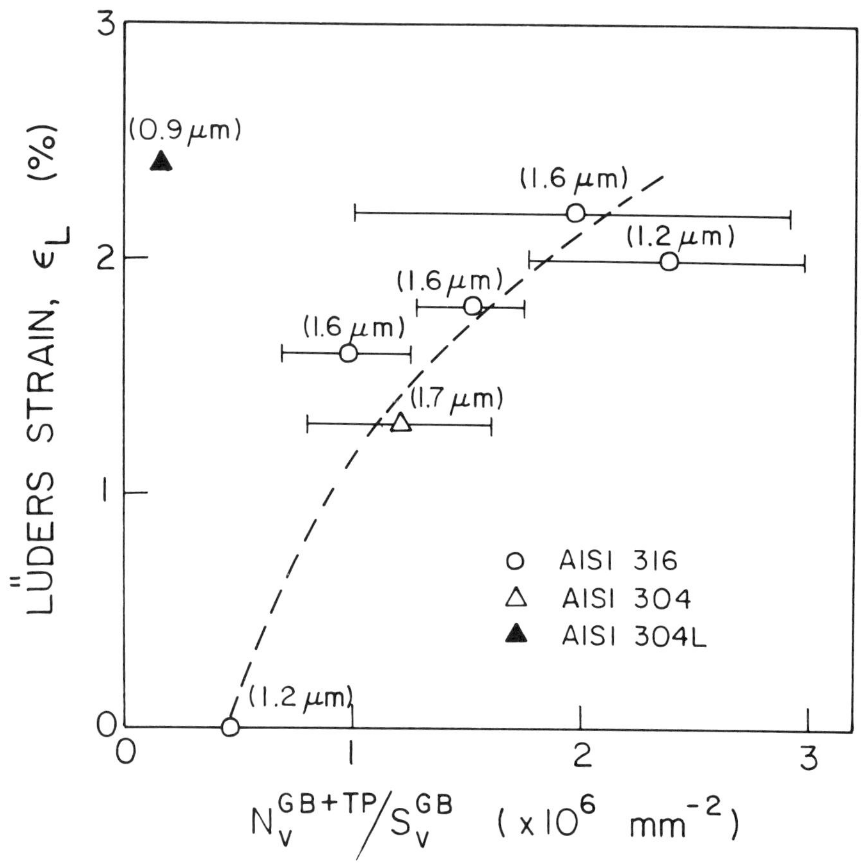

Fig. 1 Magnitude of the Lüders strain (ε_L) as a function of the number of particles per unit area of grain-boundary surface (GB) and triple points (TP). Numbers in parentheses show the mean linear intercept of grain size. S_V^{GB} the grain boundary surface area per unit volume. Source: Ref 12, 16

explained if one invokes the fundamental relationships for the Lüders strain (ε_L) in the form (Ref 12, 13):

$$\varepsilon_L \approx \frac{V_c}{NV_L} \qquad \text{(Eq 3)}$$

$$\dot{\varepsilon}_L \approx A\rho_m b\sigma_P^p \qquad \text{(Eq 4)}$$

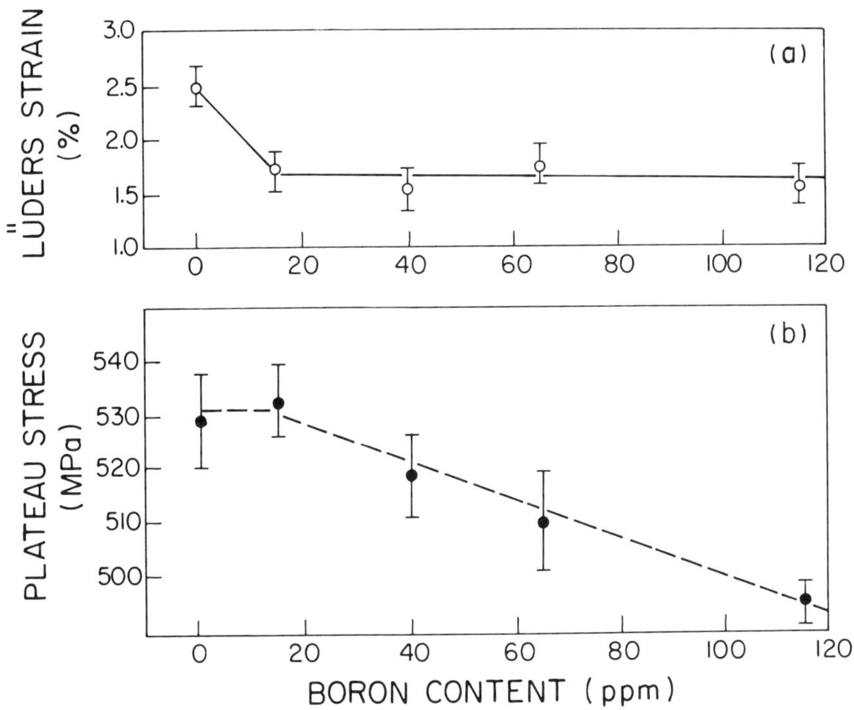

Fig. 2 Variations of (a) Lüders strain (ε_L) and (b) plateau stress (σ_P) with boron content in AISI 304L austenitic stainless steel. Bars represent confidence intervals at a 95% level of significance. Source: Ref 13

where V_c is the crosshead speed, V_L is the Lüders band front velocity, N is the number of active fronts, ρ_m is the density of mobile dislocations (generated from the sources), σ_p is the Lüders plateau stress, p is the stress dependence of dislocation velocity, b is the magnitude of Burgers vector, and A is constant. Since the Lüders front velocity (V_L) is directly proportional to $\dot{\varepsilon}_L$:

$$V_L \approx B\dot{\varepsilon}_L \tag{Eq 5}$$

By combining Eq 3, 4, and 5, one obtains:

$$\varepsilon_L \approx \frac{V_c}{NC\rho_m b\sigma_P^p} \tag{Eq 6}$$

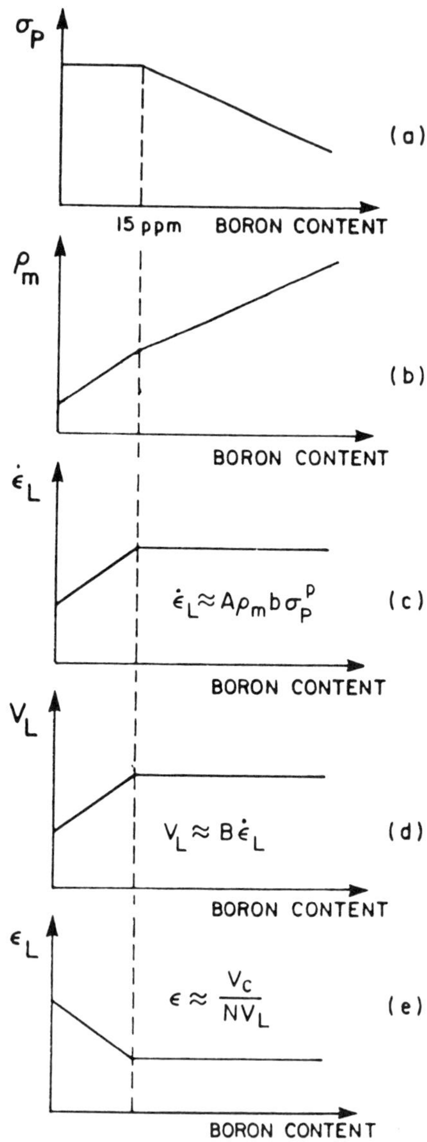

Fig. 3 Proposed schematic change of the parameters in Eq 3, 4, and 5 versus boron content leading to the experimentally observed behavior of Lüders strain versus boron content. Source: Ref 13

where C = AB is a constant. Now, if the increasing number of carbides precipitated at grain boundaries and triple points reduces the number of mobile dislocations (ρ_m) generated from the grain-boundary sources by somehow locking them, the Lüders strain will increase according to Eq 6 and Fig. 1. Of course, physically ε_L increases because V_L is reduced by decreasing the ρ_m term.

Figure 1 also implies that the ratio N_v^{GB+TP}/S_v^{GB} should be greater than about 0.5×10^6 mm^{-2} to observe some Lüders strain. However, the behavior of AISI 304L steel (0.01% C) seems to be independent of the N_v/S_v ratio. Its value for 304L is below 0.5×10^6 mm^{-2}, but the steel exhibits quite substantial Lüders strain (~2.5%). Perhaps grain boundaries formed in this steel after recrystallization (Ref 13) have an inherently limited capacity for dislocation generation, but the exact nature of this behavior is not well understood.

As the increasing number of grain-boundary carbides seems to enhance the magnitude of Lüders strain, at least in AISI 316 and 304 austenitic steels, a boron doping of austenitic steels seems to act in the opposite direction, reducing the Lüders strain (Fig. 2). Figure 2(a) shows that ε_L drops with boron content up to approximately 15 ppm and then levels off. Interestingly, the plateau stress (σ_P) (Fig. 2b) first remains constant up to approximately 15 ppm boron and then decreases continuously with increasing boron content. This behavior can be interpreted by the model based on Eq 3, 4, and 5 and shown in Fig. 3.

It is assumed in the model (Ref 13) that the boron atoms segregate at the grain boundaries, enhancing the generation of mobile dislocations (ρ_m) from grain-boundary sources (Fig. 3b). Up to about 15 ppm B the generation is not sufficient to reduce the plateau stress (σ_P) substantially, and the plateau stress remains more or less constant (Fig. 3a). Since $\dot{\varepsilon}_L$ is directly proportional to ρ_m and σ_P through Eq 4, it will continuously increase up to about 15 ppm B (Fig. 3c). At concentrations of boron above 15 ppm, the generation becomes even more enhanced because more boron is segregated at grain boundaries and σ_P starts decreasing (Fig. 3a), but the density of generated dislocations (ρ_m) continuously increases (Fig 3b). At this point it is assumed that the rate of increase in ρ_m is more or less the same as the rate of decrease in σ_P (they balance each other), and as a result $\dot{\varepsilon}_L$ given by Eq 4 remains constant above approximately 15 ppm B (Fig. 3c). Then the behavior of V_L and ε_L (Fig. 3d and e) is just a direct consequence of Eq 3 and 5.

It should be noted that Schulson *et al.* (Ref 7, 8) did not observe an effect on the Lüders strain of the addition of 750 ppm by weight of boron to stoichiometric Ni_3Al. This is not surprising because their boron content was much greater than 15 ppm, above which ε_L becomes constant (Fig. 2a). It would, however, be interesting to investigate the effect on Lüders strain of much lower boron dopings to Ni_3Al and other ordered fcc alloys. However, Schulson *et al.* (Ref 8) did observe a boron-induced weakening of the most finely grained ($\ell \leq 10$ μm) Ni_3Al samples, which is in excellent agreement with the author's findings in ultrafine-grained austenitic steels.

Grain Boundaries and Yield Strength

As was pointed out earlier, the effect of grain boundaries on yield strength is most conveniently analyzed via a common relationship between the yield strength and grain size, known as the Hall-Petch relationship:

$$\sigma_y = \sigma_0 + k_y \ell^{-n} \quad \text{(Eq 7)}$$

where ℓ is the grain size and k_y represents the average strength of grain boundaries for propagating slip during yielding. This equation with $n = 0.5$ was first proposed for the lower yield stress α-iron by Hall (Ref 17) and Petch (Ref 18). Subsequently, Armstrong *et al.* (Ref 19) extended Eq 7 ($n = 0.5$) for application to 0.2% offset and flow stresses at strains even higher than 0.2% for commercial metals and alloys without a yield point.

The Role of Twin Boundaries. One of several problems related to the experimental verification of Eq 7 is the proper meaning of the term grain size (ℓ). Two definitions are used, one of which is the mean linear intercept of grain size, and the other is true volume grain diameter (Ref 20). The most commonly used term is mean linear intercept, which is also the meaning of grain size used in all the experimental data presented in this chapter.

Another problem that arises during experimental measurements of the linear intercept is whether to count the coherent twin boundaries (CTBs) so common in both ordered and disordered fcc alloys. In counting intercepts for grain size determination in the ordered fcc alloys Zr_3Al and Ni_3Al, Schulson *et al.* (Ref 5, 7, 8) excluded the CTBs because metallography revealed that they did not appear to obstruct slip. The slip lines changed direction but did not terminate at the CTBs (Ref 7). This evidence is qualitative. More recently, a quantitative evidence that the CTBs do not constitute a strong barrier to slip has been obtained for disordered fcc alloys (Ref 21). Figure 4 shows the k_y parameter in Eq 7 ($n = 0.5$) calculated at 0.2% offset for samples of AISI 316 and 316L steels containing various numbers of CTBs per grain. It can be seen that within the experimental scatter the increasing number of twin boundaries does not measurably affect the grain-boundary-strengthening parameter ($k_{0.2}$).

On the basis of the above discussion it may be concluded that the CTBs do not constitute strong barriers to slip in either ordered or disordered fcc alloys.

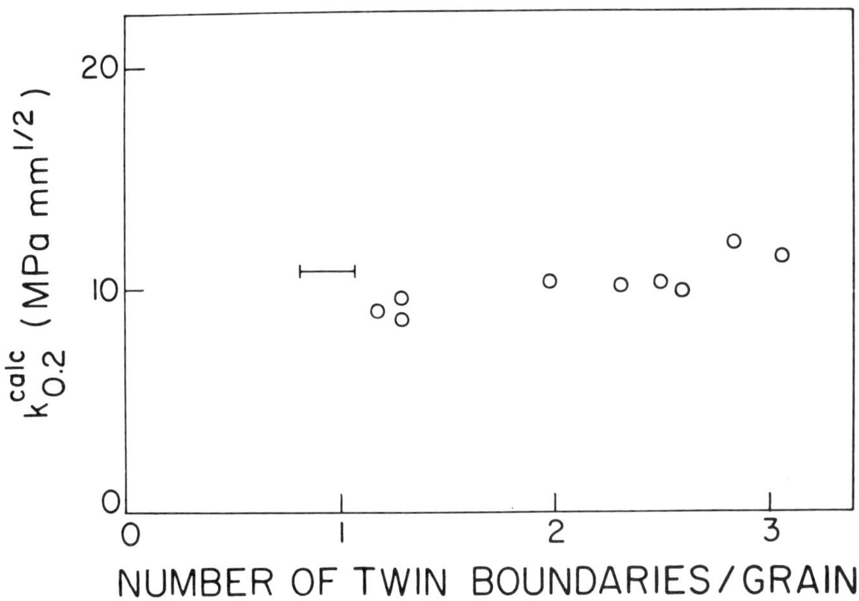

Fig. 4 Effect of coherent twin boundaries on the *k* parameter in the Hall-Petch relationship for 0.2% offset in AISI 316 and 316L austenitic steels. Source: Ref 21

The Exponent *n* in the Hall-Petch Relationship. All ordered fcc alloys seem to obey the Hall-Petch relation for the yield strength with $n = 0.5$ (Ref 2, 4-6, 22-24). One intriguing exception is the recent studies of Ni_3Al with and without boron by Schulson *et al.* (Ref 7, 8). They reported $n = 0.8$ in both materials and based their choice of exponents from 0.5 to 0.95 on the values of correlation coefficients obtained from a regression analysis. They tried to justify the $\ell^{-0.8}$ dependence theoretically by invoking the empirical Eq 1 with $m = 0.55$ for Ni_3Al and applying a well-known dislocation work-hardening model ($\sigma = \sigma_0 + \alpha Gb \rho^{0.5}$) to the Lüders strain ($\varepsilon_L$). This approach strongly implies that the $\ell^{-0.8}$ dependence is unique to all the materials that yield discontinuously, i.e., when the yield stress for the Hall-Petch relation is taken as the Lüders band propagation stress (plateau stress). It is interesting, however, that Schulson and Roy (Ref 5) in their early work on Zr_3Al, which also exhibited a discontinuous yielding with exactly the same *m* parameter in Eq 1 as that for Ni_3Al, interpreted the yield strength-versus-grain size data through a classical $\ell^{-0.5}$ dependence.

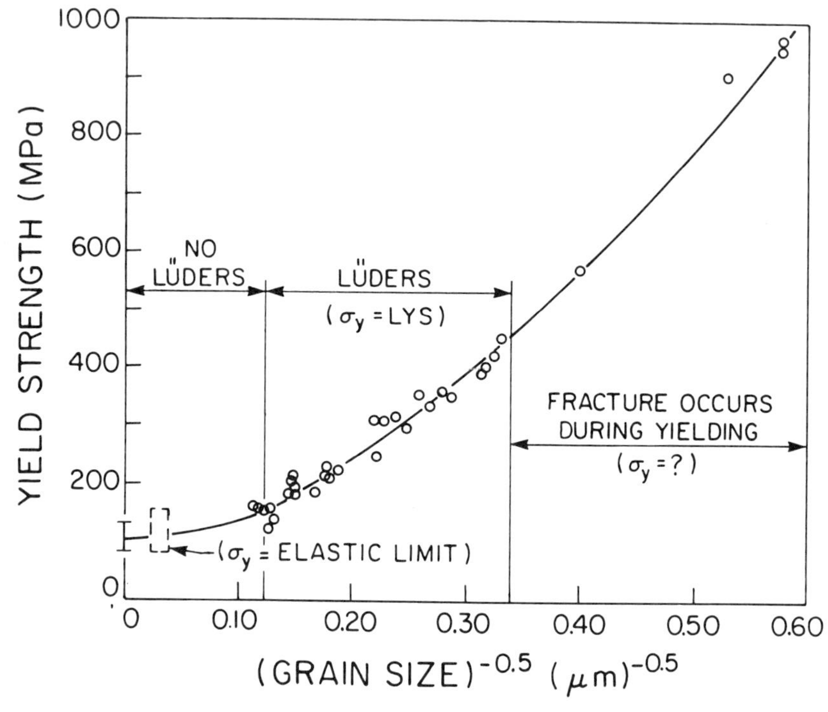

Fig. 5 Various stresses taken as the yield strength for the Hall-Petch relation Ni_3Al by Schulson *et al.* LYS, lower yield stress. Source: Ref 7

The results of Schulson *et al.* (Ref 7, 8) on Ni_3Al have subsequently been reevaluated by Takeyama and Liu (Ref 24). They have shown that most of the data presented by Schulson *et al.* can fit well with the classical Hall-Petch relation with n = 0.5. Also, Takeyama and Liu have pointed out that the specimen preparation methods, testing method (tension versus compression) and thermomechanical treatments of the specimens may have affected the results. It is also worth noting that for boron-free Ni_3Al, Schulson *et al.* (Ref 7) used a variety of stresses to plot the yield strength dependence on grain size (Fig. 5). On the one hand, at grain sizes finer than about 8 μm, a premature fracture occurred during discontinuous yielding (Ref 7), and it is not clear what stress was taken as the yield strength. On the other hand, for the coarse-grained material (~1000 μm), yielding occurred in a continuous manner, and σ_y was taken as the elastic limit.

Table 2 Chemical composition of austenitic stainless steels investigated by the present author

Steel	C	Cr	Ni	Mo	Mn	Si	P	S	N	Reference
	Composition, wt%									
316L-P	0.020	17.80	13.2	2.22	1.50	0.84	0.011	0.015	0.023	21
316L-I	0.016	16.76	11.03	2.0	1.56	0.37	0.039	0.011	0.013	21
316L	0.022	16.48	10.91	1.98	1.43	0.77	0.031	0.004	0.061	21
304	0.060	18.40	8.63	...	0.83	0.47	0.027	0.005	0.004	Present work

The Hall-Petch relationship has recently been studied by the present author and coworkers in several AISI 300-series austenitic stainless steels that can serve as examples of disordered fcc alloys. Table 2 lists their chemical compositions. An example of the Hall-Petch plot ($n = 0.5$) for 0.2% offset in AISI 304 austenitic steel is shown in Fig. 6. An excellent fit with correlation coefficient $r = 0.998$ is observed. An interesting exercise is to perform a regression analysis on the data in Fig. 6 for several exponents (n) in the Hall-Petch relation in the same manner as did Schulson *et al.* (Ref 7) for Ni_3Al. The results are listed in Table 3. It is obvious that any exponent from 0.5 to 1.0 fits the experimental data extremely well. Apparently, the correlation coefficients alone do not provide any indication of preference for the exponent.

Another aspect of the Hall-Petch analysis is presented in Fig. 7 for 316L-I and ultrafine-grained 316 (Ref 12) austenitic steels. Data points for coarse and medium grain sizes (which yielded continuously) of 316L-I steel fit well the $\ell^{-0.5}$ dependence ($r = 0.982$). However, ultrafine-grained (1.2 to 1.6 μm) samples of 316L-I and 316 exhibit much higher values of 0.2% offset than would be expected from the $\ell^{-0.5}$ dependence. Almost all of the ultrafine-grained samples (except one in Fig. 7) exhibited discontinuous yielding (Lüders type), and 0.2% offset corresponds in reality to the value of the Lüders plateau stress on the stress-strain curves (Ref 12). A very similar behavior was observed by Lloyd (Ref 9) in ultrafine-grained samples of Al-6 wt% Ni alloy that also yielded discontinuously and showed an enhanced yield strength when plotted versus $\ell^{-0.5}$. Such behavior would imply that the upward trend of the Hall-Petch plot at ultrafine-grain sizes (Fig. 7) was actually related to the stress to propagate the Lüders bands during discontinuous yielding; the Lüders stress might be greater than the 0.2% offset stress for continuous yeilding.

The present author has also performed a regression analysis on all the data in Fig. 7 for the exponents 0.8 and 1.0 in Eq 7, assuming that it is permitted to combine both 0.2% offset and lower yield stresses on the same plot. The obtained results are shown in Table 4 and plotted for $n = 1.0$ in Fig. 8. As seen,

Table 3 Results of regression analysis for various exponents (n) in the Hall-Petch relation for 304 austenitic steel from Fig. 6

n	σ_0, MPa	$k_{0.2}$, MPa mm^n	Correlation coefficient
0.5	158.4	13.36	0.998
0.7	188.4	4.09	0.999
0.8	197.7	2.34	0.999
1.0	210.8	0.789	0.998

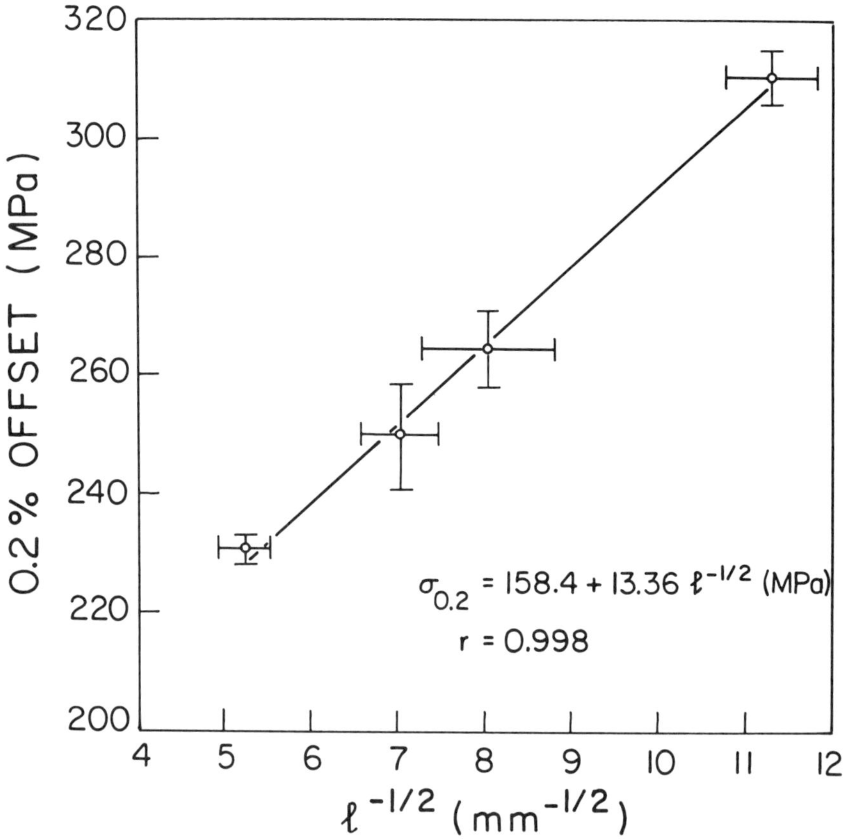

Fig. 6 Hall-Petch plot ($n = 0.5$) for 0.2% offset in AISI 304 austenitic steel. Error bars represent standard deviations. Best-fit line obtained by regression analysis is also shown: $\sigma_0 = 158$ MPa, $k_{0.2} = 13.36$ MPa mm$^{1/2}$, $r = 0.998$

Table 4 Results of regression analysis for two exponents (n) in the Hall-Petch relation for 316L-I and 316 austenitic steels from Fig. 7 (25 data points)

n	σ_0, MPa	$k_{0.2}$, MPa mm^n	Correlation coefficient
0.8	148.3	1.94	0.982
1.0	166.6	0.51	0.982

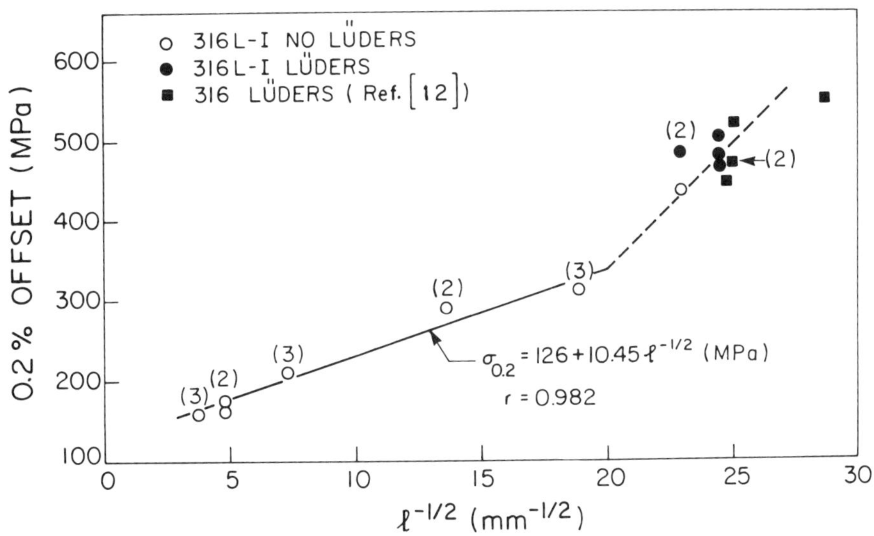

Fig. 7 Hall-Petch analysis for 316L-I and 316 austenitic steels. Best-fit line for coarse and fine grain sizes is shown for $n = 0.5$. Ultrafine-grained samples that exhibit Lüders-type yielding do not fit $\ell^{-0.5}$ dependence. The numbers in parentheses indicate that several experimental points for the yield strength are so close to each other that they cannot be separated and so are represented by one circle or square

all the data points can now be well described either by $\ell^{-0.8}$ or $\ell^{-1.0}$ dependence. The latter was obtained for Al-6 wt% Ni alloy by Lloyd (Ref 9), who also combined both types of yield stresses (0.2% offset and the lower yield stress) on the same plot. However, once again it is impossible to decide which dependence is true, on the basis of the correlation coefficient value alone, without referring to a particular model of the Hall-Petch relation (Ref 25-28). For the purpose of this work the author prefers to use a classical $\ell^{-0.5}$ dependence,

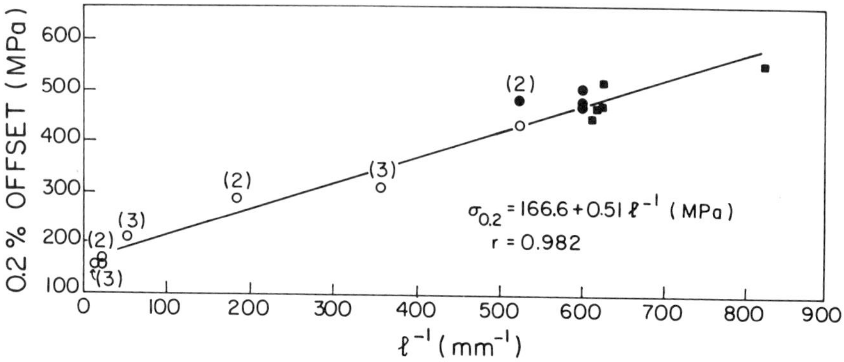

Fig. 8 Hall-Petch plot with $n = 1.0$ incorporating all the data from Fig. 7

which is used in the next section for analysis of the grain-boundary-strengthening effects.

Grain-Boundary Strengthening. The grain-boundary strengthening during yielding is conveniently represented by the magnitude of k_y parameter in Eq 7. Table 5 gives a compilation of the Hall-Petch parameters for ordered and disordered fcc alloys calculated from $\ell^{-0.5}$ dependence. The results obtained by Schulson *et al.* (Ref 7, 8) for Ni_3Al are also included as recalculated by Takeyama and Liu (Ref 24). In the disordered alloys the results obtained by Scott and Spruiell (Ref 29) for 316 austenitic steel, in which a wide range of grain sizes was obtained by rapid pulse heating after cold deformation (rapid recrystallization), are also included for comparison with other austenitic steels.

As expected and as pointed out in the literature (Ref 4-6), the average value of k_y parameter for ordered fcc alloys is much higher than that for disordered alloys, being approximately twice the k_y for austenitic steels. One exception is rapidly recrystallized 316 steel, which has $k_y = 26.88$ MPa mm$^{1/2}$, equal to that for Zr_3Al and boron-doped Ni_3Al intermetallics. Scott and Spruiell suggested that this large value of k_y might be a result of interstitial segregation to the grain boundaries and/or the special nature of boundaries created by very rapid recrystallization. No systematic studies of the segregation effects or grain-boundary nature were carried out. Obviously, a detailed explanation of why the k_y parameters are higher in ordered fcc alloys than those in disordered alloys should be based on a particular theoretical model chosen to explain the Hall-Petch relation (Ref 17-19, 25-28), but the choice seems to be to a large extent a matter of personal preference.

In Ni_3Al, the Hall-Petch slope is generally lower for boron-doped Ni_3Al, but this seems to be true only for hypostoichiometric alloys (<25 at.% Al) (Ref 24).

Table 5 The Hall-Petch parameters σ_0 and k_y in the yield strength dependence on grain size for ordered and disordered fcc alloys

Alloy	σ_0, MPa	k_y, MPa $mm^{1/2}$	Grain size range, μm	Number of data points	Correlation coefficient	Type of yield stress taken for Hall-Petch plot	Reference
Ordered							
Zr_3Al	90	24.03	1.6-65	13	...	LYS	5
Ni_3Fe	98	40.60	12.4-265	10	...	LYS + elastic limit	4
Ni_3Al(24 at.% Al) + B	163	25.93	19-216	8	...	LYS	2
Ni_3Al (24 at.% Al) + Hf + B	184	23.72	1.9-32	5	...	0.2% offset (?)	23
Ni_3Al (23 at.% Al) + Hf + B	185	26.56	15-173	7	...	LYS	24
Ni_3Al (25 at.% Al) + B	109	36.68	1.8-80	25	...	LYS	8,24
Ni_3Al (25 at.% Al)	–44	46.48	2.9-70	19	...	LYS	7,24
Ni_3Al (25 at.% Al)	0	31.94	9-25	11	...	LYS	22,24
Ni_3Al (23.9 at.% Al)	26	37.31	9-80	8	...	LYS	7,24
Ni_3Al (23.5 at.% Al) + Hf + B + Al_2O_3	350	22.45	1-35	4	...	0.2% offset	29
Disordered							
Ni_3Fe	29	30.83	12.4-265	10	...	LYS + elastic limit	4
316	105	26.88	3-50	11	...	0.2% offset	30
316L-P	152	7.84	30-80	28	0.999	0.2% offset	21
316L-I	126	10.45	2.8-73	14	0.982	0.2% offset	21 + this work
316L	184	10.46	26-230	12	0.948	0.2% offset	21
304	158	13.36	7.8-37	37	0.998	0.2% offset	This work

This would be due to the stronger segregation of boron to grain boundaries in hypostoichiometric Ni_3Al alloys rather than stoichiometric alloys (Ref 24). Schulson *et al.* (Ref 8) assumed that boron segregation increases the mobility of grain-boundary dislocations (GBDs), which enhances the slip propagation through grain boundaries and, in effect, lowers the k_y parameter. However, because the movement of GBDs is rarely observed by transmission electron microscopy, Bond *et al.* (Ref 30) and Kurzydlowski *et al.* (Ref 13) argued that the segregated boron assists the accommodation of slip, enhancing the generation of dislocations from grain-boundary sources.

Very interesting results have been obtained by Jang and Koch (Ref 31), in their investigation of the Hall-Petch relation in boron-doped Ni_3Al containing Al_2O_3 dispersoids. Surprisingly, all of the samples studied (grain sizes ranging from 1 to 35 μm) yielded in a continuous manner, and the yield strengths were defined at 0.2% offset. The obtained k_y parameter is listed in Table 5. Its value is

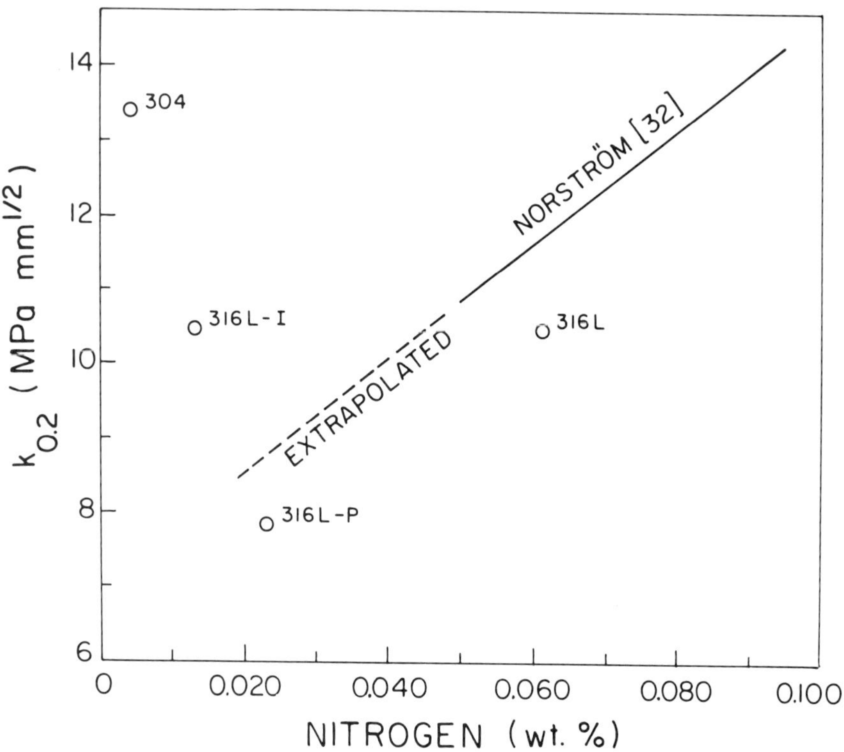

Fig. 9 The values of $k_{0.2}$ parameter for austenitic steels as a function of nitrogen content. A linear dependence reported by Norström is also shown

very similar to other k_y values for boron-doped Ni_3Al, which were obtained by taking the lower yield stresses (plateau stresses) for calculating the Hall-Petch relation. The fact that the lower yield stresses and 0.2% offset stresses give similar values of k_y implies that the grain-boundary-strengthening mechanism is similar in both discontinuously and continuously yielding Ni_3Al alloys. Thus, this mechanism may not necessarily be related to the propagation of a Lüders band at the lower yield stress, as assumed in the model proposed by Schulson *et al.* (Ref 7) to explain their $\ell^{-0.8}$ dependence. Furthermore, if the grain-boundary-strengthening mechanism were similar for both 0.2% offset and the lower yield stress, there would be nothing wrong with combining both

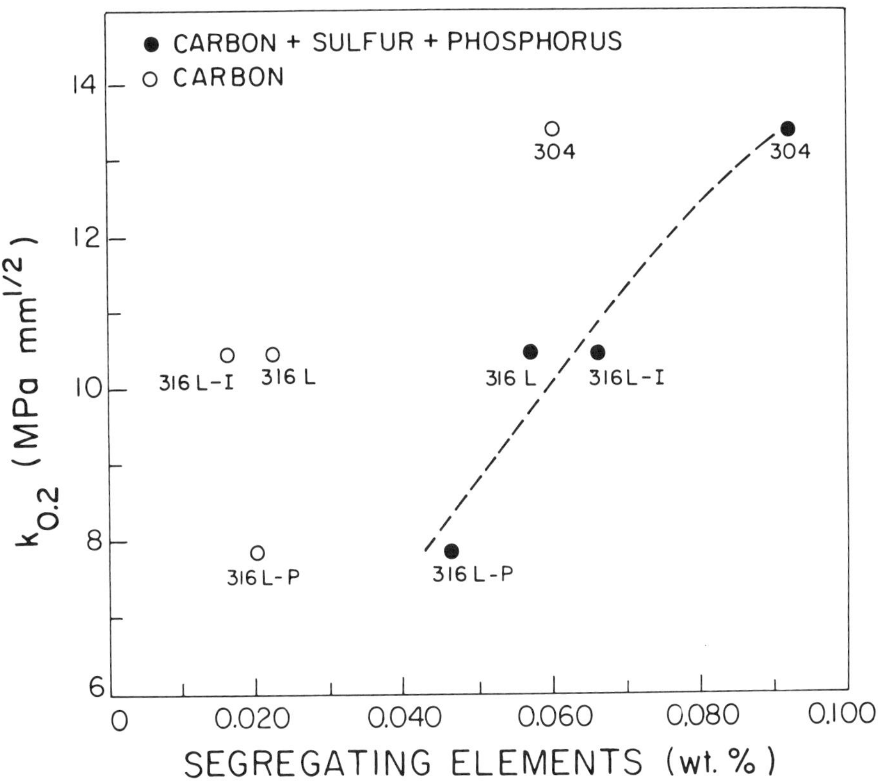

Fig. 10 Correlation of $k_{0.2}$ parameter with carbon and the combined content of carbon, sulfur, and phosphorus for austenitic steels

types of stresses to plot the Hall-Petch relation, as did Schulson *et al.* (Ref 7) (Fig. 5). Apparently, more studies are required to clarify these points.

As seen in Table 5, the $k_{0.2}$ values for austenitic stainless steels investigated by the present author vary from 7.84 to 13.36 MPa mm$^{1/2}$. These differences are too large to be merely an experimental scatter. Following Norström (Ref 32), who reported that $k_{0.2}$ in 316 steel increased linearly with the nitrogen content, the author also plotted (Fig. 9) the $k_{0.2}$ parameter as a function of nitrogen content taken from Table 2. Surprisingly, no systematic dependence on nitrogen content is observed. For the record, it must be mentioned that the 316 austenitic steel studied by Norström contained from 0.05 to 0.18 wt% N. The steels used by the author contained much lower nitrogen levels, varying from 0.004 to

0.061 wt% N (Table 2). In the next step the author tried to correlate the $k_{0.2}$ values with the carbon content (Fig. 10). Some increasing trend is visible, but a much better correlation is obtained when the $k_{0.2}$ values are plotted (Fig. 10) as a function of the combined content of carbon, sulfur, and phosphorus, taken from Table 2. This result implies some possibility of cosegregation of these elements to the grain boundaries with a resulting increase in grain-boundary resistance to slip propagation (opposite to the effect of boron).

It is interesting to note that the grain-boundary segregation of carbon, which occurs in a nonequilibrium fashion, has recently been reported in Ni_3Al (Ref 33). Although there is experimental evidence that carbon increases the yield strength of Ni_3Al (Ref 34), its exact effect on the grain-boundary strength (k_y parameter) has not been evaluated because carbon does not improve the ductility of Ni_3Al as does boron (Ref 34).

Conclusions

The conclusions that can be drawn from work reported in the literature and performed by the author are:

- Ordered fcc alloys are much more prone to develop a substantial Lüders deformation than disordered alloys, which exhibit a pronounced Lüders strain only at ultrafine-grain sizes (<2 to 4 μm)
- In one group of disordered fcc alloys such as austenitic stainless steels it has been observed that the magnitude of Lüders strain increases with an increasing number of particles (carbides) per unit area of grain-boundary surface. A locking of grain-boundary sources for mobile dislocations is suggested as the reason for such a behavior. However, a 304L austenitic steel seems to be an exception
- Boron doping up to approximately 15 ppm in a disordered fcc austenitic steel reduces the magnitude of Lüders strain, which then remains unaffected by higher boron contents. However, the plateau stress is constant up to about 15 ppm B and then decreases continuously with increasing boron content. This phenomenon is interpreted in terms of boron segregation to grain boundaries and its effect on grain-boundary sources of dislocations. There are no systematic studies of the effect of boron doping on the Lüders strain in ordered fcc alloys
- Coherent twin boundaries do not constitute strong barriers to slip in either disordered or ordered fcc alloys
- Most of the yield strength (or 0.2% offset) data versus grain size for both ordered and disordered alloys can be described by the classical Hall-Petch relation with $n = 0.5$. One group of data for Ni_3Al also fits $\ell^{-0.8}$ dependence.

Ultrafine-grained specimens of 304 and 316 austenitic steels exhibit much higher values of 0.2% offset than would be expected from the $\ell^{-0.5}$ dependence

- If the Hall-Petch relation at yielding is expressed as $\ell^{-0.5}$, the grain-boundary-strengthening parameters (k_y) for ordered fcc alloys are higher than most of those for disordered alloys, for example, austenitic steels. However, rapidly recrystallized 316 austenitic steel exhibits a k_y parameter similar to that for the ordered, boron-doped Ni_3Al
- In the ordered Ni_3Al alloys, the k_y parameter seems to have similar values for both the lower yield stress and 0.2% offset. This suggests a similar grain-boundary-strengthening mechanism for discontinuously and continuously yielding ordered fcc alloys
- In Ni_3Al, the Hall-Petch slope is lower for hypostoichiometric boron-doped alloys. This is interpreted by the enhancement of slip propagation through grain boundaries by segregated boron. No systematic studies of the effect of boron on the Hall-Petch slope in disordered fcc alloys exist
- The Hall-Petch slope in austenitic stainless steels containing up to 0.061 wt% N is independent of nitrogen content but seems to increase with increasing content of combined carbon, sulfur, and phosphorus

Acknowledgment

This work has been supported by grants from the Natural Sciences and Engineering Research Council of Canada and the Ontario Center for Materials Research, which are gratefully acknowledged. The author is grateful to Dr. K.J. Kurzydlowski for technical assistance and fruitful discussion.

References

1. K. Aoki and O. Izumi, *J. Jap. Inst. Met.*, Vol 43, 1979, p 1190
2. C.T. Liu, C.L. White, and J.A. Horton, *Acta Metall.*, Vol 33, 1985, p 213
3. F.M.C. Besag and R.E. Smallman, *Acta Metall.*, Vol 18, 1970, p 429
4. A.C. Arko and Y.H. Liu, *Metall. Trans.*, Vol 2, 1971, p 1875
5. E.M. Schulson and J.A. Roy, *Acta Metall.*, Vol 26, 1978, p 29
6. E.M. Schulson, *Int. Met. Rev.*, Vol 29, 1984, p 195
7. E.M. Schulson, T.P. Weihs, D.V. Viens, and I. Baker, *Acta Metall.*, Vol 33, 1985, p 1587
8. E.M. Schulson, T.P. Weihs, I. Baker, H.J. Frost, and J.A. Horton, *Acta Metall.*, Vol 34, 1986, p 1395
9. D.J. Lloyd, *Met. Sci.*, Vol 14, 1980, p 193
10. H. Westengen, in *Strength of Metals and Alloys, Proc. 6th ICSMA*, Vol 1, 1982, p 461
11. R.A. Varin and K. Tangri, *Mater. Sci. Eng.*, Vol 72, 1985, p 177

12. R.A. Varin, B. Mazurek, and D. Himbeault, *Mater. Sci. Eng.*, Vol 94, 1987, p 109
13. K. Kurzydlowski, R.A. Varin, and K. Tangri, *Z. Metallkunde*, Vol 80, 1989, p 469
14. E.O. Hall, *Yield Point Phenomena in Metals and Alloys*, Macmillan, 1970
15. R.A. Varin, K.J. Kurzydlowski, and K. Tangri, *Mater. Sci. Eng.*, Vol 85, 1987, p 115
16. R.A. Varin, *Proceedings of International Symposium on Advanced Structural Materials*, D.S. Wilkinson, Ed., Pergamon Press, 1989, p 235
17. E.O. Hall, *Proc. Phys. Soc. London*, Vol B64, 1951, p 747
18. N.J. Petch, *J. Iron Steel Inst.*, Vol 174, 1953, p 25
19. R.W. Armstrong, I. Codd, R.M. Douthwaite, and N.J. Petch, *Philos. Mag.*, Vol 7, 1962, p 45
20. A.W. Thompson, *Metallography*, Vol 5, 1972, p 366
21. R.A. Varin and K.J. Kurzydlowski, *Mater. Sci. Eng. A.*, Vol 101, 1988, p 221
22. S. Hanada, S. Watanabe, and O. Izumi, *J. Mater. Sci.*, Vol 21, 1986, p 302
23. P.S. Khadkikar, K. Vedula, and B.S. Shabel, *Metall. Trans. A*, Vol 18A, 1987, p 425
24. M. Takeyama and C.T. Liu, *J. Mater. Res.*, Vol 3, 1988, p 665
25. J.C.M. Liu and Y.T. Chou, *Metall. Trans.*, Vol 1, 1970, p 1145
26. M.M. Hutchinson and R.T. Pascoe, *Met. Sci. J.*, Vol 5, 1972, p 90
27. J.P. Hirth, *Metall. Trans.*, Vol 3, 1972, p 3047
28. R.W. Armstrong, in *Yield, Flow and Fracture of Polycrystals*, T.N. Baker, Ed., Applied Science Publishers, 1983, p 1
29. J.A. Scott and J.E. Spruiell, *Metall. Trans.*, Vol 5, 1974, p 255
30. G.M. Bond, I.M. Robertson, and H.K. Birnbaum, *J. Mater. Res.*, Vol 2, 1987, p 436
31. J.S.C. Jang and C.C. Koch, *Scr. Metall.*, Vol 22, 1988, p 677
32. L.-A. Norström, *Met. Sci.*, Vol 11, 1977, p 208
33. S.S. Brenner and Hua Ming-Jian, *Scr. Metall. Mater.*, Vol 24, 1990, p 667
34. N. Masahashi, T. Takasugi, and O. Izumi, *Acta Metall.*, Vol 36, 1988, p 1823

12

Dislocation Reactions at Interfacial Boundaries in Ti_3Al + TiAl Two-Phase Alloys and Their Relation to Mechanical Properties

Linruo Zhao and K. Tangri
Metallurgical Sciences Laboratories, Department of Mechanical Engineering
University of Manitoba, Winnipeg, MB, Canada R3T 2N2

The nature and distribution of dislocations at the different interfaces in the Ti_3Al + TiAl lamellar structure were studied by transmission electron microscopy (TEM). Dislocation reactions occurring during deformation at these interfaces and their relation to mechanical properties of the Ti_3Al + TiAl alloy were also investigated. In the as-cast and homogenized state, two types of interfaces, i.e., Ti_3Al/TiAl interfaces and TiAl/$TiAl_T$ twin boundaries in twin-related TiAl variants, are predominantly formed in this lamellar structure. After heat treatment is used to refine the microstructure, three types of interfaces, Ti_3Al/TiAl interfaces, antiphase domain boundaries (APBs) in Ti_3Al laths, and cross twin boundaries in TiAl laths, are present. The Ti_3Al/TiAl interfaces contain two types of intrinsic interfacial dislocations (IIDs), i.e., dissociated dislocations (IID-1) and 1/6⟨112] partials (IID-2), which are controlled by the interfacial misfit. The defect structures on TiAl twin boundaries in twin-related TiAl variants are characterized by triple-twinning partials 1/6⟨112], whereas those on cross twin boundaries are characterized by individual twinning partials 1/6⟨112]. The antiphase vectors of the APBs were determined to be $1/6\langle 11\bar{2}0\rangle$ and $1/6\langle 10\bar{1}0\rangle$. Planar slip during compression, both in Ti_3Al and TiAl laths, is partly interrupted by the presence of these interfaces. Dislocation tangles are formed at and are also near all of the interfaces. Some of the dislocation reactions occurring on the interfaces result in sessile dislocation segments, which in turn block the movement of interfacial dislocations. It is suggested that the improvement of ductility in the presence of lamellar structure is due to the more uniform deformation of Ti_3Al and TiAl laths. The enhancement in strain hardening is predominantly attributed to the large incidence of interaction between lattice dislocations and these interfaces.

Titanium aluminide alloys based on $Ti_3Al(\alpha_2)$ and $TiAl(\gamma)$ are currently receiving a great deal of attention for their application as high-temperature structural materials in the aerospace industry. These alloys possess attractive properties, including high melting temperature, low density, good elevated-temperature strength and modulus, and high resistance to oxidation and creep, but are somewhat deficient in ductility at room-to-intermediate temperatures (Ref 1, 2). In spite of a lack of proper understanding of the mechanism of brittleness, significant progress in improving the ductility of these alloys has been made in recent years through the control of composition and/or microstructure (Ref 3-5). The former is known to affect the types of dislocations involved in the deformation of Ti_3Al (Ref 6) and TiAl (Ref 4), the tetragonality (*c/a* ratio) of TiAl (Ref 3-5), and the chemical bonds of TiAl (Ref 7). The microstructure, i.e., the phases present and the size and morphology of grains, is dependent on both the composition and the processing history (Ref 2). Blackburn and Smith (Ref 8) have shown that the TiAl alloy containing 30 vol% lamellar structure consisting of Ti_3Al and TiAl laths possesses the best balance of properties in high-temperature strength and creep resistance and room-temperature ductility. It was argued that the improvement of the ductility of the two-phase alloys with lamellar structure is due to the segregation of interstitial elements to the Ti_3Al phase, which in turn either causes the dislocations 1/2⟨110] in TiAl to be significantly more glissile at room temperature (presumably because of a reduction in the effects of covalence) (Ref 9) or changes the deformation mode of TiAl by altering the core configurations of dislocations and their energies (Ref 4). It was further argued that this improvement is also due to a more fine-scale and uniform deformation of Ti_3Al and TiAl laths in the lamellar structure (Ref 3, 9) and to the difficulty of forming significant pileups of dislocations because of the abundance of dislocation sources in this microstructure (Ref 9). More recently, intensive studies have been carried out on the defect structure of various interfacial boundaries present in the lamellar structure because it was expected that the high density of interfaces played a significant role in deformation (Ref 10-12).

The present investigation was carried out to identify the dislocation reactions occurring at the interfacial boundaries during room-temperature deformation of the wholly lamellar structure and the relation of the dislocation reaction to mechanical behavior. It is shown in this chapter that all of the different types of interfaces interact with lattice dislocations in Ti_3Al and TiAl. The dislocation structure within the laths, as well as the defect structure of the interfaces, is dependent on this interaction, which in turn alters the mechanical behavior of this two-phase alloy.

Experimental Procedure

An ingot of Ti-44 at.% Al alloy was prepared by consumable electrode arc melting in vacuum and then homogenized at 1000 °C in vacuum for 7 days. The size of the primary grains with Ti_3Al + TiAl lamellar morphology ranged from 0.5 to 1.0 mm. After heat treatment (Ref 13) at 1150 °C in vacuum for 72 h and cooling in air, and then at 1000 °C for 4 h and cooling in air, the grain size of the alloy was refined to about 50 μm, but Ti_3Al and TiAl laths, on average, became coarser from the original thickness of approximately 0.5 to 1.3 μm. This treatment produced a completely lamellar structure, although the composition of this alloy deviates to the Al-rich side from the eutectoid stoichiometry (Ref 14). Samples with the dimensions 5 × 5 × 10 mm were deformed in compression at room temperature and a strain rate of 1×10^{-4} s^{-1} to about 8.0% true plastic strain (Ref 13).

To prepare TEM foils, samples were cut from the as-cast and homogenized ingot, the heat-treated sample, and the deformed sample parallel to the compression axis. The TEM foils were made by grinding and twin-jet polishing using an electrolyte of 64% methanol, 31% butanol, and 50% perchloric acid at –40 to –60 °C and 20 to 30V. The structure of various interfaces before and after deformation was studied through bright-field (BF) and weak-beam dark-field (WBDF) observations using a JEOL 2000FX microscope. To identify these defect structures quantitatively, they were imaged near low-index crystal zones and with several reflections (g) and then analyzed according to their image contrast and traces.

Results and Discussion

The Structure of the Interfaces in the Undeformed State. The completely lamellar structure of the alloy after homogenization consists of twin-related TiAl variants and Ti_3Al laths where the latter are much narrower in thickness. It has been shown that the TiAl variants are always present adjacent to Ti_3Al laths (Ref 11). In this state of the alloy only two types of interfacial boundaries predominate, i.e., Ti_3Al/TiAl interfaces and the TiAl/$TiAl_T$ twin boundaries. The former type has orientation relationships $(0001)\alpha_2//\{111\}\gamma$ and $\langle 11\bar{2}0\rangle\alpha_2//\langle 1\bar{1}0\rangle\gamma$, and the latter belongs to $\langle 112]\{11\bar{1}\}\gamma$ order twins. The fine defect structure of these interfaces was examined under WBDF operations.

Figures 1(a) and (b) show two types of intrinsic interfacial dislocation present on the same Ti_3Al/TiAl interfaces (arrows), which were imaged with different reflections. Contrast analysis near $[1\bar{1}1]\gamma$ and $[5\bar{2}1]\gamma$ zones indicated that the dislocation arrays shown in Fig. 1(a) have a dissociated structure (thus are IID-1):

$$\tfrac{1}{2}\langle 110] \rightarrow \tfrac{1}{6}\langle 211] + (\text{APB}' + \text{SISF}') + \tfrac{1}{6}\langle 12\bar{1}]$$

where APB′ and SISF′ refer to the antiphase domain boundary and the superlattice intrinsic stacking fault formed on the interfaces (Ref 11). These dislocations align predominantly along the $\langle 1\bar{1}2]$ direction. Other dislocations shown in Fig. 1(b) were identified as $1/6\langle 112]$ partials (thus are IID-2) with a line direction close to $\langle 110]\gamma$. This result is in disagreement with the TEM studies carried out by Feng *et al.* (Ref 15) in which only a single array of parallel dislocations on a Ti_3Al/TiAl interface was observed in Ti-45 at% Al alloy. Calculations (Ref 11) were made to determine the spacing of the IIDs required to account for the interfacial misfits that were resolved into $\langle 110]\gamma$ and $\langle 112]\gamma$ directions. A comparison between the calculations and the experimental measurements showed that these two types of IID arrays are able to accommodate most of the interfacial misfits.

Figure 2 shows the defect structure of TiAl/$TiAl_T$ twin boundaries in the twin-related variants. This structure, rather than being characterized by the presence of individual partials on deformation twin boundaries, as in TiAl (Ref 16), is characterized by triple dislocations (arrow) that were identified as $1/6\langle 112]$ twinning partials through contrast analysis. It was suggested (Ref 11) that the triplet structure is formed by the dissociation of $1/2\langle 112]$ superdislocations onto three successive $\{11\bar{1}\}\gamma$ twinning planes, followed by the movement of the triple partials, which results in a twin plate lying adjacent to a Ti_3Al lath. This process is driven by thermal stresses generated primarily during casting.

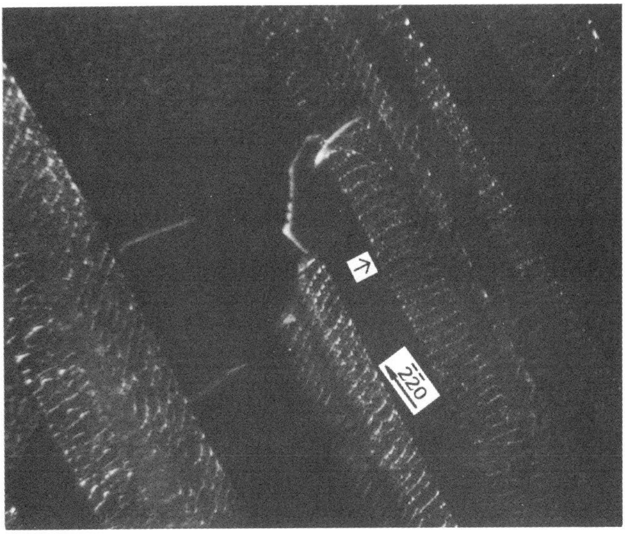

a

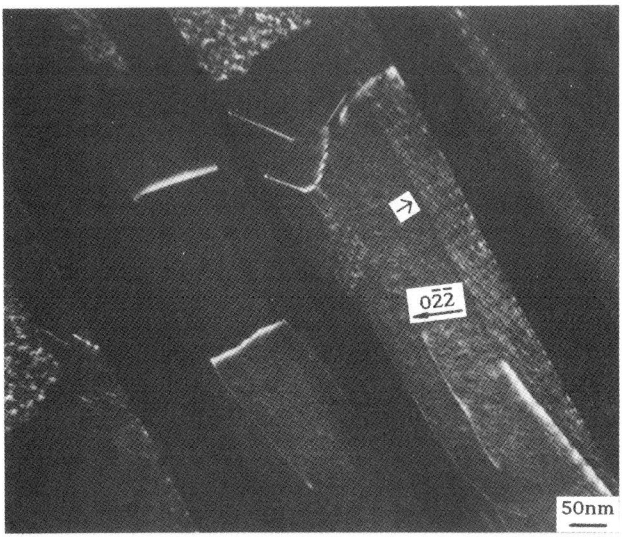

b

Fig. 1 Two types of intrinsic interfacial dislocations (IIDs) present on the same Ti_3Al/TiAl interfaces in the homogenized alloy. (a) Weak-beam dark-field (WBDF) image taken with $g/3g_{\bar{2}\bar{2}0}$ near $[1\bar{1}1]\gamma$. Only the IID-1 dislocations are in contrast. (b) WBDF image taken with $g/3g_{0\bar{2}\bar{2}}$ near $[1\bar{1}1]\gamma$. Another type of dislocation, i.e., IID-2, is visible

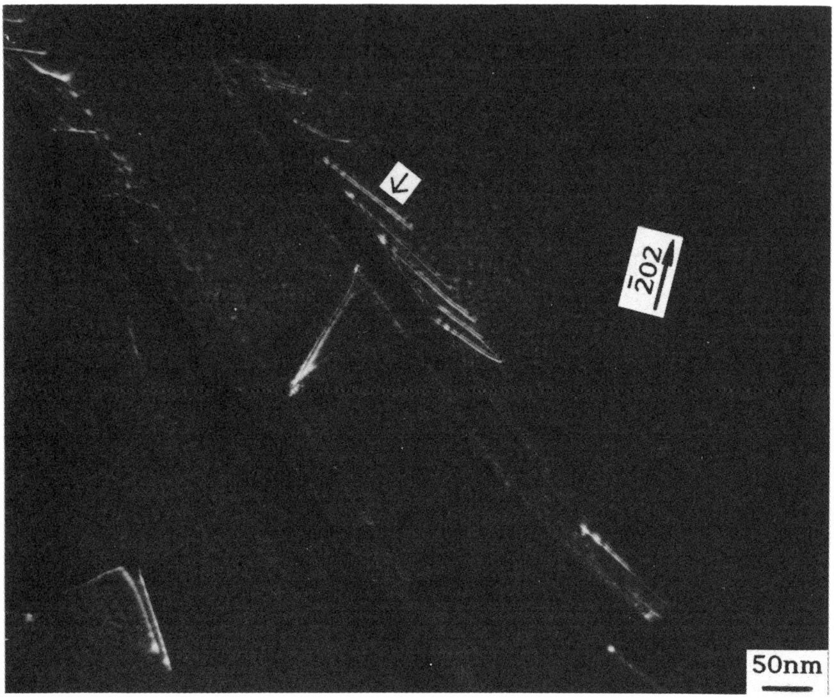

Fig. 2 Triple partials 1/6⟨112] appearing on a TiAl/TiAl$_T$ twin boundary in twin-related TiAl variants in the homogenized alloy. The WBDF image was taken with g/3g$_{\bar{2}02}$ near $[1\bar{1}1]\gamma$

The heat treatment used for the refinement of the microstructure produced several changes in the as-homogenized structure: while the primary grain size was significantly reduced, the lamellar structure within the primary grains became coarser; in particular, the Ti_3Al laths increased in thickness, and many of the TiAl twin-related variants disappeared by growth. Thus, a more equilibrated state of the microstructure is achieved by this heat treatment. Detailed TEM studies (Ref 12) on both Ti_3Al/TiAl interfaces and TiAl/TiAl$_T$ twin boundaries in TiAl twin-related variants have indicated that the defect structure on these interfaces remains essentially unchanged in nature but has a state of greater equilibrium. For instance, more hexagonal networks of IIDs were observed on Ti_3Al/TiAl interfaces formed by the interaction of IID-1 and IID-2 arrays during heat treatment. A significant change in the microstructure was

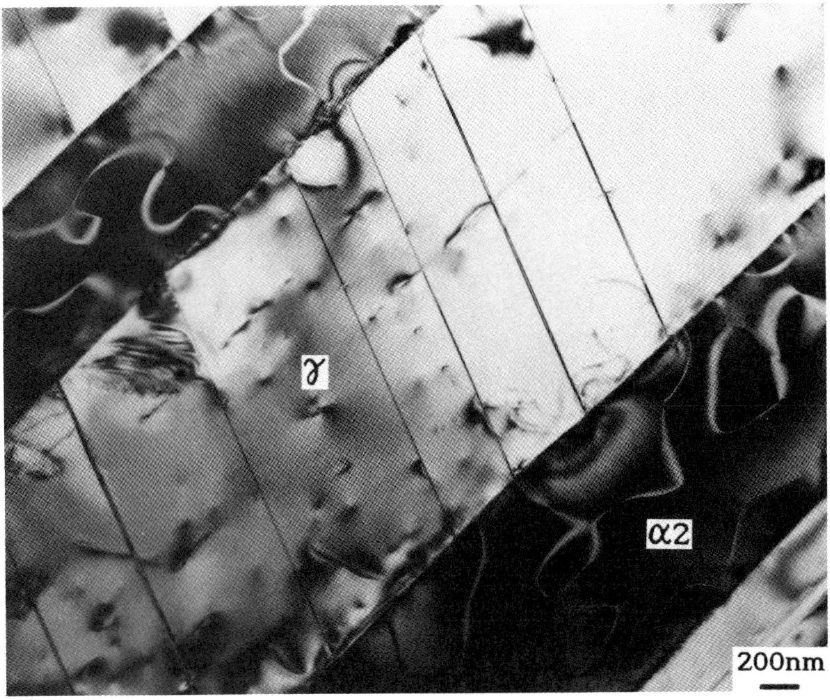

Fig. 3 Ti_3Al + TiAl lamellar structure after heat treatment. Three types of interfaces are distinguishable: Ti_3Al/TiAl interfaces, TiAl/$TiAl_T$ twin boundaries in TiAl(γ) laths, and APBs in $Ti_3Al(\alpha_2)$ laths

the formation of APBs with a maze pattern in every Ti_3Al lath and the generation of cross twin plates in some of the TiAl laths (Fig. 3). These APBs were characterized by contrast analysis as pure-APB $1/6\langle 11\bar{2}0\rangle$ and APB + SF $1/6\langle 10\bar{1}0\rangle$. Individual dislocations, rather than triple dislocations, on the cross twin boundaries (Fig. 4) were also identified as $1/6\langle 112]$ twinning partials. In the section below, the role of these interfaces in the deformation behavior of this alloy is considered.

Dislocation Reactions at the Interfaces during Deformation. It was seen that after compression deformation at room temperature, the lattice dislocation density in most of the TiAl laths is much higher than that in Ti_3Al laths. This implies that the TiAl laths in this two-phase alloy are more deformable.

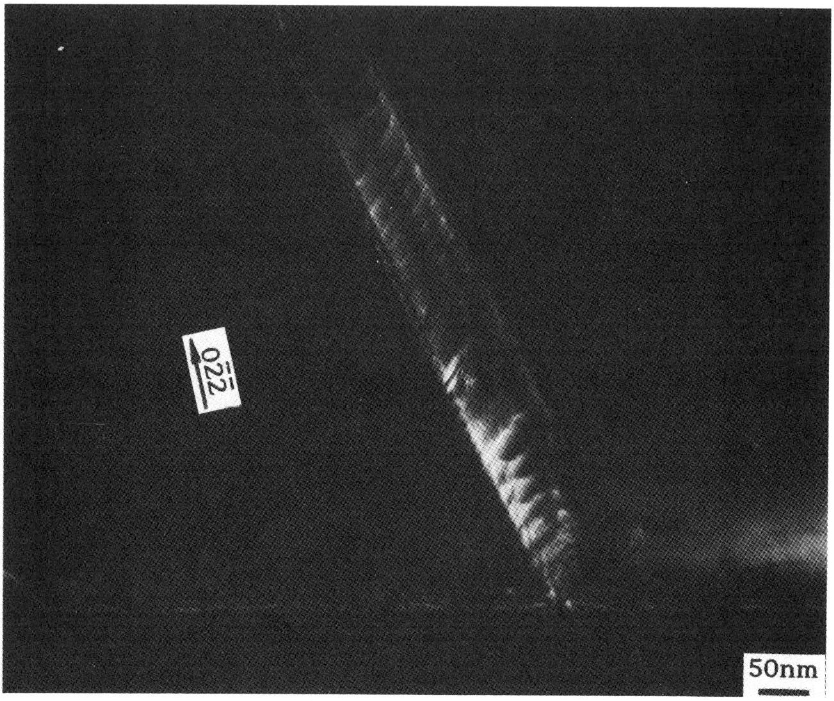

Fig. 4 Individual partials 1/6⟨112] present on a cross twin boundary in the heat-treated alloy. The WBDF image was taken with $g/3g_{0\bar{2}\bar{2}}$ near [100]γ.

Dislocation tangles on the TiAl side are preferentially formed at, as well as near, Ti_3Al/TiAl interfaces, as shown in Fig. 5(a) and (b), which were taken from two opposite interfaces of one TiAl(γ) lath. In Fig. 5(a) the IIDs are locally out of contrast (arrows). It is suggested that this is due to the interaction of the "easy" slip-type lattice dislocations $1/2[\bar{1}10]$ (b_L) in the tangles on the TiAl side with the IID-1 dislocations that have combined under stresses to form perfect interfacial dislocations $1/2[110](b_{IID\text{-}1})$. That is:

$$\tfrac{1}{2}[\bar{1}10](b_L) + \tfrac{1}{2}[110]\,(b_{IID\text{-}1}) \rightarrow 1[010]\,(b_{int})$$

where the resulting 1[010] dislocations are sessile on the interface $(0001)\alpha_2/(\bar{1}11)\gamma$ and invisible under the reflection $g_{20\bar{2}}$. The structure produced by this dislocation reaction on the opposite interface, as shown in Fig. 5(b), resembles that in Fig. 5(a). Here, instead of being out of contrast, the same kind

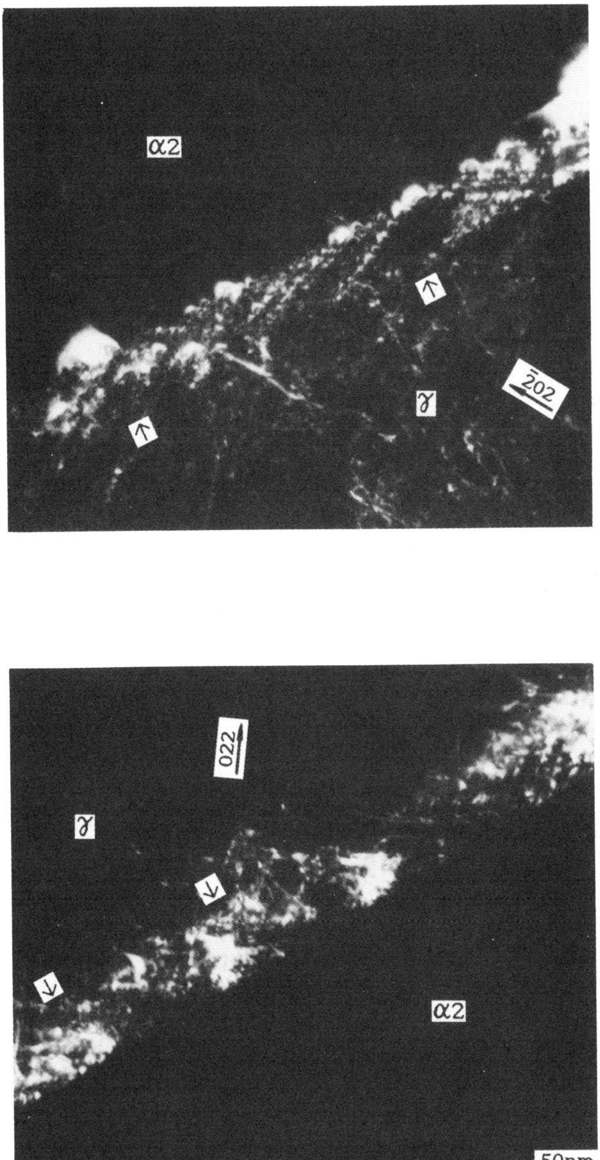

a

b

Fig. 5 Dislocation structures on, as well as near, Ti_3Al/TiAl interfaces of the alloy deformed in compression. (a) WBDF image taken with $g/3g_{\bar{2}02}$ near $[1\bar{1}1]$. (b) WBDF image taken with $g/3g_{022}$ near $[1\bar{1}1]\gamma$

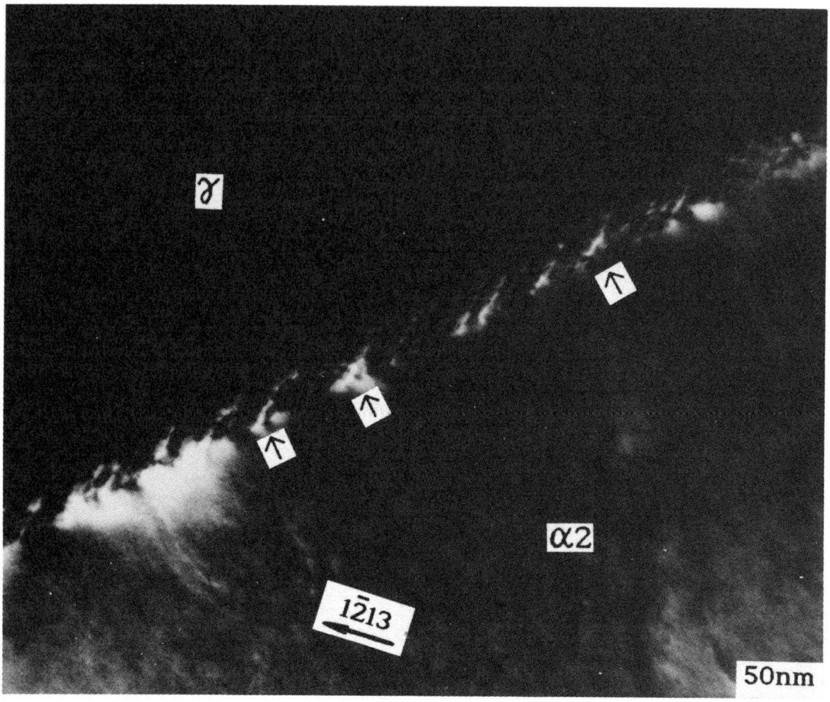

Fig. 6 Ti_3Al/TiAl interface with irregularly arranged IIDs. The WBDF image was taken with g/3g$_{1\bar{2}13}$

of interaction between 1/2[$\bar{1}$10] (b_L) and 1/2[110] ($b_{IID\text{-}1}$) results in increased contrast under the reflection g_{022} (arrows). The IID configuration was revealed by imaging Ti_3Al/TiAl interfaces with reflections of the Ti_3Al phase as the high-density tangles in TiAl laths were out of contrast. Figure 6 shows an interface in which both IID-1 and IID-2 arrays are visible, the latter arranged almost parallel to the interface trace. Two features regarding the IID-1 configuration should be noted. First, some of the IID-1 segments have a significantly increased contrast due to the reaction between lattice dislocations and IIDs described above. Second, the spacing of the IIDs becomes irregular, compared to that in the undeformed state (Fig. 1a), because of blocking of the unreacted and thus glissile IIDs by the reacted and thus sessile IIDs (arrows), as shown in Fig. 7.

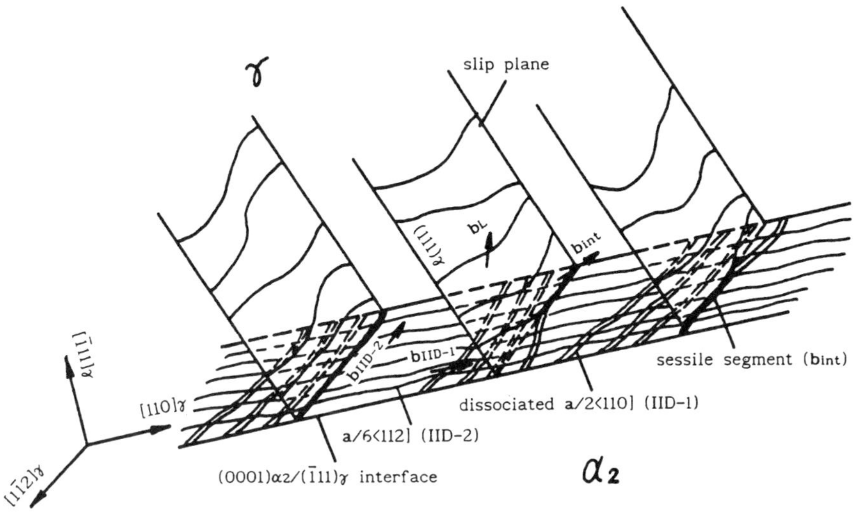

Fig. 7 The blocking of IIDs by the sessile products of dislocation reactions at $Ti_3Al/TiAl$ interfaces

Dislocation structures associated with the presence of thermal APBs in Ti_3Al laths are shown in Fig. 8, where two sets of dislocation segments (S-1 and S-2) with different line directions predominate. Trace analysis near the $[0001]\alpha_2$ zone showed that these segments lie on the prism planes $(10\bar{1}0)\alpha_2$ for S-1 and $(01\bar{1}0)\alpha_2$ for S-2. The Burgers vectors of S-1 and S-2 segments were determined to be $a_0'/3[1\bar{2}10]$ and $a_0'/3[\bar{2}110]$, respectively, i.e., a-type superdislocations. Contrast experiment also confirmed that these segments are screw dipoles consisting of superdislocation pairs of $1/6\langle 11\bar{2}0\rangle$, which bound pure APB faults. Moreover, in this lamellar structure a low incidence of nonbasal $\langle c + a\rangle$ and c-type slips takes place during deformation.

An important feature of this defect structure in the presence of the thermal APBs is that many short dislocation segments and even tangles are formed in the APB areas (arrows in Fig. 8). The interior of the antiphase domains is relatively free of dislocations. This implies that these thermal APBs act as slip barriers to the moving superdislocations, $a_0'/3\langle 11\bar{2}0\rangle$, during deformation. As shown in Fig. 9, a superdislocation of Burgers vector (b) gliding through an APB (p) will leave behind an APB ledge described by p′:

$$p' = p + b$$

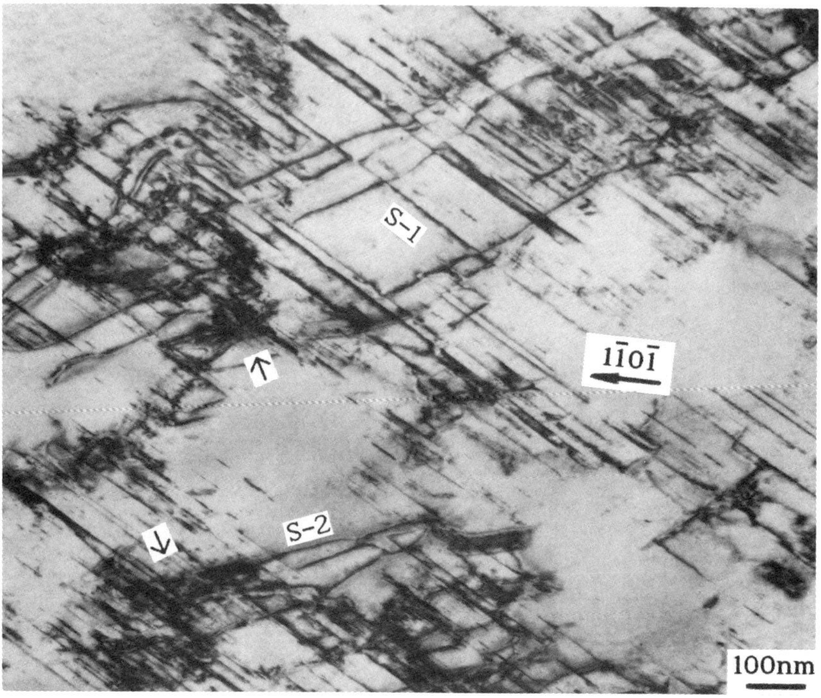

Fig. 8 Dislocation structure associated with the APBs in a Ti_3Al lath. The bright-field (BF) image was taken with $g_{1\bar{1}0\bar{1}}$ near $[1\bar{2}13]\alpha_2$

where p is the antiphase vector. Considering the dissociated structure of $a_0'/3\langle 11\bar{2}0\rangle$ superdislocations, the interaction between dislocations $1/6\langle 11\bar{2}0\rangle$ and APBs $1/6\langle 11\bar{2}0\rangle$ or $1/6\langle 10\bar{1}0\rangle$ results in, for most of the incidents, $p' = a_0'/2\langle 10\bar{1}0\rangle$, $a_0'/3\langle 10\bar{1}0\rangle$, $1/6\langle 10\bar{1}0\rangle$, $a_0'/3\langle 11\bar{2}0\rangle$, and $a_0'/6\langle 1\bar{3}20\rangle$, which are energetically unfavorable compared to the antiphase vector of the original APBs. Resistance of the APBs against moving dislocations originates in these reactions. As a consequence, dislocation tangles are formed by the intersection of the superdislocations activated on different slip planes and stopped at APBs.

Many cross twins are produced in the TiAl laths during compression deformation (Fig. 10a). This is attributed to the presence of Ti_3Al/TiAl interfaces where the twinning partials $1/\langle 112]$ are emitted from one of the interfaces, for example, at A in Fig. 10(b), and travel across the TiAl lath to form cross twin plates (arrow). Diffraction analysis has shown that these twin plates always

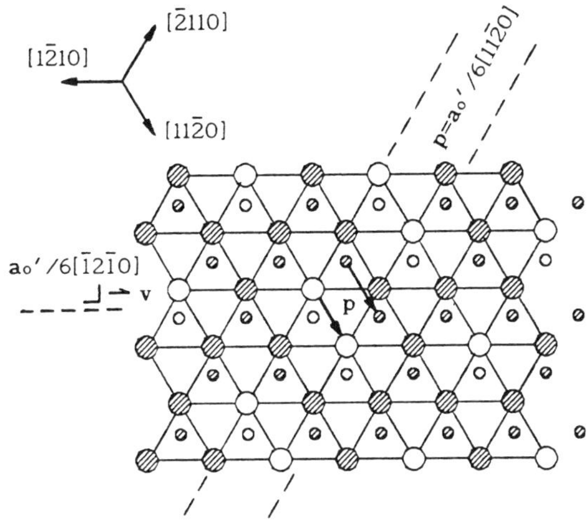

a

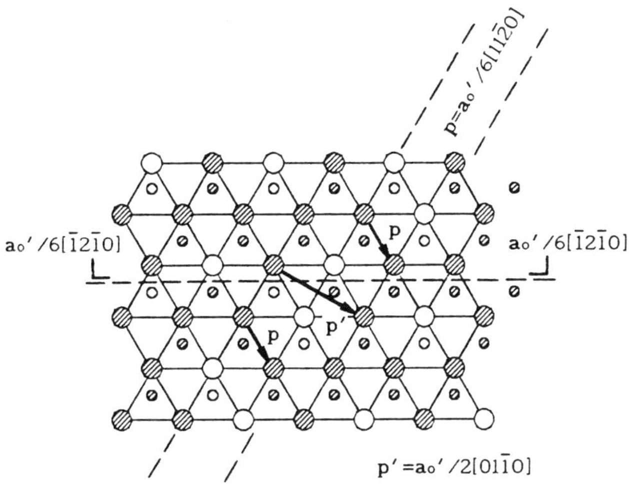

b

Fig. 9 The glide of a lattice dislocation through a thermal APB in the Ti_3Al phase. (a) Before the dissociated superdislocation $a_0'/6[\bar{1}2\bar{1}0]$ meets the APB $a_0'/6[11\bar{2}0]$. (b) After the dislocation glides through this APB intersect the $Ti_3Al/TiAl$ interfaces in a common direction, $\langle 101]\gamma$, as shown in Fig. 10(a). Thus, it is suggested that the generation of these twinning partials on the interfaces is associated with the interfacial ledge structure (Ref 10), which has a line direction altered from $\langle 110]\gamma$ to close to $\langle 101]\gamma$ after IID networks are formed (Ref 12).

Figures 11(a) and (b) show the details of the dislocation structure associated with the presence of twin boundaries. Many of the dislocations in Fig. 11(a) and (b) are seen along the traces of slip planes $(1\bar{1}\bar{1})\gamma$ and $(11\bar{1})\gamma$, respectively. The dislocation segments between slip traces shown in Fig. 11(a) are actually located on $(11\bar{1})\gamma$ planes that have an orientation inclined to the electron beam. Contrast experiments carried out near $[1\bar{1}2]\gamma$ and $[101]\gamma$ zones showed that most of the lattice dislocations are $1/2\langle 110]$ easy-slip dislocations. It should be noted that the planar slip mode in TiAl laths is often interrupted by the presence of the cross twin plates. The fine structure resulting from the reaction of the lattice dislocations on the twin boundaries is also seen in Fig. 11(a) and (b) (arrows). Dislocation tangles, for example, at A in Fig. 11, are formed at these interaction sites. The reaction between $1/2\langle 110]$ lattice dislocations and $1/6\langle 112]$ twinning partials during room-temperature deformation produces sessile dislocation segments on the twin boundary planes since their Burgers vectors lie outside these planes. These segments, in turn, hinder the glide of twinning partials, producing distortion in them, as seen in Fig. 11(a) and (b) (arrows).

The Role of the Interfaces in Deformation. As analyzed in the above sections, three types of interfaces in Ti_3Al + TiAl lamellar structure, i.e., $Ti_3Al/TiAl$, $TiAl/TiAl_T$, and APB, are intimately involved in compression deformation at room temperature. Mechanical tests have shown that TiAl alloys with a lamellar structure exhibit both improved ductility (Ref 8) and higher strain-hardening behavior (Ref 17). It is suggested that the improvement in ductility is due primarily to the refinement of the structure resulting from the presence of Ti_3Al and TiAl laths, cross twins in TiAl, and the APBs in Ti_3Al, which in turn produces a more homogeneous deformation of the material. A further contribution to improved ductility comes from the frequent interruption of planar slip by the various interfaces present which makes a further contribution to the uniformity of deformation. As mentioned earlier, Ti_3Al and TiAl laths undergo different degrees of deformation, necessitating the accommodation of the sharp strain gradients around the interfaces. This is largely accomplished by the activation of multislip in the interface vicinity, as well as by the movement of IIDs along the interfaces. However, the formation of sessile segments on some of the interfaces, as described in the section above,

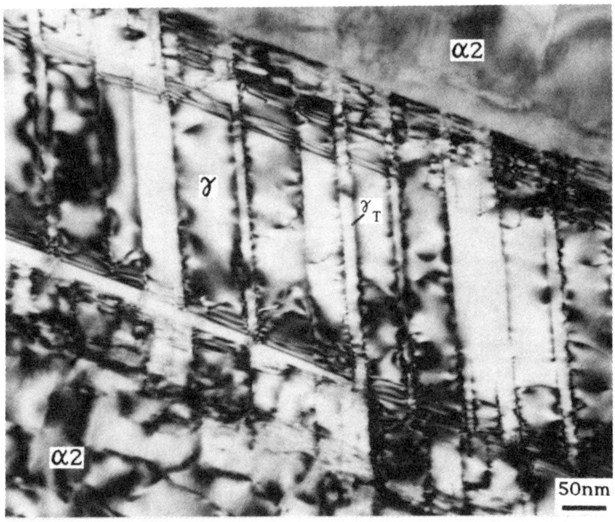

a

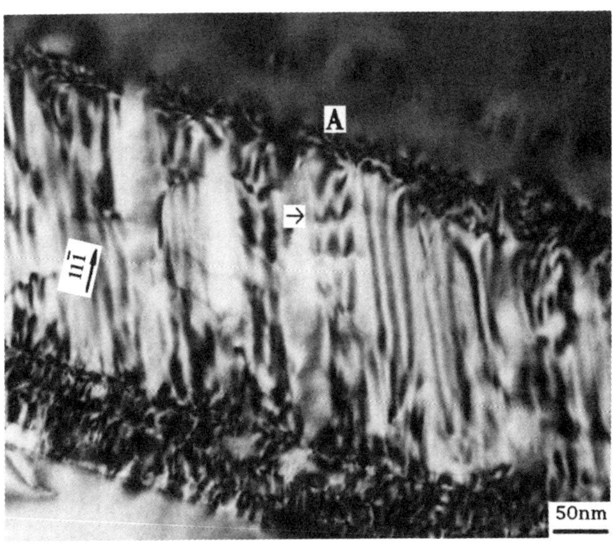

b

Fig. 10 Formation of cross twin plates associated with Ti_3Al/TiAl interfaces. (a) BF image taken in the edge direction of the twin plates [101]γ. (b) BF image taken with $g_{11\bar{1}}$ in the inclined direction [112]γ

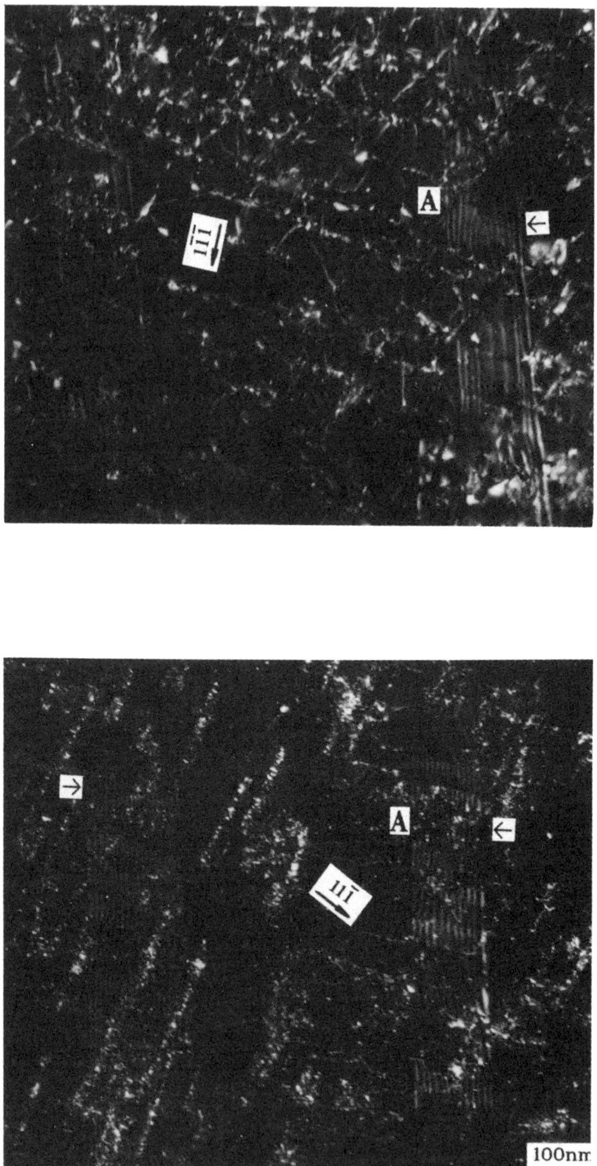

a

b

Fig. 11 Dislocation structures associated with the presence of cross twin boundaries. (a) WBDF image taken with $g/3g_{1\bar{1}\bar{1}}$ near $[1\bar{1}2]\gamma$. (b) WBDF image taken with $g/3g_{11\bar{1}}$ near $[101]\gamma$

may hinder the glide of IIDs and thus interfere with the process of accommodation. As a consequence, crack formation along the interfaces may take place, as observed in a previous investigation (Ref 17). Similarly the formation of sessile dislocation segments on cross twin boundaries, as described in the section above may hinder the glide of twinning partials, thereby contributing to crack formation on these twin boundaries.

It is proposed that the enhanced strain-hardening characteristics of this structure result from the large incidence of interactions between the lattice dislocations and the numerous interfaces present. Some of these interactions result in the formation of sessile configurations of dislocations, which make a further contribution to the strengthening of this structure.

Summary

The dislocation structures at the different interfaces in the Ti_3Al + TiAl lamellar structure were studied by TEM. In the homogenized state, there are predominantly only two types of interfacial boundaries present, Ti_3Al/TiAl interfaces and the TiAl/$TiAl_T$ twin boundaries in twin-related TiAl variants. The Ti_3Al/TiAl interfaces contain two types of intrinsic interfacial dislocations that are controlled by the interfacial misfit. The dislocation structure on the twin boundaries is characterized by triple twinning partials $1/6\langle 1\bar{1}2]$. In the heat-treated state, the structure of the Ti_3Al/TiAl interfaces remains essentially unchanged except for the formation of more IID networks. Two types of new interfaces, i.e., APBs in Ti_3Al laths and cross twin boundaries in TiAl laths, are formed. The former possess antiphase vectors $1/6\langle 11\bar{2}0\rangle$, as well as $1/6\langle 10\bar{1}0\rangle$, and the latter are characterized by individual twinning partials $1/6\langle 112]$ rather than the triplets.

Interactions between lattice dislocations and these interfaces take place during compression at room temperature. The planar slip in both Ti_3Al and TiAl laths is often interrupted because of the presence of the interfaces, and the dislocation tangles are formed near and at these interfaces, which contribute to the uniformity of the deformation. Furthermore, many cross twins in TiAl laths are generated from the Ti_3Al/TiAl interfaces, further contributing to a more uniform deformation. It is suggested that the improvement of ductility in the presence of the lamellar structure is due to the more homogeneous deformation and to the enhanced strain-hardening characteristics of this structure that result from a large incidence of interactions between lattice dislocations and the numerous interfaces present. The differences in the degree of deformation between Ti_3Al and TiAl laths are accommodated by both the multislip in the interface vicinity and the glide of the IIDs. However, the formation of sessile dislocation segments on Ti_3Al/TiAl interfaces and TiAl cross twin boundaries may hinder the glide of the interfacial dislocations and consequently contribute to crack formation at these interfaces.

Acknowledgment

This research was supported by the Natural Sciences and Engineering Research Council of Canada and the Office of Research Administration at the University of Manitoba. Technical assistance from D. Mardis and J. Van Dorp is gratefully acknowledged. Thanks are also due the Beijing Institute of Aeronautical Materials for supplying the alloy.

References

1. H.A. Lipsitt, in *High-Temperature Ordered Intermetallic Alloys, Proc. MRS Symp.*, Vol 39, 1985, p 351-364
2. Y.-W. Kim, *J. Met.*, Vol 41, 1989, p 24
3. T. Kawabata, T. Tamura, and O. Izumi, in *High-Temperature Ordered Intermetallic Alloys III, Proc. MRS Symp.*, Vol 133, 1989, p 329-334
4. E.L. Hall and S.C. Huang, in *High-Temperature Ordered Intermetallic Alloys III, Proc. MRS Symp.*, Vol 133, 1989, p 693-698
5. T. Tsujimoto and K. Hashimoto, in *High-Temperature Ordered Intermetallic Alloys III, Proc. MRS Symp.*, Vol 133, 1989, p 391-396
6. S.M.L. Sastry and H.A. Lipsitt, *Metall. Trans.*, Vol 8A, 1977, p 1543
7. M. Morinaga, J. Saito, N. Yukawa, and H. Adachi, *Acta Metall. Mater.*, Vol 38, 1990, p 25
8. M.J. Blackburn and M.P. Smith, Technical Report AFWAL-TR-82-4086, Air Force Wright Aeronautical Laboratories, 1982
9. S.A. Court, J.P.A. Löfvander, M.A. Stucke, P. Kurath, and H.L. Fraser, in *High-Temperature Ordered Intermetallic Alloys III, Proc. MRS Symp.*, Vol 133, 1989, p 675-680
10. G.J. Mahon and J.M. Howe, *Metall. Trans.*, Vol 21A, 1990, 1655
11. L. Zhao and K. Tangri, *Acta. Metall. Mater.*, in print 1991
12. L. Zhao and K. Tangri, *Philos. Mag. A*, in print 1991
13. P.P. Rao and K. Tangri, *Mater. Sci. Eng.*, Vol 132A, 48 (1991)
14. C. McCullough, J.J. Valencia, H. Mateos, C.G. Hevi, R. Mehrabian, and K.A. Rhyne, *Scr. Metall.*, Vol 22, 1988, p 1131
15. C.R. Feng, K.J. Michel, and C.R. Crowe, *Scr. Metall.*, Vol 23, 1989, p 1707
16. D. Shechtman, M.J. Blackburn, and H.A. Lipsitt, *Metall. Trans.*, Vol 5, 1974, p 1373
17. P.P. Rao and K. Tangri, unpublished work

13

Intergranular Fracture and Microalloying Effects in Ordered Intermetallics

C.T. Liu and E.P. George
Metals and Ceramics Division
Oak Ridge National Laboratory
P.O. Box 2008
Oak Ridge, Tennessee 37831-6115

Grain boundaries in strongly ordered intermetallic alloys are intrinsically brittle at ambient temperatures. Both experimental studies and theoretical calculations indicate that a high ordering energy and large electronegativity difference between the constituent atoms in A_3B compounds are the major causes of brittle grain boundaries. Boron tends to segregate strongly to grain boundaries and suppresses intergranular fracture in many $L1_2$ and $B2$ intermetallic alloys, and carbon shows the same segregation behavior when added to $L1_2$-Ni_3Si. The suppression of grain-boundary fracture may or may not be accompanied by a significant increase in the tensile ductility of ordered intermetallics. The improvement in ductility in intermetallics depends on whether the grain-boundary weakness is the sole cause of brittleness (for example, in Ni_3Al) or just part of the problem (for example, in FeAl and NiAl).

There is a great deal of current interest in ordered intermetallic alloys (Ref 1-7), particularly aluminides and silicides because of their potential for becoming new-generation structural materials for high-temperature applications.

These intermetallics contain sufficient amounts of aluminum and silicon to form, in oxidizing environments, thin films of oxides that often are compact and protective (Ref 8). Ordered intermetallics have relatively low densities and high melting points. Their strength and stiffness do not degrade rapidly with increasing temperature. In many cases, the yield strength of intermetallics increases with test temperature (Ref 1-6, 9-15) rather than decreases, as is common for conventional metallic alloys. A combination of good high-temperature strength and low material density with excellent corrosion resistance makes aluminides and silicides attractive for structural applications at elevated temperatures in corrosive environments (Ref 6).

Ordered intermetallics, however, generally have low ductility and undergo brittle fracture (Ref 1-7, 16-22), which severely restricts their use as engineering materials. Many intermetallics with high crystal symmetries exhibit severe brittleness that originates at grain boundaries. A classic example is Ni_3Al with the $L1_2$ structure (Ref 23-26). Single crystals of Ni_3Al are highly ductile, whereas polycrystals are very brittle at ambient temperatures, even though there are 12 independent slip systems. Other nickel-based intermetallics having brittle grain boundaries (Ref 27-32) include Ni_3Ga, Ni_3Si, and Ni_3Ge. Some *B*2 intermetallics such as FeAl (Ref 33-38) and NiAl (Ref 24, 39-41) also show brittle intergranular fracture, with low ductility at room temperature.

Recently, considerable effort has been devoted to the study of grain-boundary structure and properties in ordered intermetallics (Ref 1-5). In particular, intergranular fracture and the beneficial effect of boron additions have been investigated extensively in Ni_3Al, a model material for the study of grain-boundary fracture (Ref 25, 26, 42, 43). Consequently, a great deal of knowledge has been gained and significant progress has been made on the improvement of ductility and toughness in ordered intermetallics, using physical metallurgy principles. This chapter summarizes recent progress in these areas, with emphasis, first, on understanding the intrinsic nature of intergranular brittleness in intermetallics and, second, on grain-boundary engineering by microalloying with boron and carbon.

Intrinsic Grain-Boundary Brittleness

Brittle intergranular fracture has been observed in many polycrystalline materials. In conventional metals and alloys, grain boundaries are commonly ductile, and intergranular fracture is generally caused by embrittlement involving the segregation of harmful impurities, such as O, S, and P, to grain boundaries (Ref 44). In earlier work, trace impurities were also suspected to be a possible cause for the embrittlement of grain boundaries in ordered inter-

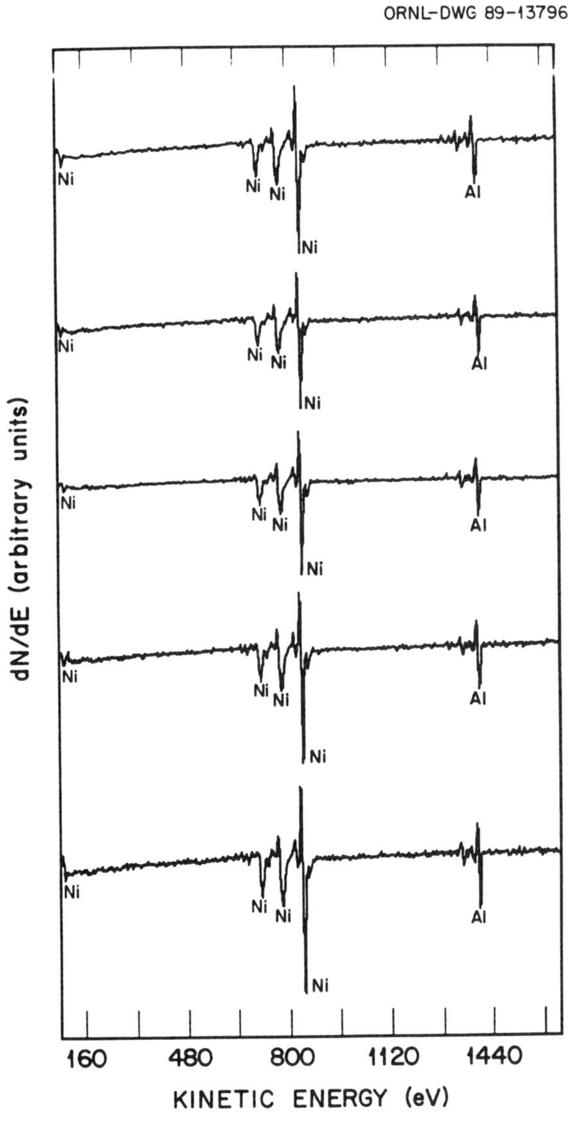

Fig. 1 Auger spectra from grain boundaries in high-purity NiAl. Only peaks of the constituent elements, Ni and Al, are present in the spectra. Source: Ref 41

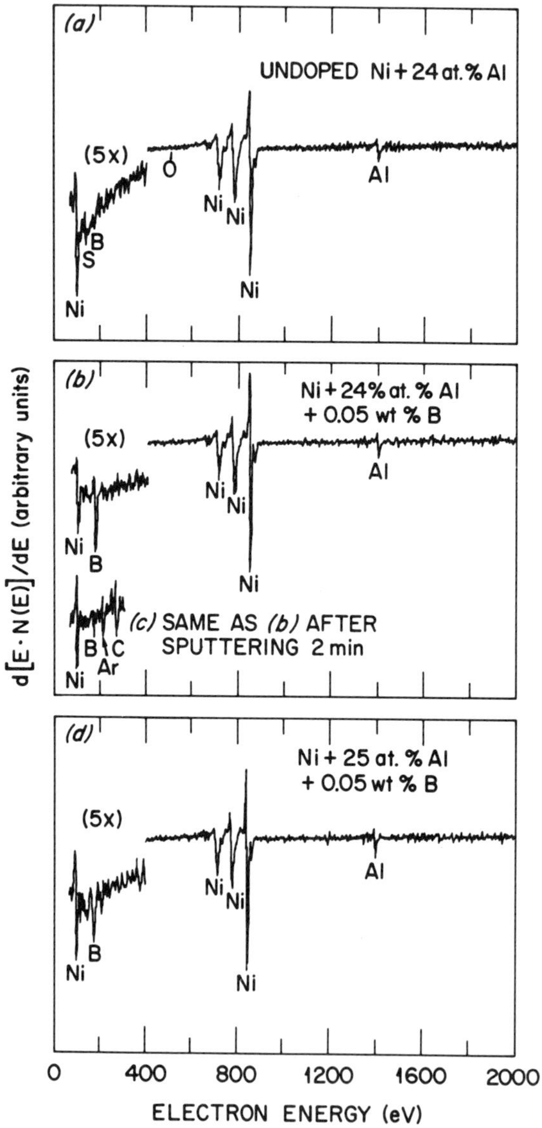

Fig. 2 Auger spectra from intergranularly fractured region in Ni_3Al specimens. (a) Undoped Ni-24 at.% Al. (b) Ni-24 at.% Al doped with 500 wt ppm B. (c) Same as (b) after sputtering from 2 min. (d) Ni-25 at.% Al doped with 500 wt ppm B. Source: Ref 26

Table 1 Effect of aluminum concentration and ordering energy on the fracture mode and tensile elongation of Fe_3Al-FeAl alloys, tested at room temperature in dry oxygen

Alloy, at.%	Critical ordering temperature, °C	Fracture mode(a)	Elongation, %
Fe-28Al	760	TF	12.0
Fe-36.5Al	1110	Mainly GBF	17.6
Fe-40Al	1200	GBF	3.2
Fe-43Al	1260	GBF	~0(b)

(a) TF, transgranular fracture; GBF, grain-boundary fracture. (b) Fracture prior to macroscopic yielding.
Source: Ref 48-51

metallics. In fact, sulfur was found on the grain boundaries of polycrystalline Ni_3Al (Ref 45).

However, later Auger work showed that grain boundaries in sufficiently pure ordered intermetallics are for the most part clean and free from impurities (Ref 26, 41, 46-49). This is indicated in Fig. 1 and 2(a), showing Auger spectra obtained from grain boundaries in $B2$-NiAl (Fig. 1) and $L1_2$-Ni_3Al (Fig. 2a). All peaks are associated with nickel and aluminum, and no appreciable amounts of impurities are detected at the grain boundaries. These high-purity intermetallic alloys nevertheless fracture intergranularly, implying that the grain boundaries are intrinsically brittle. Other intermetallics showing intrinsic grain-boundary fracture at ambient temperatures include Ni_3Si, Ni_3Ga, Ni_3Ge, and NiAl (Ref 27-41).

Considerable effort has been devoted to understanding intrinsic grain-boundary brittleness in ordered intermetallics. Table 1 attempts to correlate the fracture mode in Fe_3Al-FeAl (Ref 48-51) with critical ordering temperature (T_c) and aluminum concentration. (Since iron aluminides exhibited environmental embrittlement involving moisture in air, all tests reported here were performed in a neutral environment, i.e., dry oxygen.) The T_c is generally related to the ordering energy in intermetallics (Ref 16). In the iron aluminides, the ordering energy increases substantially with increasing aluminum concentration, giving higher T_cs for higher aluminum alloys (Ref 33-38, 52, 53). For a relatively weakly ordered Fe_3Al with $T_c = 760$ °C, the iron aluminide exhibited no grain-boundary problem and fractured transgranularly (Fig. 3a). For the strongly ordered FeAl aluminides with T_c above 1100 °C, all the specimens failed by grain-boundary separation (Fig. 3b). However, the total elongation, which somewhat reflects the degree of grain-boundary brittleness, is sensitive to aluminum concentration or T_c. The elongation decreases from 17.6 to approximately 0% when aluminum concentration is increased from 36.5 to

a

b

Fig. 3 Fractographs of iron aluminides fractured at room temperature in dry oxygen. (a) Fe_3Al (28 at.% Al). (b) FeAl (43 at.% Al). Marker length, 100 μm

Table 2 Correlation of critical ordering temperature and chemical parameters with fracture behavior of $L1_2$-Ni_3X alloys

Alloy	Critical ordering temperature, °C	Valency difference	Electronegativity difference	Fracture mode
Ni_3Fe	516	0.2	...	TF
Ni_3Mn	440	0.9	−0.36	TF
Ni_3Al	1395	3.0	−0.30	GBF
Ni_3Ga	1210	3.0	−0.10	GBF
Ni_3Si	1035	4.0	−0.01	GBF
Ni_3Ge	1132	4.0	0.10	GBF

43%. These results indicate that the ordering energy, or T_c, plays a major role in grain-boundary brittleness in the iron aluminides.

Table 2 correlates T_c, valency difference, and electronegativity difference with fracture mode in $L1_2$-ordered alloys Ni_3X (Ref 30), where X = Fe, Mn, Al, Ga, Si, or Ge. Weakly ordered alloys such as Ni_3Fe and Ni_3Mn, with T_c less than 600 °C, are ductile and show transgranular fracture at ambient temperatures. Strongly ordered alloys such as Ni_3Al, Ni_3Ga, Ni_3Si, and Ni_3Ge, with T_c greater than 1000 °C, are all brittle and fracture intergranularly. Examination of the fracture behavior in Ni_3X thus reveals that T_c, or the ordering energy, is a key parameter affecting intrinsic grain-boundary fracture in ordered intermetallics.

An explanation of the intrinsic grain-boundary brittleness in strongly ordered intermetallics was put forward by Vitek and his coworkers (Ref 54-56) on the basis of grain-boundary structures calculated by atomic simulations. In the case of weakly ordered $L1_2$ polycrystals (for example, $T_c < 800$ °C), grain boundaries are generally analogous to those in fcc materials, where the atom arrangement at the boundary is not dissimilar from that in the bulk. In the case of strongly ordered $L1_2$ compounds, however, columns of atomic-sized cavities are formed in stoichiometric boundaries in order to preserve the chemical order up to the boundaries. Figure 4 compares $\Sigma = 5$ (310) boundary in fcc and $L1_2$ materials, calculated by pair and embedded-atom-type potentials, respectively (Ref 54). The intrinsic grain-boundary brittleness of strongly ordered $L1_2$ compounds can be attributed to those large defects, which could act as nuclei for intergranular cracks, thereby weakening the grain boundaries and resulting in intrinsic grain-boundary brittleness. By considering slip grain-boundary interactions, King and Yoo (Ref 57) showed that considerably fewer dislocation reactions are permissible in those grain boundaries that are required to maintain chemical order than in those that are disordered. Since slip

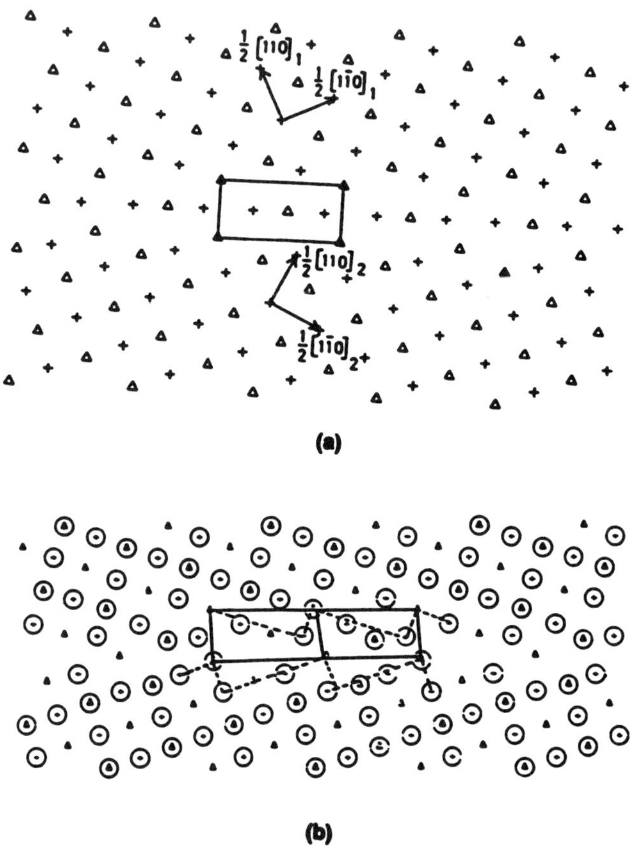

Fig. 4 Atomic structure of $\Sigma = 5$ (310) tilt boundary. (a) In fcc material. (b) In strongly ordered $L1_2$ compounds. Source: Ref 54

compatibility between adjacent grains requires dislocation reactions at grain boundaries, their work also suggested that strongly ordered grain boundaries are intrinsically more brittle than compositionally disordered ones.

In addition to the ordering energy, other chemical factors (see Table 2) have been considered as possible causes for the intrinsic grain-boundary brittleness in intermetallics. Takasugi and Izumi (Ref 27, 28) first initiated a systematic study of the effects of various metallurgical and chemical factors on the

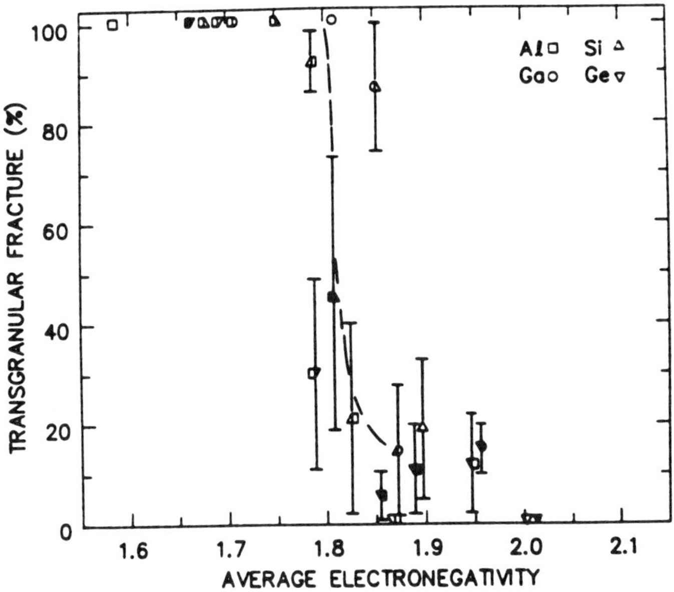

Fig. 5 The bend ductility of binary and pseudobinary alloys based on Ni_3X (X = Al, Si, Ga, and Ge) as a function of average electronegativity of the X species. Source: Ref 31

grain-boundary cohesive strength of A_3B alloys (including Ni_3X with the $L1_2$ structure). They found that those A_3B alloys that had large valency differences between their constituent atoms were more prone to brittle intergranular fracture than those with relatively small valency differences. Other factors, such as the different atom sizes of the constituent atoms, were also important, but to a lesser degree. In later work Taub *et al.* (Ref 29-31) found that a better correlation is obtained if the electronegativity difference between the constituent atoms is used rather than the valency difference. They were able to rank correctly the rapidly solidified Ni_3X alloy ribbons in terms of fracture mode and bend ductility using their electronegativity scheme. The ranking of the grain-boundary cohesive strength of Ni_3X in the order $Ni_3Fe > Ni_3Mn > Ni_3Al > Ni_3Ga > Ni_3Si > Ni_3Ge$ is in good agreement with experimental data tabulated by Taub *et al.* (Ref 29, 30). In addition, the fracture mode of binary and pseudobinary intermetallic alloys based on Ni_3X (X = Al, Ga, Si, or Ge) doped with boron (produced by a melt-spinning technique) can be well correlated with the average electronegativity of the X atom (Fig. 5). The alloys showed a sharp transition from transgranular to intergranular fracture when the average electronegativity was larger than 1.8 to 1.9.

In sum, both experimental studies and atomic-simulation calculations indicate that the grain boundary in strongly ordered intermetallics is intrinsically brittle. Other chemical factors, such as the electronegativity difference between the constituent atoms in A_3B compounds, also influence grain-boundary cohesion.

Grain-Boundary Engineering by Microalloying

As discussed in the section above, the grain boundary in strongly ordered intermetallics is intrinsically brittle at ambient temperatures. Because of the intrinsic nature of this brittleness, the grain-boundary problem cannot be solved by material purification. Instead, an effective solution involves microalloying with beneficial dopants that have a strong tendency to segregate to grain boundaries and suppress intergranular fracture (Ref 58). This chapter reviews recent research on microalloying with the interstitial elements boron and carbon, which were found to be most effective in suppressing intergranular fracture in a number of $L1_2$ and $B2$ intermetallics.

Boron Effect in $L1_2$-Ni_3Al (24 at.% Al). Boron was found to be most effective in improving the room-temperature tensile ductility of Ni_3Al containing less than 25 at.% Al (Ref 25, 26, 42, 43). Figure 6 is a plot of room-temperature tensile ductility as a function of boron addition in Ni_3Al (24% Al). Microalloying with boron sharply increases ductility and effectively suppresses intergranular fracture. A tensile ductility as high as 50% was achieved in Ni_3Al by microalloying with 0.1 wt% B. The striking effect of boron on the ductility of Ni_3Al, first discovered by Aoki and Izumi (Ref 42), occurs over a wide range of concentrations of boron in solid solution (the solubility of boron ~1.5 at.%) (Ref 26). Both Auger microprobe studies (Fig. 2) (Ref 26) and imaging atom probe studies (Ref 59-61) have shown that boron segregates strongly to grain boundaries in Ni_3Al, although the extent of segregation varies from boundary to boundary. Figure 7 is an atom probe image, showing a high-angle ($\Sigma = 27$) grain boundary in Ni_3Al densely covered with boron atoms (bright spots) (Ref 60).

Considerable effort has been spent in the attempt to understand the ductilizing effect of boron in Ni_3Al, but as yet a unique mechanism cannot be identified. There is evidence in support of both currently favored theories: the boron-enhanced cohesive strength theory (Ref 26, 62-68) and the boron-facilitated slip transfer theory (Ref 69-73). The first theory is supported by the observation of the unique segregation behavior of boron (Ref 26, 63, 64) and TEM study of *in situ* deformation (Ref 68), as well as atomic simulation and first-principles quantum-mechanical calculations (Ref 65-67). The second

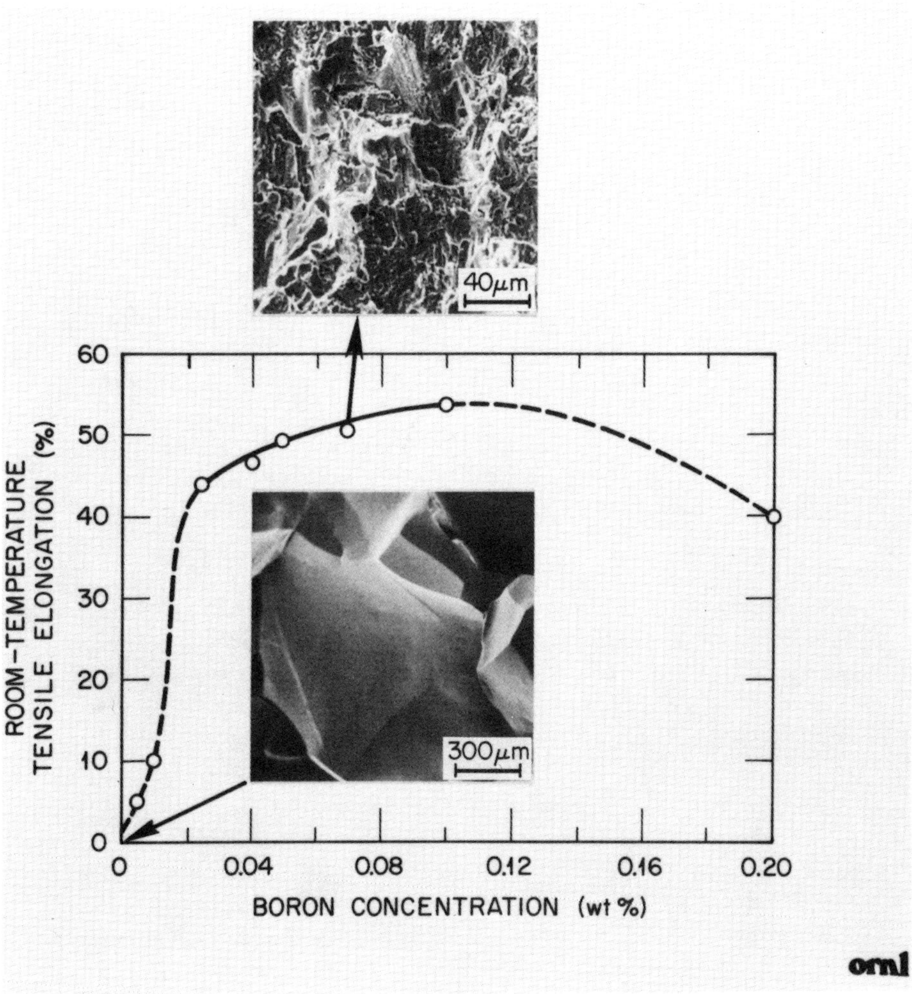

Fig. 6 Effect of boron additions on tensile elongation and fracture behavior of Ni_3Al (24 at.% Al) tested at room temperature

theory is supported by Hall-Petch analyses (Ref 69-70) and TEM studies of dislocation grain-boundary interaction in Ni_3Al (Ref 71-73). However, TEM studies of dislocation-spreading kinetics conducted in 1989 showed that the introduction of boron to Ni_3Al changes the properties of the grain boundary,

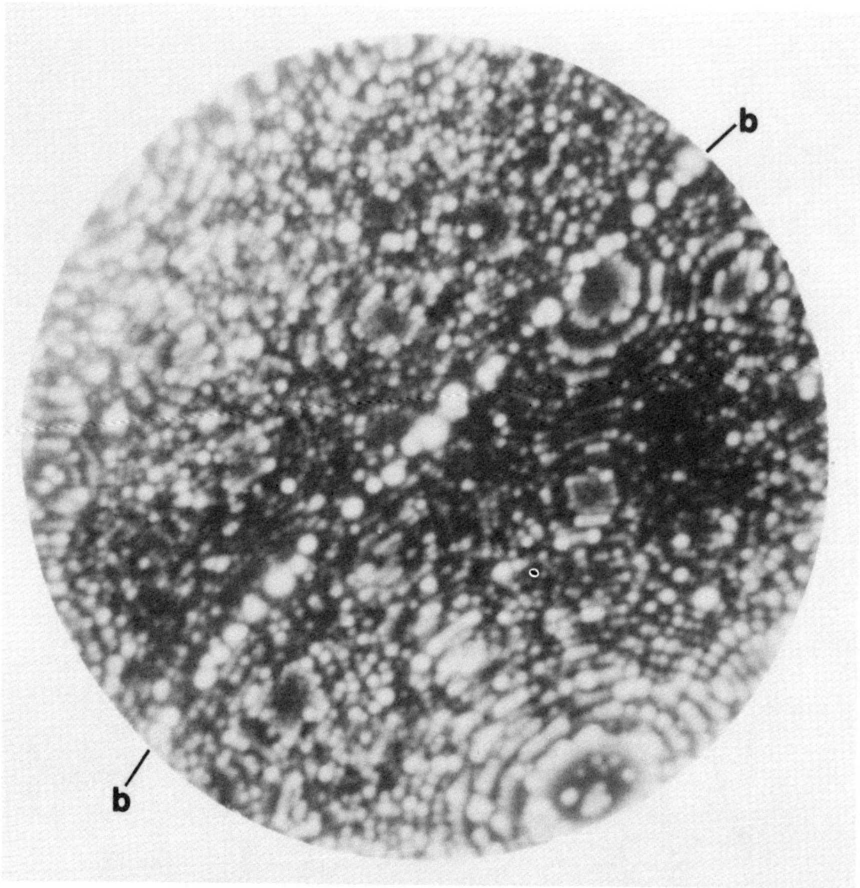

Fig. 7 Field ion image of a high-angle ($\Sigma = 27$) grain boundary in B-doped Ni_3Al showing extensive boron segregation along the boundary marked b-b. Source: Ref 60

which results in a substantial decrease in the mobility of grain-boundary dislocations (Ref 74), rather than the increase suggested by the second theory.

The beneficial effect of boron has also been suggested to be the result of chemical disorder of the grain boundary in boron-doped Ni_3Al (Ref 57). This can be rationalized by considering that the chemical disorder may increase the grain-boundary cohesion and facilitate slip transfer across the boundary. Considerable effort has been expended in recent years to determine whether boron does, in fact, disorder grain boundaries in Ni_3Al. Unfortunately, direct experimental observation of grain-boundary structures by high-resolution TEM

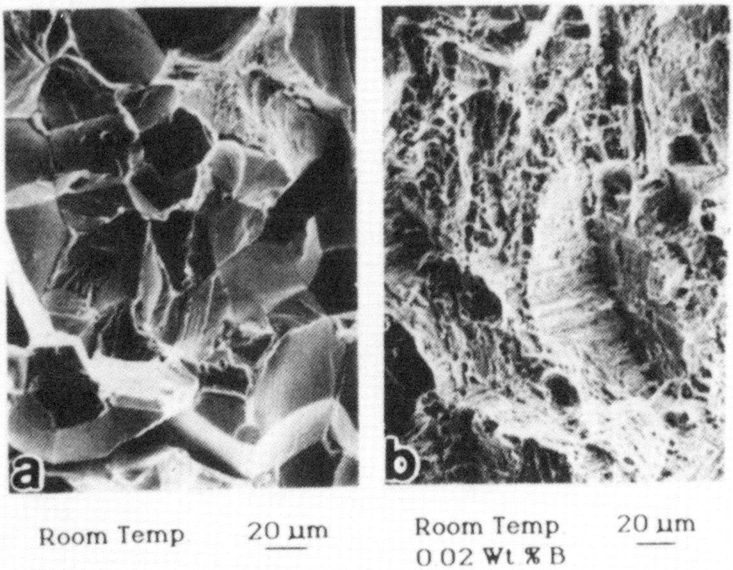

Fig. 8 Scanning electron micrograph of fracture surface of Ni-18.9Si-3.2Ti (at.%). (a) Room-temperature fracture. (b) Room-temperature fracture of the alloy doped with 0.02 wt.% B. Source: Ref 78

is quite tricky, and there was no consensus among the reports published at this writing (Ref 75-77).

Boron Effect in Ni_3X (X = Al, Ga, Si, and Ge). In addition to Ni_3Al, other Ni_3X alloys including Ni_3Ga, Ni_3Si, and Ni_3Ge exhibit brittle intergranular fracture at room temperature (Ref 27-31). Taub *et al.* (Ref 29) first reported that boron is effective in suppressing grain-boundary fracture and ductilizing Ni_3Ga and Ni_3Si with hypostoichiometric compositions (<25% X) but that it is not effective with Ni_3Ge. The beneficial effect of boron in Ni_3Si alloys has since been confirmed by Oliver and White (Ref 78, 79) and Takasugi *et al.* (Ref 80). Auger studies showed a strong segregation of boron to grain boundaries in Ni_3Si alloys (Ref 78). As shown in Fig. 8, boron doping changed the fracture path from predominantly brittle intergranular fracture to completely ductile transgranular at room temperature.

In order to rationalize the ductilizing effect of boron in Ni_3X (mainly, the reason that boron cannot ductilize Ni_3Ge), Taub *et al.* studied the bend ductility and fracture behavior of boron-doped binary and pseudobinary Ni_3X alloy

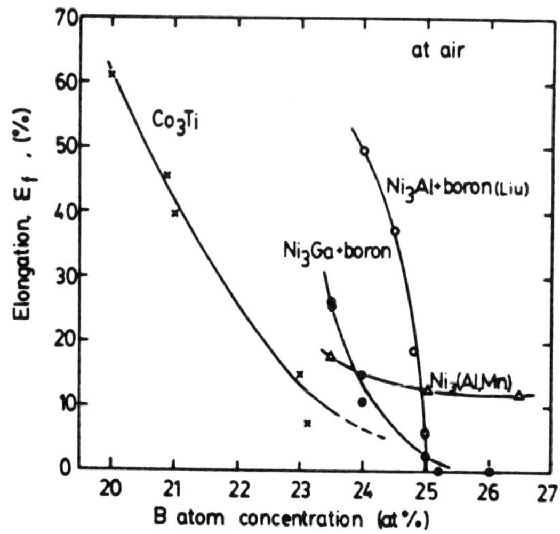

Fig. 9 Plot of room-temperature elongation of A_3B compounds as a function of B atom concentration. Source: Ref 81

ribbons (Ref 29-31). As shown in Fig. 5, the fracture mode of all B-doped Ni_3X alloys can be well correlated with the average electronegativity of the X atom (including Ge). The electronegativity difference between the constituent atoms (i.e., Ni and X atoms) can generally be regarded as a simple measure of the transfer of electrons between the atoms and is thereby related to atomic bonding. Compared with Al, Ga, and Si, Ge atoms (electronegativity = 2.01) are more electronegative with respect to Ni atoms (1.91); consequently, germanium has a greater tendency to pull electron charge out of Ni-Ni bonds, thereby reducing the cohesive strength and promoting intergranular fracture in Ni_3Ge. The grain boundary in Ni_3Ge is relatively weaker than that in Ni_3Al, Ni_3Si, or Ni_3Ga, and boron is considered less effective in ductilizing Ni_3Ge and some Ni_3(Ge, X) alloys containing large levels of Ge. Further research is required to extend the observed electronegativity effect to other alloy systems.

Alloy Stoichiometric Effect. As mentioned in the foregoing two subsections, boron is effective in ductilizing Ni_3Al, Ni_3Ga, and Ni_3Si with hypostoichiometric compositions ($X < 25\%$) but ineffective in hyperstoichiometric compositions ($X > 25\%$). Recently Takasugi *et al.* (Ref 81) compiled data to show the stoichiometric effect observed in boron-doped Ni_3Al, boron-doped Ni_3Ga, Co_3Ti, and Ni_3(Al, Mn) (Fig. 9). Among these alloy systems, the room-temperature ductility of boron-doped Ni_3Al is most sensitive to deviations

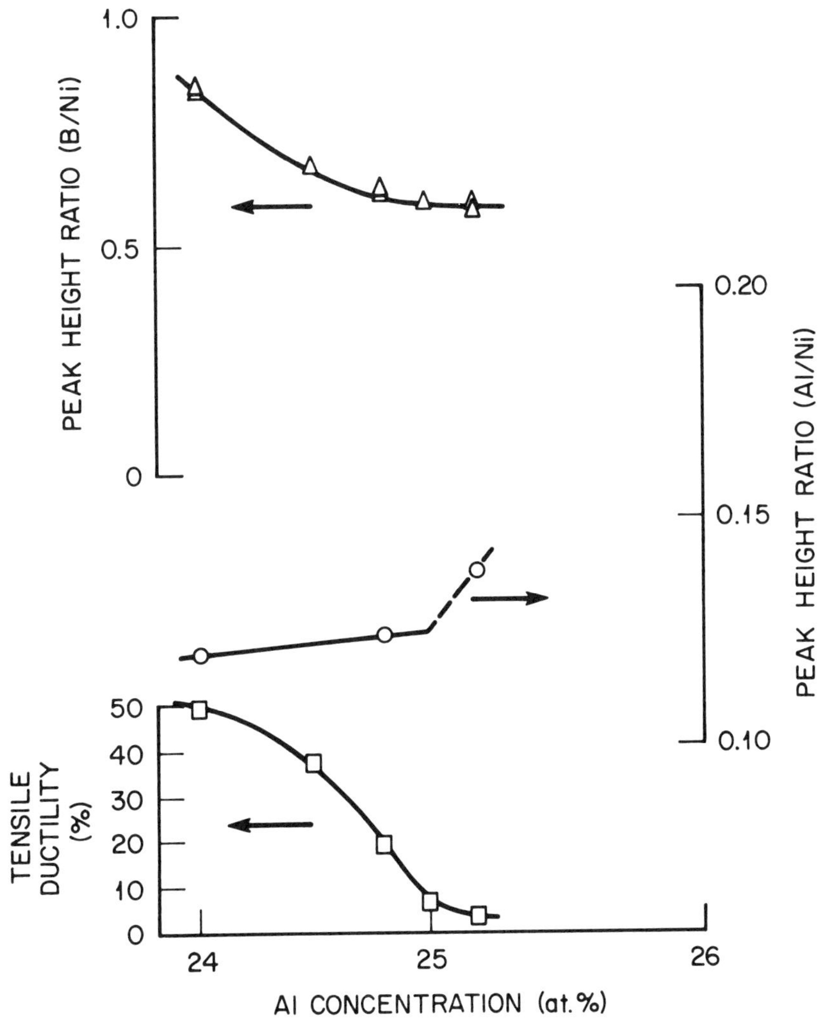

Fig. 10 Effect of stoichiometry on grain-boundary segregation (in terms of Auger peak height ratio) and room-temperature tensile elongation of boron-doped Ni_3Al containing 24 to 25.2% Al. Source: Ref 86

from the stoichiometric composition, whereas the ductility of $Ni_3(Al, Mn)$ is least sensitive. As the bulk aluminum concentration in Ni_3Al is increased from 24 to 25.2 at.%, the tensile ductility decreases sharply from 50 to 3% (Ref 26) (Fig. 10). Accompanying the ductility decrease is a change in the failure mode from transgranular to mixed mode fracture and then to mainly intergranular fracture. The same change in fracture mode was observed in Co_3Ti and B-doped Ni_3Ga (Ref 81). In the case of $Ni_3(Al, Mn)$ alloys, manganese was reported to occupy both aluminum and nickel lattice sites in the $L1_2$ structure (Ref 82). Alloying Ni_3Al with manganese thus smears out the alloy stoichiometric effect; as a result, the ductility of the alloy $Ni_3(Al, Mn)$ is less sensitive to alloy stoichiometry (Fig. 9).

The alloy stoichiometric effect is one of the most fascinating features in the study of the ductility and fracture behavior of intermetallic compounds. Takasugi *et al.* have correlated the stoichiometric effect to the electronic environment in the grain-boundary region (Ref 81) in heteroelectric A_3B compounds. Off-stoichiometry generally results in the enrichment of A or B atoms at grain boundaries. The enrichment of A atoms in hypostoichiometric compositions is expected to create a more homogeneous electronic environment in grain-boundary planes, resulting in higher grain-boundary cohesion and thus higher alloy ductility. On the other hand, the depletion of A atoms in hyperstoichiometric compositions creates a more heterogeneous electronic environment in grain-boundary planes, leading to lower grain-boundary cohesion and alloy ductility.

In a study of grain-boundary chemistry in Ni_3Al as a function of aluminum and boron concentration, Liu *et al.* (Ref 26) observed that alloy stoichiometry strongly affects boron segregation to the grain boundaries. As shown in Fig. 10, the level of boron at the boundaries decreases with increasing bulk aluminum above 24%. A systematic study of boron segregation in Ni_3Al alloys clearly shows that the grain boundary is saturated with a much lower level of boron in 25.2% Al alloy (Ref 83) (in Fig. 11, the saturation boron level = 200 wt ppm) than in the 24% Al alloy (the saturation boron level < 1000 wt ppm) (Ref 84). These observations indicate that alloy stoichiometry may affect the grain-boundary structure and thus boron segregation. The grain-boundary properties in Ni_3Al are found to be sensitive to the level of boron segregation. The boundary remains brittle when the amount of boron segregation is insufficient (Ref 85).

Auger studies also suggest that the bulk aluminum concentration affects the level of aluminum at the grain boundaries in Ni_3Al (Ref 86, 87). As shown in Fig. 10, the grain-boundary aluminum shows a sharp increase when the

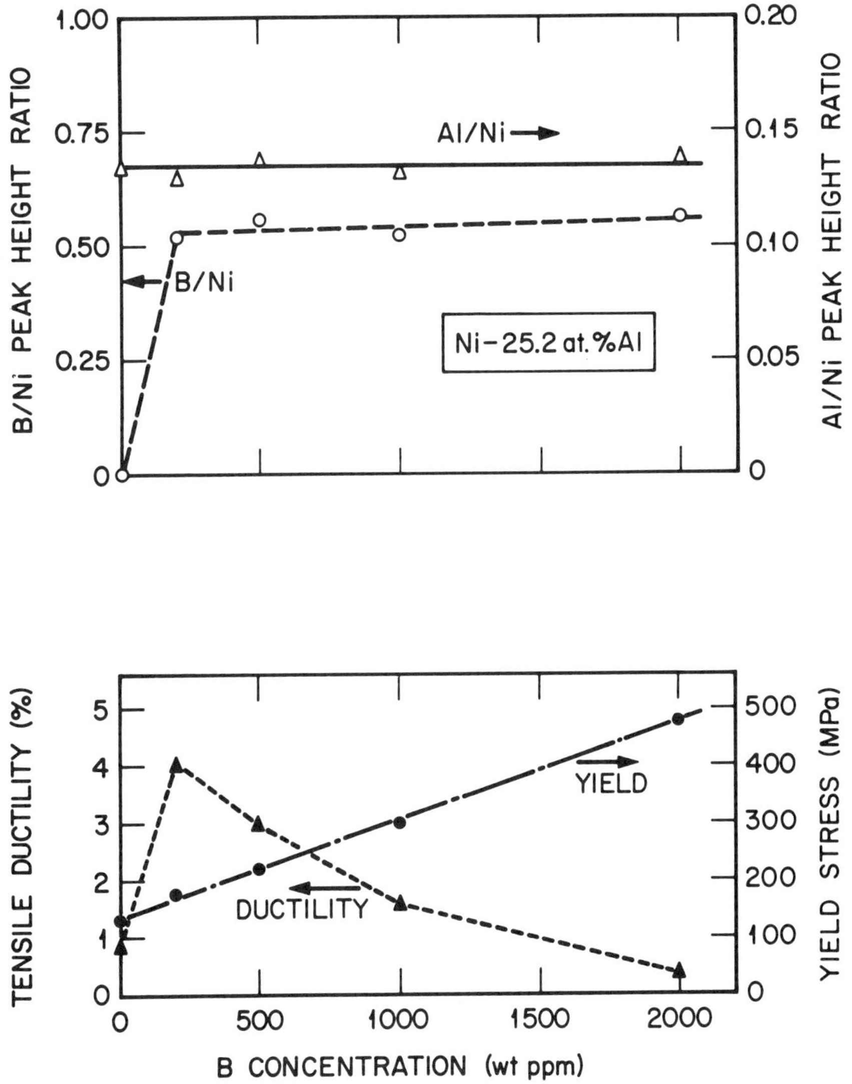

Fig. 11 Plot of B/Ni and Al/Ni AES peak height ratio obtained from intergranularly fractured surface, and tensile properties of Ni-25.2Al as a function of boron concentration. Source: Ref 83

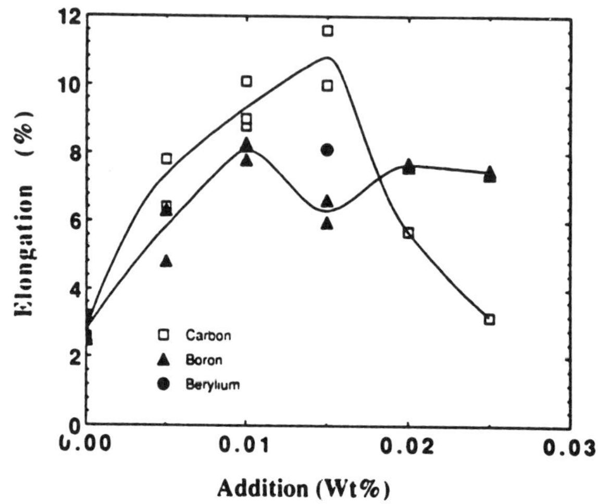

Fig. 12 The room-temperature ductility of Ni-22.5Si microalloyed with B, C, and Be. Source: Ref 79

bulk aluminum surpasses 25%. According to Takasugi *et al.* (Ref 81) the aluminum enrichment increases the heterogeneous electronic environment in the boundary and reduces cohesion. Atomic-simulation calculations based on an embedded atom method also confirm the reduction in grain-boundary cohesion by the enrichment of aluminum at the boundary (Ref 67). Alloy stoichiometry not only changes grain-boundary chemistry, it also strongly affects the solid-solution hardening behavior in Ni_3Al, with a distinctly higher hardening effect of boron in the 25% Al alloys (Ref 83) (Fig. 11 shows roughly a fourfold increase in the yield strength of Ni-25.2Al by alloying with 1 at.% B). Thus, the dramatic effect of alloy stoichiometry on room-temperature ductility observed in B-doped Ni_3Al is a complicated phenomenon involving boron segregation to the grain boundary, aluminum enrichment at the boundary, and solid-solution hardening in bulk by boron (Ref 83).

Microalloying Effect of Boron and Carbon in $L1_2$-Ni_3Al and -Ni_3Si. Boron and carbon are chemically similar to each other, and both occupy interstitial positions in Ni_3Al (Ref 88) and Ni_3Si (Ref 89). However, the doping effect of boron and carbon is quite different in these two alloys.

In the case of Ni_3Al (24% Al), boron has a dramatic ductilizing effect (Ref 25, 26, 42, 43), whereas carbon has no appreciable beneficial effect (Ref 90, 91). At this writing, only limited data are available regarding carbon segregation

Table 3 Prediction of cleavage strength of ordered intermetallics from total-energy quantum-mechanical calculations and atomic simulations

Alloy	Cleavage strength, GPa	Ideal cleavage energy, J/m^2	Reference
Al_3Sc(a)	19	3.5	94
Ni_3Al(b)	30	6.0	95
Ni_3Al(c)	27.5-26.5	...	67
NiAl(a)	29	...	96
FeAl(b)	35	6.5	95

(a) Cleavage strength is essentially independent of crystallographic planes (110) or (100). (b) Cleavage on (001) planes. (c) Cleavage on (001) and (011) planes

to grain boundaries in Ni_3Al. Huang *et al.* (Ref 90) observed an Auger peak corresponding to carbon enrichment at grain boundaries in Ni_3Al alloy ribbons produced by rapid solidification. Using atom probe imaging, Brenner *et al.* (Ref 92) found nonequilibrium segregation of carbon to high-angle grain boundaries in Ni_3Al specimens produced by melting and casting, followed by cold work and recrystallization. In the latter case, the carbon enrichment factor was only about 4 at the boundaries. In contrast, boron exhibited equilibrium-type segregation with a segregation factor larger than 20 (Ref 26). The lack of strong segregation of carbon to the grain boundary may be the reason that carbon doping does not impart ductility to Ni_3Al at ambient temperatures.

In the case of Ni_3Si alloys, boron and carbon both tend to suppress grain-boundary fracture and improve the tensile ductility of the silicide with hypo-stoichiometric compositions (Ref 78-80). Figure 12 shows the beneficial effect of boron and carbon doping in Ni_3Si containing 22.5 at.% Si (Ref 79). The room-temperature ductility of the silicide increases with increasing carbon and peaks at 0.05 wt% C. Above that level, the ductility decreases sharply, indicating the importance of carbon level control for maximum beneficial effect. The beneficial effect of boron, on the other hand, is less sensitive to the boron level, and peak ductility appears to exist over a wider range of boron concentration. It should be pointed out that boron and carbon produce only a moderate improvement in ductility in Ni_3Si (3 to 8-11%, at room temperature), while boron dramatically improves the ductility in Ni_3Al (from 1 to 50%) (Ref 26). The different behavior can be rationalized by taking into consideration the electronegativity of silicon and aluminum atoms (see Table 2). Because of the larger electronegativity difference, the grain boundary in Ni_3Si is more brittle than that in Ni_3Al. Consequently, boron is less effective in ductilizing Ni_3Si.

Boron and Carbon Effects in *B*2 Intermetallics NiAl and FeAl. Polycrystalline NiAl with equiatomic composition exhibits grain-boundary fracture

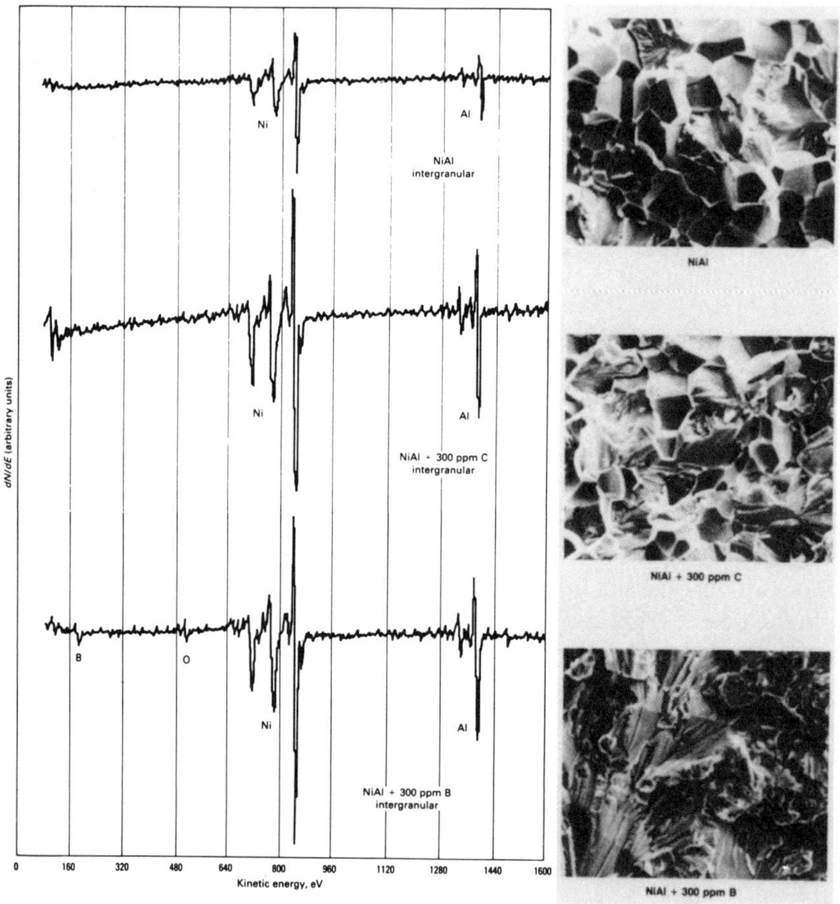

Fig. 13 Results of Auger analyses showing the effect of microalloying on the grain-boundary composition and room-temperature fracture mode of NiAl. Top to bottom: unalloyed NiAl, NiAl doped with 300 wt ppm C, and NiAl doped with 300 wt ppm B. 400×. Source: Ref 41

with a low tensile ductility of about 2% (Ref 39-41). Recently, detailed Auger analyses have shown that the grain boundaries in NiAl are extremely clean and free of impurities (Ref 41) (Fig. 1 and 13a), indicating intrinsically brittle boundaries in the aluminide. Microalloying with boron in the parts per million range can completely suppress the intergranular fracture (Ref 41), and Auger analyses (Ref 41, 93) have confirmed that the beneficial effect of boron in NiAl, like that in Ni_3Al, is due to its strong segregation to the grain boundaries (Fig. 13c). In contrast, carbon does not segregate to the grain boundaries and consequently does not suppress intergranular fracture (Fig. 13b).

Surprisingly, the suppression of grain-boundary fracture is not accompanied by an increase in the ductility of NiAl (Ref 41, 93). For instance, doping with 300 wt ppm B actually reduces the plastic strain from approximately 2% to almost zero at room temperature. This is very different from the dramatic improvement in tensile ductility obtained by boron doping in Ni_3Al (Ref 26). Of particular interest is why boron causes embrittlement instead of ductilization in NiAl. First, because of the strong solid-solution-hardening effect of boron in B2-NiAl, the addition of excess boron, even 300 wt ppm (0.12 at.%), leads to embrittlement (Ref 93). The hardening potency of boron is estimated to be as high as 4500 MPa per atomic percent of boron in solid solution. Second, the weak grain boundary is not the sole problem for brittleness in NiAl. The other problem involves brittle cleavage fracture observed in boron-doped NiAl (Ref 41, 93) (Fig. 13). This cleavage may not be simply a result of poor cleavage strength, as indicated in Table 3, which compares the cleavage strength and ideal cleavage energy of a number of intermetallic compounds as obtained from first-principles quantum-mechanical calculations by Fu *et al.* (Ref 94-96) and from atomic simulations by Chen *et al.* (Ref 67). The cleavage strength of NiAl is significantly higher than that of $L1_2$-Al_3Sc, which is a model material for the study of $L1_2$ trialuminides, which show cleavage fracture with no appreciable plastic deformation at room temperature (Ref 97-102). Actually, the cleavage strength of NiAl is comparable to that of $L1_2$-Ni_3Al, which does not have a major cleavage problem. Of course, brittle cleavage in NiAl could be the result of low cleavage energy rather than poor cleavage strength; additional studies are needed to verify this possibility. Another possibility is that since NiAl shows $\langle 100 \rangle$-type slip instead of the $\langle 111 \rangle$ slip commonly observed in bcc materials, high stress concentrations due to lack of sufficient deformation modes could be the major reason for brittle cleavage in B-doped NiAl. Further studies are required to promote $\langle 111 \rangle$ slip in NiAl by macroalloying (Ref 103, 104).

Table 4 Effect of boron doping on room-temperature tensile properties of Fe-40Al tested in air and dry oxygen

Boron doping	Strength, MPa		Ductility, %	Fracture mode
	Yield	Ultimate		
Air test				
No boron	390	405	1.2	GBF
300 wt ppm B	391	577	4.3	TF
Oxygen test(a)				
No boron	402	537	3.2	GBF
300 wt ppm B	392	923	16.8	TF

(a) Oxygen pressure = 6.7×10^{-4} Pa. Source: Ref 48, 49

When tested in air at room temperature, *B*2-FeAl aluminides containing 38% or more Al showed brittle grain-boundary fracture with limited ductilities (Ref 33-38). Auger analyses showed the grain boundaries in the iron aluminides to be clean and free of impurities (Ref 48, 49), indicating the intrinsic nature of grain-boundary brittleness, as in *B*2-NiAl and $L1_2$-Ni_3X. Boron was added to FeAl (40% Al) to solve the grain-boundary problem, and Auger studies showed a strong segregation of boron to the boundaries, with a segregation factor of about 43 (Ref 48). Tensile properties obtained from air tests (Table 4) show that boron doping in FeAl (40% Al), as expected, suppresses intergranular fracture at room temperature but unexpectedly produces only a limited increase in tensile ductility. The B-doped FeAl specimens exhibited brittle cleavage fracture.

Although weak grain boundaries are part of the brittleness problem in FeAl aluminides with 38% or more Al, they are not the only problem. Recent studies have shown that environmental embrittlement is a major cause of low tensile ductility and brittle fracture at ambient temperatures in iron aluminides (Fe_3Al-FeAl) (Ref 48-51), as had been shown for many other ordered intermetallics (Ref 105-109) containing reactive elements such as Al, Si, and Ti. In the case of the iron aluminides, the embrittlement involves the reaction of aluminum atoms at the crack tips with moisture in the air. This reaction generates atomic hydrogen that drives into the metal at the crack tips, promoting brittle crack propagation mainly along cleavage planes, resulting in brittle cleavage fracture. In order to minimize environmental embrittlement, the aluminides are required to be tested in neutral environments such as vacuum or dry oxygen (Ref 48-51). As shown in Table 4, boron-doped Fe-40Al alloy exhibited extensive ductility when tested in oxygen at room temperature. It should be noted that the undoped Fe-40Al showed low ductility when tested in oxygen because of the grain-boundary problem.

In 1990 Gaydosh and Nathal studied the tensile ductility of FeAl aluminides with and without 0.5 at.% B in various environments at room temperature (Ref 110). The aluminides were produced by powder metallurgy, which had much finer grain sizes (19 to 24 μm) than conventionally prepared materials (~200 μm) (Ref 48, 49). The fine-grained Fe-40Al exhibited substantial plastic deformation prior to intergranular fracture in neutral environments (i.e., vacuum or oxygen), even when no boron was added to the aluminide. This indicates that the grain-boundary problem in iron aluminides can be alleviated to some extent by grain refinement.

Summary and Remarks

Many ordered intermetallic alloys having high crystal symmetries (such as $L1_2$ and $B2$) exhibit brittle fracture and low ductility at ambient temperatures, and the brittleness originates at grain boundaries. Auger analyses showed that the grain boundaries are clean and free of impurities, indicating the intrinsic nature of grain-boundary brittleness. Both experimental studies and theoretical calculations indicate that a high ordering energy and large electronegativity difference (or valency difference) between constituent atoms in A_3B alloys are the major causes of intrinsically brittle grain boundaries.

Because of the intrinsic nature of brittleness, the grain-boundary problem cannot be solved by material purification. Instead, an effective solution to this problem is to control grain-boundary chemistry through microalloying. This chapter has provided a comprehensive review of the current research on microalloying with the interstitial elements, boron and carbon, which are commonly added in the parts per million range to intermetallic alloys, as well as to conventional alloys. Boron is found to have a strong tendency to segregate to grain boundaries and to suppress intergranular fracture in $L1_2$-Ni_3Al, Ni_3Ga, and Ni_3Si with hypostoichiometric compositions; and in $B2$-NiAl and FeAl. The same beneficial effect has been observed for carbon added to Ni_3Si, but not for carbon added to Ni_3Al or NiAl. This is because carbon does not segregate strongly to grain boundaries in the nickel aluminides, as evidenced from Auger analyses and atom probe studies.

The suppression of grain-boundary fracture may or may not be accompanied by a significant increase in the ductility of ordered intermetallics. In the case of Ni_3Al, Ni_3Ga, and Ni_3Si in which grain-boundary weakness is the major cause of brittleness, the elimination of intergranular fracture leads to a substantial increase in tensile ductility at room temperature. Boron is most effective in ductilizing Ni_3Al (24% Al), and a high ductility of 50% has been achieved by microalloying with 1000 wt ppm B. In the case of NiAl and FeAl, in which weak grain boundaries are part of the brittleness problem but not the sole problem, the suppression of grain-boundary fracture by boron doping does not result in significant improvement in tensile ductility. Once other embrittlement factors are alleviated by metallurgical means, the beneficial effect of microalloying is revealed. This is clearly demonstrated in FeAl (40% Al) doped with boron. B-doped FeAl showed transgranular fracture but limited tensile ductility at room temperature because of environmental embrittlement involving moisture in air. When this embrittlement was alleviated by testing in a neutral environment (i.e., dry oxygen), B-doped FeAl showed extensive deformation without failure.

Alloy stoichiometry strongly affects the grain-boundary properties of ordered intermetallics, particularly those connected with the effects of microalloying. Boron is effective in ductilizing Ni_3X (X = Al, Ga, and Si) at hypostoichiometric compositions ($X < 25\%$) but is ineffective at hyperstoichiometric compositions ($X > 25\%$). Takasugi *et al.* (Ref 81) have correlated the stoichiometric effect to electronic environments in the grain-boundary region in heteroelectric A_3B intermetallics. In a study of grain-boundary chemistry in Ni_3Al as a function of aluminum and boron concentrations, Liu *et al.* (Ref 26, 83) found that alloy stoichiometry strongly affects boron segregation to the grain boundaries, aluminum concentration at the boundaries, and solid-solution hardening in the bulk by boron. All these factors influence the intergranular fracture and tensile ductility of Ni_3Al.

Grain-boundary chemistry can be controlled not only by microalloying but also by macroalloying with major elements. Takasugi and Izumi (Ref 27, 28) observed that the ductility and fracture behavior of Ni_3Al were improved by alloying with iron and manganese, which reduced the heterogeneous electronic environments in the grain-boundary planes. A number of other metallurgical means have been employed to control grain-boundary properties in ordered intermetallics, as well as conventional metals and alloys, including refinement of grain structure, control of grain geometry, control of grain-boundary character, and precipitation of second-phase particles. These means have proved to be useful in certain intermetallic alloys. For instance, grain refinement is quite effective in improving the ductility of FeAl aluminides at room temperature (Ref 110, 111). A brief discussion of grain-boundary engineering in ordered intermetallics has been given elsewhere (Ref 86, 112).

Acknowledgment

The authors are grateful to C.L. Fu for valuable discussion and to J.A. Horton and P.F. Tortorelli for reviewing the manuscript. This research was sponsored by the Division of Materials Sciences, U.S. Department of Energy, under contract DE-AC05-84OR21400 with Martin Marietta Energy Systems, Inc.

References

1. C.T. Liu and J.O. Stiegler, *Science*, Vol 226, 1984, p 636
2. *High-Temperature Ordered Intermetallic Alloys, Proc. MRS Symp.*, Vol 39, 1985
3. *High-Temperature Ordered Intermetallic Alloys II, Proc. MRS Symp.*, Vol 81, 1987
4. *High-Temperature Ordered Intermetallic Alloys III, Proc. MRS Symp.*, Vol 133, 1989
5. *High-Temperature Aluminides and Intermetallics, Proceedings of TMS/ASM Symposium*, S.H. Whang, C.T. Liu, D.P. Pope, and J.O. Stiegler, Ed., The Minerals, Metals & Materials Society, 1990
6. C.T. Liu, J.O. Stiegler, and F.H. Froes, Ordered Intermetallics, in *Properties and Selection: Nonferrous Alloys and Special-Purpose Materials*, Vol 2, 10th ed., *Metals Handbook*, ASM International, 1990, p 913-942
7. A.I. Taub and R.L. Fleischer, *Science*, Vol 243, 1989, p 616
8. E.A. Aitken, in *Intermetallic Compounds*, J.H. Westbrook, Ed., John Wiley, 1967, p 491-516
9. S. Takeuchi and E. Kuramoto, *Acta Metall.*, Vol 21, 1973, p 415
10. P.H. Thornton, R.G. Davies, and T.L. Johnston, *Metall. Trans.*, Vol 1, 1970, p 207
11. S.M. Copley and B.H. Kear, *Trans. TMS-AIME*, Vol 239, 1967, p 977
12. D.P. Pope, *Philos. Mag.*, Vol 25, 1972, p 917
13. V. Paidar, D.P. Pope, and V. Vitek, *Acta Metall.*, Vol 32, 1984, p 435
14. M.H. Yoo, *Scr. Metall.*, Vol 20, 1986, p 915
15. M.H. Yoo, J.A. Horton, and C.T. Liu, *Acta Metall.*, Vol 36, 1988, p 2935
16. N.S. Stoloff and R.G. Davies, *Prog. Mater. Sci.*, Vol 13 (No. 1), 1966, p 1
17. *Ordered Alloys—Structural Applications and Physical Metallurgy*, in *Proceedings of the Third Bolton Landing Conference*, B.H. Kear, C.T. Sims, N.S. Stoloff, and J. H. Westbrook, Ed., 1970
18. *Mechanical Properties of Intermetallic Compounds,* J.H. Westbrook, Ed., Wiley, 1959
19. *Intermetallic Compounds,* J.H. Westbrook, Ed., Wiley, 1967
20. L.E. Tanner, *et al.*, "Mechanical Behavior of Intermetallic Compounds," Report AST-TDR62-1087, parts 1-3, Manlabs, Inc., 1963-1964
21. H.A. Lipsitt, D. Schechtman, and R.E. Schafrik, *Metall. Trans.*, Vol 11A, 1980, p 1369
22. K. Aoki and O. Izumi, *Acta Metall.*, Vol 27, 1979, p 807
23. K. Aoki and O. Izumi, *Trans. Jpn. Inst. Met.*, Vol 19, 1978, p 203
24. E.M. Grala, *Mechanical Properties of Intermetallic Compounds*, J.H. Westbrook, Ed., Wiley, 1960, p 358-404

25. C.T. Liu and C.C. Koch, in *Proceedings of a Public Workshop on Trends in Critical Materials Requirements for Steels of the Future: Conservation and Substitution Technology for Chromium*, NBSIR-83-2679-2, National Bureau of Standards, June 1983
26. C.T. Liu, C.L. White, and J.A. Horton, *Acta Metall.*, Vol 33, 1985, p 213-219
27. T. Takasugi and O. Izumi, *Acta Metall.*, Vol 33, 1985, p 1247-1258
28. T. Takasugi, O. Izumi, and N. Masahashi, *Acta Metall.*, Vol 33, 1985, p 1259
29. A.I. Taub, C.L. Briant, S.C. Huang, K.M. Chang, and M.R. Jackson, *Scr. Metall.*, Vol 20, 1986, p 129-134
30. A.I. Taub and C.L. Briant, in *High-Temperature Ordered Intermetallic Alloys II, Proc. MRS Symp.*, Vol 81, 1987, p 343
31. A.I. Taub and C.L. Briant, *Acta Metall.*, Vol 35, 1987, p 1597-1603
32. D. Farkas and V. Rangarajan, *Acta Metall.*, Vol 35, 1987, p 353
33. M.A. Crimp, K.M. Vedula, and D.J. Gaydosh, in *High-Temperature Ordered Metallic Alloys, Proc. MRS Symp.*, Vol 39, 1985, p 499
34. D.J. Gaydosh, S.L. Draper, and M.V. Nathal, *Metall. Trans.*, Vol 20A, 1989, p 1701
35. M.A. Crimp and K.M. Vedula, *Mater. Sci. Eng.*, Vol 78, 1986, p 193
36. M.G. Mendiratta, S.K. Ehlers, and D.K. Chatterji, in *Proceedings of the NBS Symposium on Rapid Solidification Processing: Principles and Technologies*, National Bureau of Standards, 1983, p 420
37. I. Baker and D.J. Gaydosh, *Mater. Sci. Eng.*, Vol 96, 1987, p 147
38. I. Baker and D.J. Gaydosh, in *High-Temperature Ordered Metallic Alloys, Proc. MRS Symp.*, Vol 39, 1985, p 315
39. A.G. Rozner and R.J. Wasilewski, *J. Inst. Met.*, Vol 94, 1966, p 169
40. K.H. Hahn and K. Vedula, *Scr. Metall.*, Vol 23, 1989, p 7
41. E.P. George and C.T. Liu, *J. Mater. Res.*, Vol 5, 1990, p 754
42. K. Aoki and O. Izumi, *Nippon Kinzoku Gakkaishi*, Vol 43, 1979, p 1190
43. A.I. Taub, S.C. Huang, and K.M. Chang, *Metall. Trans.*, Vol 15A, 1984, p 399
44. D.F. Stein and L.A. Heldt, *Interfacial Segregation*, W.C. Johnson and J.M. Blakely, Ed., ASM International, 1979, p 239-260
45. C.L. White and D.F. Stein, *Metall. Trans.*, Vol 9A, 1978, p 13
46. T. Takasugi, E.P. George, D.P. Pope, and O. Izumi, *Scr. Metall.*, Vol 19, 1985, p 551
47. T. Ogura, S. Hanada, T. Masumoto, and O. Izumi, *Metall. Trans.*, Vol 16A, 1985, p 441
48. C.T. Liu and E.P. George, *Scr. Metall.*, Vol 24, 1990, p 1285-1290
49. C.T. Liu and E.P. George, in *High-Temperature Ordered Intermetallic Alloys IV, Proc. MRS Symp.*, 1991, to be published
50. C.T. Liu, E.H. Lee, and C.G. McKamey, *Scr. Metall.*, Vol 23, 1989, p 875
51. C.T. Liu, C.G. McKamey, and E.H. Lee, *Scr. Metall.*, Vol 24, 1990, p 385
52. J.A. Horton, C.T. Liu, and C.C. Koch, in *High-Temperature Alloys: Theory and Design*, J.O. Stiegler, Ed., American Institute of Mining, Metallurgical, and Petroleum Engineers, 1984, p 309
53. C.G. McKamey, J.A. Horton, and C.T. Liu, in *High-Temperature Ordered Metallic Alloys, Proc. MRS Symp.*, Vol 39, 1985, p 321

54. V. Vitek, S.P. Chen, A.F. Voter, J.J. Kruisman, and J.T.M. De Hosson, in *Grain Boundary Chemistry and Intergranular Fracture*, G.S. Was, Ed., Trans Tech Publications, 1989
55. J.J. Kruisman, V. Vitek, and J.T.M. De Hosson, *Acta Metall.*, Vol 36, 1989, p 2729
56. V. Vitek, J.J. Kruisman, and J.T.M. De Hosson, in *Interfacial Structure, Properties and Design, Proc. MRS Symp.*, 1988, p 139-144
57. A.H. King and M.H. Yoo, in *High-Temperature Ordered Metallic Alloys, Proc. MRS Symp.*, Vol 39, 1985, p 99
58. C.T. Liu and C.L. White, in *High-Temperature Ordered Metallic Alloys, Proc. MRS Symp.*, Vol 39, 1985, p 365-380
59. M.K. Miller and J.A. Horton, *J. Phys.*, Vol C7, 1986, p 263
60. J.A. Horton and M.K. Miller, in *High-Temperature Ordered Metallic Alloys, Proc. MRS Symp.*, Vol 39, 1985, p 105
61. D.D. Sieloff, S.S. Brenner, and M.G. Burke, in *High-Temperature Ordered Metallic Alloys, Proc. MRS Symp.*, Vol 39, 1985, p 87
62. J.R. Rice, in *The Effect of Hydrogen on the Behavior of Metals*, American Institute of Mining, Metallurgical, and Petroleum Engineers, 1976, p 455
63. C.T. Liu, C.L. White, C.C. Koch, and E.H. Lee, in *Proceedings of the Symposium on High-Temperature Materials and Chemistry II*, Electrochemical Society, Inc., 1983, p 32
64. C.L. White, R.A. Padgett, C.T. Liu, and S.M. Yalisove, *Scr. Metall.*, Vol 18, 1984, p 1417
65. G.S. Painter and F.W. Averill, *Phys. Rev. Lett.*, Vol 58, 1987, p 234
66. M.E. Eberhart and D.D. Vvedinsky, *Phys. Rev. Lett.*, Vol 58, 1987, p 61
67. S.P. Chen, A.F. Voter, R.C. Albers, A.M. Boring, and P.J. Hay, *Scr. Metall.*, Vol 23, 1989, p 217
68. G.M. Bond, I.M. Robertson, and H.K. Birnbaum, *J. Mater. Res.*, Vol 2, 1987, p 436
69. E.M. Schulson, T.P. Wiehs, D.V. Viens, and I. Baker, *Acta Metall.*, Vol 33, 1985, p 1587
70. P.S. Khadkikar, K. Vedula, and B.S. Shale, *Metall. Trans.*, Vol 18A, 1987, p 425
71. I. Baker, E.M. Schulson, and J.A. Horton, *Acta Metall.*, Vol 35, 1987, p 1533
72. E.M. Schulson, I. Baker, and H.J. Frost, in *High-Temperature Ordered Metallic Alloys, Proc. MRS Symp.*, Vol 39, 1985, p 195
73. E.M. Schulson, T.P. Wiehs, I. Baker, H.J. Frost, and J.A. Horton, *Acta Metall.*, Vol 34, 1986, p 1395
74. W.A. Swiatnicki and M.W. Grabski, *Acta Metall.*, Vol 37, 1989, p 1307
75. R.A.D. Mackenzie and S.L. Sass, *Scr. Metall.*, Vol 22, 1988, p 1807
76. M.J. Mills, *Scr. Metall.*, Vol 23, 1989, p 2061
77. J.E. Krzanowski, *Scr. Metall.*, Vol 23, 1989, p 1219
78. W.C. Oliver and C.L. White, in *High-Temperature Ordered Intermetallic Alloys II, Proc. MRS Symp.*, Vol 81, 1987, p 241-246
79. W.C. Oliver, in *High-Temperature Ordered Intermetallic Alloys III, Proc. MRS Symp.*, Vol 133, 1989, p 397
80. T. Takasugi, M. Nagashima, and O. Izumi, *Acta Metall.*, Vol 38, 1990, p 747
81. T. Takasugi, N. Masahashi, and O. Izumi, *Acta Metall.*, Vol 35, 1987, p 381

82. S. Ochiai, Y. Oya, and T. Suzuki, *Acta Metall.*, Vol 32, 1984, p 289-298
83. D.P. Pope and C.T. Liu, in *Superalloys, Supercomposites and Superceramics*, J.K. Tien and T. Caulfield, Ed., Academic Press, 1989, p 583-624
84. A. Choudhury, "The Intergranular Segregation of Boron in Substoichiometric Ni_3Al," Ph.D. thesis, The University of Tennessee, 1987
85. A. Choudhury, C.L. White, and C.R. Brooks, *Scr. Metall.*, Vol 20, 1986, p 1061
86. C.T. Liu, in *Science of Advanced Materials*, H. Weidersich and M. Meshii, Ed., ASM International, 1990, p 423-459
87. E.P. George, C.T. Liu, and R.A. Padgett, *Scr. Metall.*, Vol 23, 1989, p 979
88. N. Masahashi, T. Takasugi, and O. Izumi, *Acta Metall.*, Vol 36, 1988, p 1815
89. T. Takasugi, in *High-Temperature Ordered Intermetallic Alloys IV, Proc. MRS Symp.*, 1991, to be published
90. S.C. Huang, C.L. Briant, K.-M. Chang, A.I. Taub, and E.L. Hall, *J. Mater. Res.*, Vol 1, 1986, p 60
91. N. Masahashi, T. Takasugi, and O. Izumi, *Acta Metall.*, Vol 36, 1988, p 1823
92. S.S. Brenner and Hua Ming-Jian, *Scr. Metall.*, Vol 24, 1990, p 667-670
93. E.P. George, C.T. Liu, and J.J. Liao, *Alloy Phase Stability and Design, Proc. MRS Symp.*, Vol 186, 1991, to be published
94. C.L. Fu, *J. Mater. Res.*, Vol 5, 1990, p 971
95. C.L. Fu and G.S. Painter, *J. Mater. Res.*, Vol 6, 1991, in press
96. C.L. Fu and M.H. Yoo, in *High-Temperature Ordered Intermetallic Alloys IV, Proc. MRS Symp.*, 1991, to be published
97. J.H. Schneibel and P.M. Hazzledine, in *High-Temperature Ordered Intermetallic Alloys IV, Proc. MRS Symp.*, 1991, to be published
98. E.P. George, W.D. Porter, H.M. Henson, W.C. Oliver, and B.F. Oliver, *J. Mater. Res.*, Vol 4, 1989, p 78
99. E.P. George, W.D. Porter, and D.C. Joy, in *High-Temperature Ordered Intermetallic Alloys III, Proc. MRS Symp.*, Vol 133, 1989, p 311-315
100. J.H. Schneibel and W.D. Porter, in *High-Temperature Ordered Intermetallic Alloys III, Proc. MRS Symp.*, Vol 133, 1989, p 335-340
101. C.D. Turner, W.O. Powers, and J.A. Wert, *Acta Metall.*, Vol 37, 1989, p 2635-2644
102. K.S. Kumar and J.R. Pickens, *Scr. Metall.*, Vol 22, 1988, p 1015
103. D.B. Miracle, S. Russell, and C.C. Law, in *High-Temperature Intermetallic Alloys II, Proc. MRS Symp.*, Vol 81, 1987, p 225
104. R. Darolia, D.F. Lahrman, R.D. Field, and A.J. Freeman, in *High-Temperature Ordered Intermetallic Alloys II, Proc. MRS Symp.*, Vol 81, 1987, p 113
105. T. Takasugi and O. Izumi, *Acta Metall.*, Vol 34, 1986, p 607
106. N. Masahashi, T. Takasugi, and O. Izumi, *Metall. Trans.*, Vol 19A, 1988, p 353
107. O. Izumi and T. Takasugi, *J. Mater. Res.*, Vol 3, 1988, p 426
108. A.K. Kuruvilla, S. Ashok, and N.S. Stoloff, in *Proc. Third Int. Cong. Hydrogen Met.*, Vol 2, 1982, p 629
109. A.K. Kuruvilla and N.S. Stoloff, *Scr. Metall.*, Vol 19, 1985, p 83
110. D.J. Gaydosh and M.V. Nathal, *Scr. Metall.*, Vol 24, 1990, p 1281-1284
111. S. Strothers and K. Vedula, in *Proc. Powder Metall. Conf.*, Vol 43, 1987, p 597
112. C.T. Liu, *Interfacial Structure, Properties and Design, Proc. MRS*, Vol 122, 1988, p 429-441

14

Types of Grain-Boundary Sliding in Zinc Bicrystals: Phenomenology and Mechanisms

R.Z. Valiev and V.G. Khairullin
Institute for Metals Superplasticity Problems
USSR Academy of Sciences, Ufa 45000l USSR
A.D. Sheikh-Ali
Aircraft Engineering Institute, Ufa 450000 USSR

Research on grain-boundary sliding (GBS) in zinc bicrystals of different types, distinguished by their geometry and straining conditions, has been done by the present authors. In this chapter a classification of four basic types of GBS, specifying their phenomenological and structural distinctions, is proposed. Experimental evidence for the dislocation nature of sliding is presented. It is assumed that grain-boundary dislocations responsible for GBS in high-angle random boundaries are dislocations with alternating Burgers vectors.

Grain-boundary sliding (GBS), i.e., a shift of grains against one another along their common boundary, plays an important role during the high-temperature deformation and failure of polycrystals. In recent years GBS has been the subject of detailed investigation, but its nature requires further clarification. Among the principal unresolved problems are whether there is more

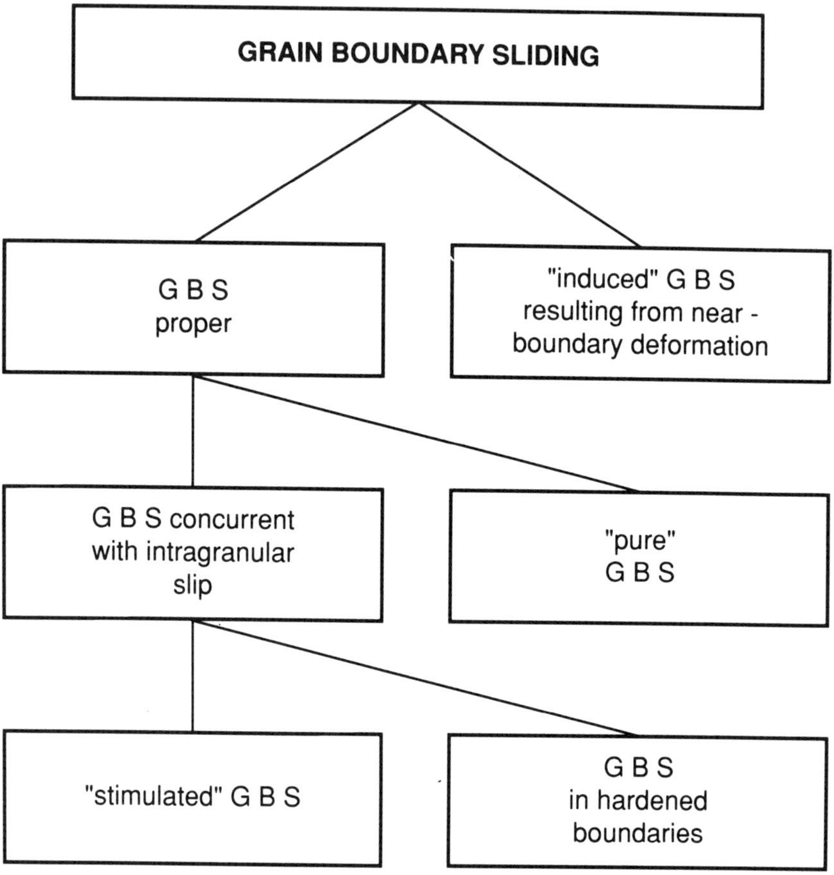

Fig. 1 Types of grain-boundary sliding

than one type of GBS; if so, what are their atomic mechanisms; and what are the features of GBS with a curved boundary and triple points.

To answer these questions, several years ago the authors launched a comprehensive research program on the bicrystals and tricrystals of zinc. Zinc has one preferential basal slip plane, which allows one to distinguish between effects produced by the coupling of various deformation processes in the boundaries and within the grains. Based on results reported in Ref 1 to 8 and other data published in the literature, the following classification of the types of GBS are proposed (Fig. 1).

First, a distinction should be made between GBS proper and so-called induced GBS. The difference is that GBS proper is promoted by the tangential stresses applied in the boundary plane and is effected through the shift of grains along the boundary; i.e., grains shift against one another as the deformation takes place along the boundary proper. In this case the area of grain-boundary shear is comparable to the crystallographic width of the boundary. In the case of induced GBS, an important role is played by local stresses generated by the intragrain slip, which is incomplete near the boundary. GBS operates here as an accommodation process and can be accompanied by shear in a broad near-boundary zone.

GBS proper can be subdivided into two types: "pure" GBS, i.e., sliding without intragrain slip (IS), and GBS accompanied by IS. In the latter case "stimulated" GBS should be distinguished from GBS along the "hardened" boundaries.

The objective of this chapter is, first, to discuss the phenomenological features of the various types of GBS and, second, to reveal the dislocation nature of grain-boundary sliding proper.

Experimental Procedure

Bicrystalline samples were grown from 99.97% pure zinc using the improved Bridgeman method (Ref 1, 2). Bicrystals had a symmetric-tilt boundary $\langle 10\bar{1}0\rangle$, but different misorientation angles ($90^\circ \pm 1^\circ$, $30^\circ \pm 1^\circ$, and $86^\circ \pm 1^\circ$) and different orientations of the grain plane against the applied stresses, depending on the problem investigated.

Tensile tests were performed in the temperature range of 323 to 553 K. Sliding was measured by the offset of marker lines (scratches) on the polished sample surface using an optical microscope with an accuracy of ±0.3 μm. The values of the activation energy of sliding (Q) and strain rate sensitivity coefficient ($m = 1/n$) where n is the sensitivity of the GBS rate ($\dot{s}$) to the stress of shear along the boundary (τ) were derived from the expression $\dot{s} \sim \tau^n$.

Results and Discussion

Regularities of Different Types of GBS. The geometry of the zinc bicrystals used to study sliding proper is shown in Fig. 2 (type 1 and 2 samples) (Ref 1, 2, 5). Type 1 samples (Fig. 2a, b) had basal planes parallel and normal to the tensile axis, which eliminated shear stresses along these planes and, hence, prevented the development of intragrain slip. In the type 2 sample (Fig. 2c), basal planes had a 15° deviation from the position of the basal planes in the type 1 samples, with reference to the tensile axis, causing shear stresses to arise

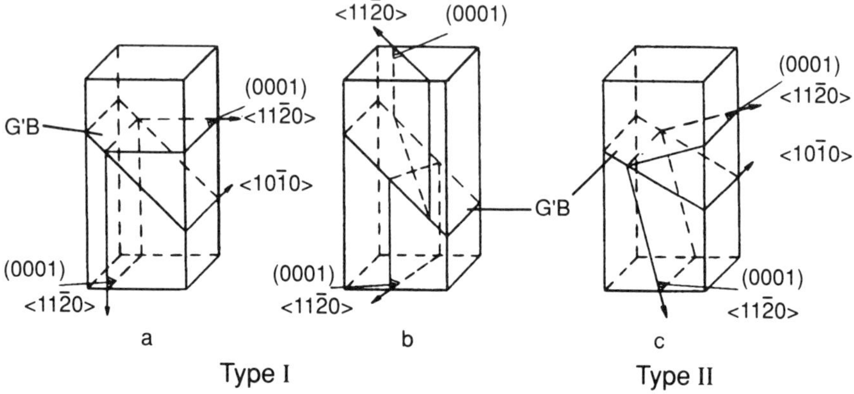

Fig. 2 Geometry of the bicrystals under study (a and c) with a 90° ⟨10$\bar{1}$0⟩ tilt boundary and (b) with a 90° ⟨$2\bar{2}$01⟩ twist boundary

during tensile straining. All the studied boundaries were, according to a well-known Brandon criterion, random high-angle intercrystalline boundaries.

Thus, the geometry of the type 1 samples allowed pure sliding, i.e., GBS without IS; the type 2 samples were used to study GBS occurring simultaneously with IS.

The study of strained type 1 bicrystals indicated (Ref 1) that during sliding, the shear of neighboring crystals took place in a narrow boundary area without any noticeable curving of the surface marker lines or boundary migrations (Fig. 3a) and with no slip lines. Further evidence for the absence of IS was obtained through the study of etching pits. All these observations point to pure GBS.

This type of sliding is characterized by the nonuniformity of GBS along the boundary (see also the section below). This is the reason, when plotting the strain rate-to-time (*s-t*) curves, that the dependence of an average value of GBS (based on five marker lines) on testing time became evident. Figure 4 shows that all the curves plotted at different shear stresses (τ) at the 90° ⟨10$\bar{1}$0⟩ boundary are monotonous, with the sliding rate decreasing continuously. These curves illustrate the dependence of the GBS rate on shear stresses along the boundary (for $t \leq 5$ h). At low stresses ($\tau \leq 2$ MPa), this dependence is linear and, when extrapolated, it yields the value of the threshold stress equal to 0.5 MPa. Similar regularities were observed in a bicrystal with the 90° ⟨$2\bar{2}$01⟩ twist

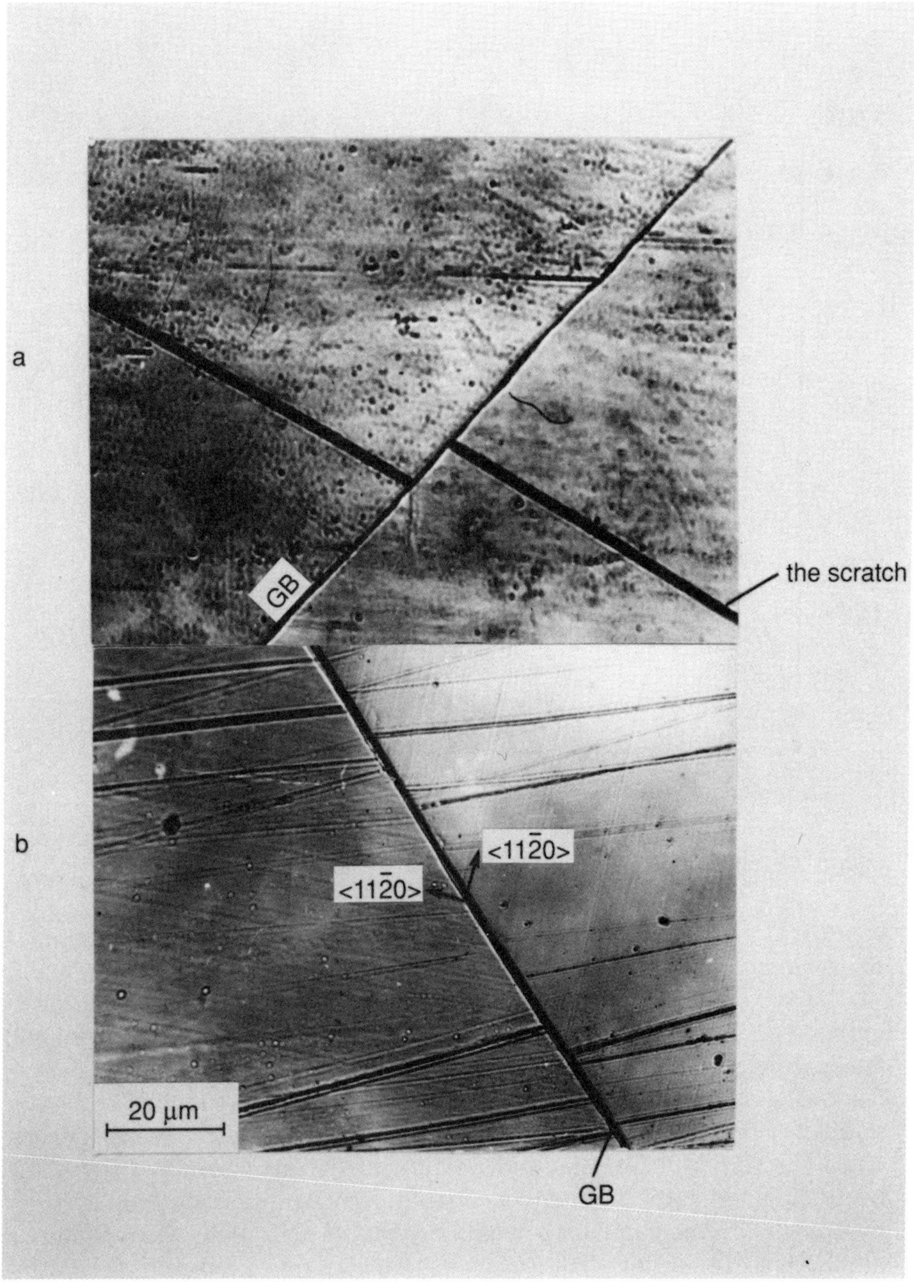

Fig. 3 Grain-boundary sliding (b) with and (a) without intragrain slip in the 90° ⟨10$\bar{1}$0⟩ tilt boundary at 323 K

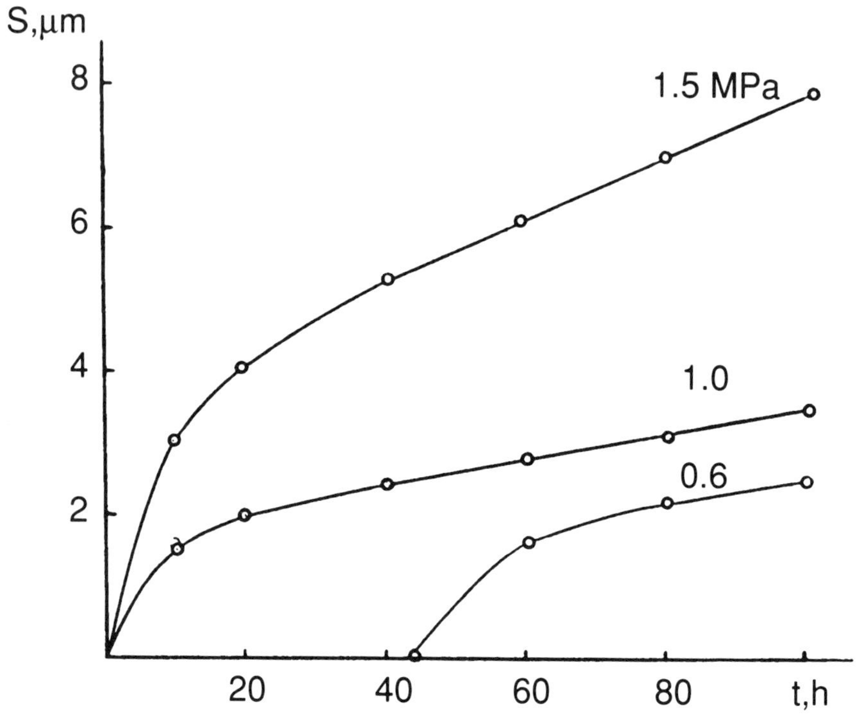

Fig. 4 Sliding-time curves for type 1 bicrystals with a 90° ⟨10$\bar{1}$0⟩ tilt boundary

boundary. Various parameters of pure GBS in the bicrystals studied are presented in Table 1.

The type 2 samples also preserved a narrow boundary area while shear was in progress. There was no curving of marker lines in the boundary vicinity, although lines of basal intragrain slip were also observed (Fig. 3b) (Ref 2). Figure 5 shows the dependence of the average GBS value on time in these samples at two shear stresses (curves 1 and 2). This figure also presents the *s-t* curve for the type 1 bicrystal under similar conditions (curve 3). In type 2 bicrystals, where GBS develops concurrently with IS, the variation of the rate of sliding is not monotonous, in contrast to that of pure sliding. Two regions are clearly distinguished on the *s-t* curve: the initial region, corresponding to a rapid, almost linear growth of sliding with time, and a region where this growth is insignificant (Ref 2).

Table 1 Parameters of different types of GBS: rate of sliding (s) at $\tau = 1.2$ MPa, threshold stress (τ_0), activation energy (Q), and strain rate sensitivity coefficient $m = 1/n$ where n is taken from the equation $\dot{s} \sim t^n$

Type of grain-boundary sliding	Grain boundary	$\dot{s} \cdot 10^5$, μm/s	Threshold stress (τ_0), MPa	Activation energy (Q), J/mol · K	Strain rate sensitivity coefficient, (m)	Reference
"Pure"	90° $\langle 10\bar{1}0 \rangle$ tilt,	3.1	0.5	42	0.9	1
	90° $\langle 2\bar{2}01 \rangle$ twist	10	0.9	56	0.9	1
"Stimulated"	90° $\langle 10\bar{1}0 \rangle$ tilt	70	0.4	55	0.5	2
Along the "hardened" boundary	90° $\langle 10\bar{1}0 \rangle$ tilt	2.7	...	76	...	2
"Induced"	90° $\langle 10\bar{1}0 \rangle$ tilt	...	...	...	...	8

It is clear that when GBS develops concurrently with IS, the rate of sliding in the initial region is remarkably higher than that of pure sliding. This indicates that IS has a strong stimulating effect on GBS. Under actual deformation conditions (T = 50 °C and τ = 0.9 MPa), the intragrain shear strain that develops concurrently with the sliding along the boundary approaches 10% (γ = 0.1) in both crystallites of the bicrystal by the end of the first region of the s-t curve. Assuming that all the dislocations providing for the deformation run into the boundary, the change in the misorientation angle of the adjacent crystals is evaluated to be $\Delta\theta \cong \gamma$ (rad) = 5.7° (Ref 2). This value is very similar to that observed experimentally in the end of the first region, i.e., 6°, which confirms the authors' assumption that intragrain (lattice) dislocations run into the boundary and are probably absorbed there, since in zinc the absorption of dislocations in the boundaries occurs at temperatures above 293 K (Ref 7).

Quite a different pattern of GBS development is observed in type 2 samples in the second region of the s-t curve. The rate of sliding here decreases sharply and is comparable to the rate of pure GBS. Changes in the microhardness indicate considerable hardening of the boundary region. In this case, the boundary probably fails to be an active sink for intragrain dislocations, and the dislocations accumulate near the boundary. The lesser capability to absorb dislocations is probably due to the attainment of an ultimate density of dislocations in the boundary by the end of the first region of the s-t curve (Ref 7).

The parameters of GBS concurrent with IS are given in Table 1. As in the case of pure GBS, the sliding activation energy in the first region of the s-t curve is quite similar to the activation energy of the grain-boundary diffusion, which is 46 kJ/mol · K in zinc (volume diffusion activation energy is 90 kJ/mol · K) (Ref 9, 10). However, the strain rate sensitivity coefficient (m) approaches the

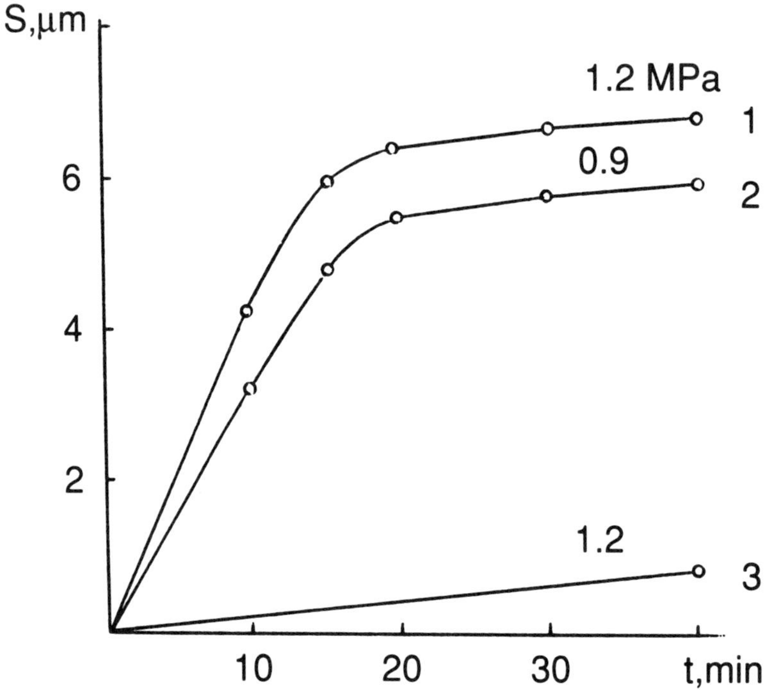

Fig. 5 Sliding-time curves for pure sliding (3) and sliding concurrent with intragrain slip (1 and 2)

value of 0.5. Meanwhile, the GBS illustrated by the second region of the *s-t* curve (Fig. 5) has all the parameters presented in the table. This supposes the development of two physically distinguishable types of GBS concurrent with IS. These two types are called here stimulated sliding and GBS along the hardened boundary.

The considerable differences observed in the parameters of different types of GBS and their structural manifestations indicate that their physical nature and atomic mechanisms are also different. In this regard the problem of the contribution of dislocations to the development of sliding is of crucial significance.

Dislocation Nature of GBS Proper. The question of whether GBS is related to the motion of dislocations along a crystallographically ordered boundary or is determined by shear along a grain boundary that has been rendered amor-

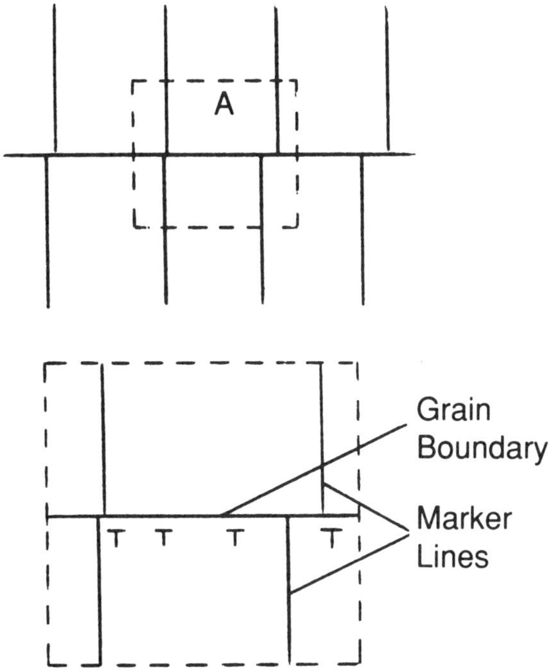

Fig. 6 Diagram of a nonuniform shift of marker lines during pure GBS. On the bottom is an enlargement of section A

phous is still open to discussion (Ref 11, 12). Among the authors' experimental results, three seem to be the most important in proving the dislocation nature of GBS.

Nonuniformity of Sliding. It has been mentioned previously that pure sliding is characterized by the nonuniformity of sliding along the boundary. As shown in Fig. 6, the different values of GBS in neighboring regions of the boundary can be viewed as the result of the incompleteness of grain-boundary shears. The resulting plastic inconsistencies can be compensated for by the generation of lattice dislocations into adjacent crystals. However, the measurements of the density of dislocations indicate that its value is constant throughout the deformation (Ref 2); i.e., the observed nonuniformity of sliding is not related to the generation of lattice dislocations. Nevertheless, the incompleteness of grain-boundary shears (like the introduction of dislocations into crystals) can be interpreted as the result of the formation of pileups of grain-boundary dislocations in the boundary plane (see Fig. 6). It is probably the motion of grain-boundary dislocations generated within the boundary that determines the nature of pure GBS.

Relationship between Sliding and Slip. It is commonly accepted that during sliding along a boundary that has become amorphous, a strict relationship between sliding and slip is not fulfilled but that it must be fulfilled when the dislocation mechanism of grain-boundary sliding is realized. This problem was studied on type 2 bicrystals (Ref 2) under the concurrent development of GBS and IS. As mentioned earlier, the evaluation indicates that in the first region of the curve of these bicrystals most dislocations run into the boundary.

Using a relationship $s = \rho \cdot b \cdot L$ where L is the length of the dislocation run and ρ is the density of dislocations that have run into the boundary, the contribution of trapped lattice dislocations to the development of sliding can be evaluated. It is important that in this relationship $\rho \cdot b = \gamma$ be the value of shear deformation. L can also be the distance between the slip lines in crystals. This is important because the intersections of slip lines with the boundary hinders the development of grain boundary shear. According to the experiment, at $T = 323$ K and $\tau = 1.2$ MPa, the value of intragrain shear deformation (γ) was 0.1, and the value of L was $7 \cdot 10^{-3}$ cm. From this was obtained $s = 5.4$ μm, which actually coincides with the experimental value of s, which was found to be 6 μm. Also, the relationship $s = \gamma \cdot L$ was established as valid under other experimental conditions. It should be mentioned that in regard to zinc bicrystals Ref 13 reported that the value of sliding (s) is related to the intragrain shear (γ) as $s \sim \gamma a$ where a is the coefficient dependent on the misorientation of crystals. Thus, the results obtained provide sufficiently convincing evidence for the existence of a certain relationship between GBS and IS, the nature of which is related to the dislocation character of sliding. This relationship may have a very simple physical meaning. It is well known that at elevated temperature lattice dislocations that run into the boundary may dissociate into mobile grain-boundary dislocations. If sliding is considered a process of motion of these grain-boundary dislocations, the relationship between GBS and IS is automatically fulfilled.

Relation between GBS and Boundary Migration. Zinc bicrystals make possible the experimental observation of the relationship between GBS and migration, and results of the experiments can be used to support the dislocation approach to the process of sliding (Ref 6). Zinc bicrystals with symmetric-tilt boundaries $30^\circ \pm 1^\circ \langle 10\bar{1}0 \rangle$ and $86^\circ \pm 1^\circ \langle 10\bar{1}0 \rangle$ were used. These grain boundaries are near coincidence with $\Sigma = 15$ and $(c/a)^2 = 7/2$ and belong to the constrained coincidence site lattice related boundaries (Ref 14). The deformation scheme of the

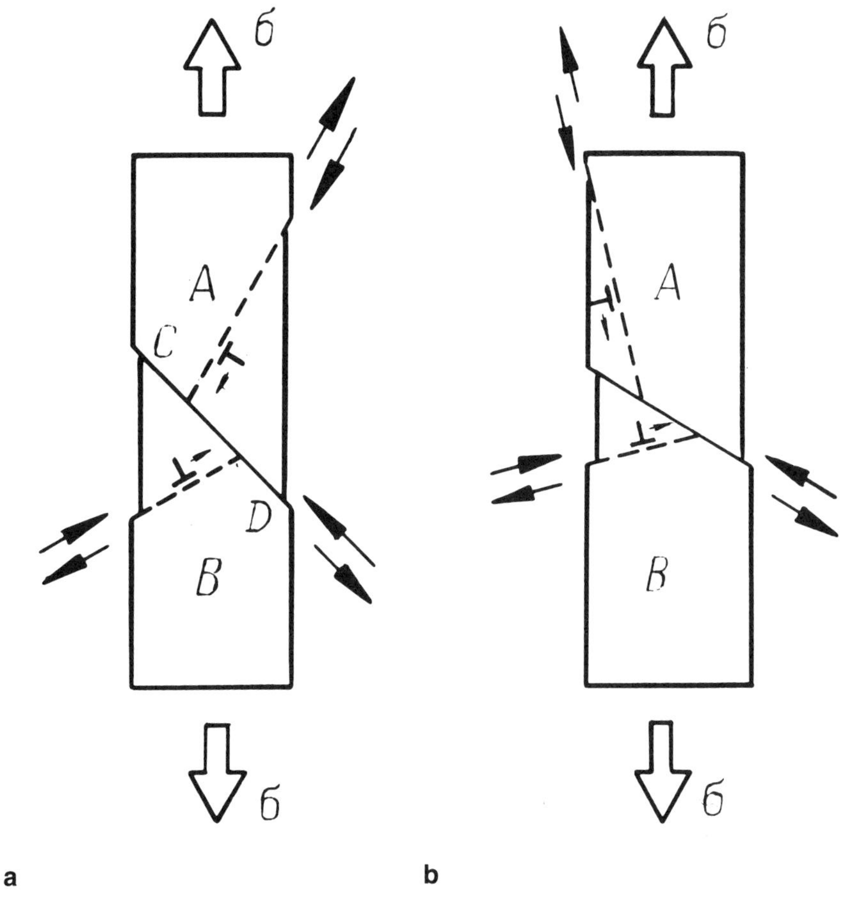

Fig. 7 Geometry of intragrain slip in zinc bicrystals containing (a) 30° $\langle 10\bar{1}0 \rangle$ and (b) 86° $\langle 10\bar{1}0 \rangle$ tilt boundaries

bicrystalline samples and the geometry of IS are shown in Fig. 7, where it can be seen that in type 1 bicrystals crystallographic relationships of slip planes in adjacent grains must be similar in direction, whereas in type 2 bicrystals the slip planes must rotate in different directions. The orientations of the Burgers vectors of the edge lattice dislocations gliding in a crystal are determined by the direction of intragrain shears and crystallographic rotations, which are interrelated. Thus, during the deformation of these bicrystals, various combinations of lattice dislocations appear in grain boundaries.

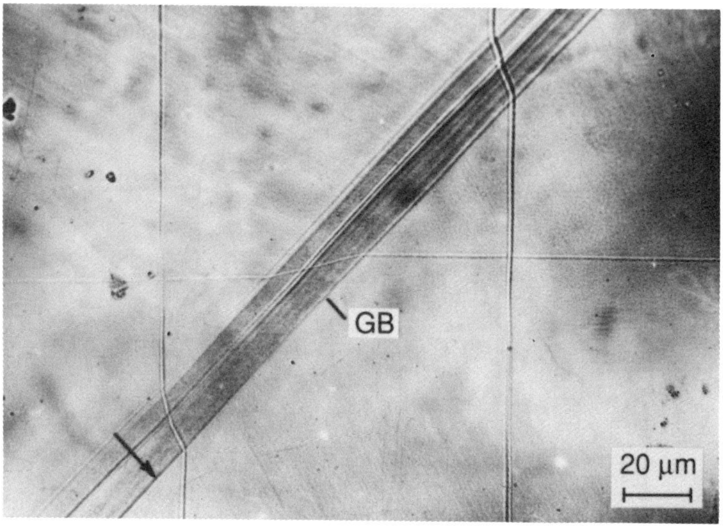

a

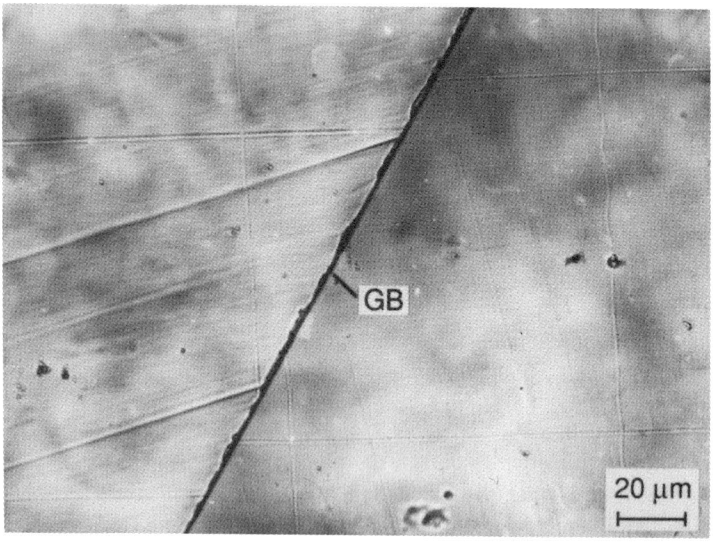

b

Fig. 8 Interaction of sliding and migration observed on (a) 30° ⟨10$\bar{1}$0⟩ and (b) 86° ⟨10$\bar{1}$0⟩ boundaries in zinc. $\tau_b = 0.3$ MPa, $T = 553$ K. The arrow in (a) indicates the direction of grain-boundary migration, and the solid lines show (0001) traces. The tensile axis is horizontal

The Schmid factors for basal slips in both bicrystal 1 grains were the same ($\psi_A = \psi_B = 0.43$), and in bicrystal 2 grains they differed 1.3 times ($\psi_A = 0.29$; $\psi_B = 0.22$). The samples were tensile strained under permanent load at 553 K (80% of melting point). The shear stress along the grains varied in the range of 0.2 to 0.3 MPa. Figure 8 shows typical portions of bicrystal surfaces after deformation. In bicrystal 1 grains, concurrent with GBS, an intensive full-length migration of grain boundaries takes place, which is confirmed by the deviation of marker lines in the migration zone from their original direction. In bicrystals of this type, boundaries migrate in the direction of grain B (Fig. 7a and arrow in 8a). In the boundary area, thin basal slip lines, reaching as far as the boundary line, are observed in both grains. With selected magnification (500×) these lines are seen to be located uniformly along the grain boundary, with the average spacing between them being 1.6 μm.

In the case of type 2 bicrystals, the effects of GBS and boundary migration are also observed (Fig. 8b). However, the average value of boundary migration here is considerably lower than in the former case. Type 1 bicrystals display a certain correlation between the m' (migration) and s (sliding) values: the m' value, as well as s, grows from point D to point C; the m'/s ratio is 2.0 units on the average. In type 2 bicrystals, there is no correlation between m' and s, and their ratio in different regions of the grain boundary varies from 0.05 to 0.2 units. The experimental results obtained indicate that in zinc bicrystals having different IS geometries, the development of GBS can be accompanied by intensive boundary migration that is greater in value than the GBS value and by weak boundary migration that constitutes only a decimal fraction of the value of shear along grain boundaries. The results are interpreted by using the approach suggested in Ref 15 based on the crystallogeometric analysis of the interaction of lattice dislocations with grain boundaries and on the mechanism of grain-boundary migration. According to the latter, an elementary act of boundary migration, being an elementary act of GBS, is realized through grain-boundary dislocations with the Burgers vector gliding component and a core that is crystallographically related to the grain-boundary ledge and that moves along the grain boundary (Ref 16-18).

First consider the type 1 bicrystal deformation. Suppose that basal dislocations enter a 30°$[10\bar{1}0]$ boundary and dissolve into dislocations with Burgers Vectors b and $2b$ (Fig. 9a and b). As a result of the dissociation of lattice dislocations that have run into the grain boundary from different grains, slipping grain-boundary dislocations of the same sign, with similar ledges, are produced (Fig. 9b). Subjected to external shear stresses along the grain boundary, the grain-boundary dislocations move towards point C, as shown in Fig.

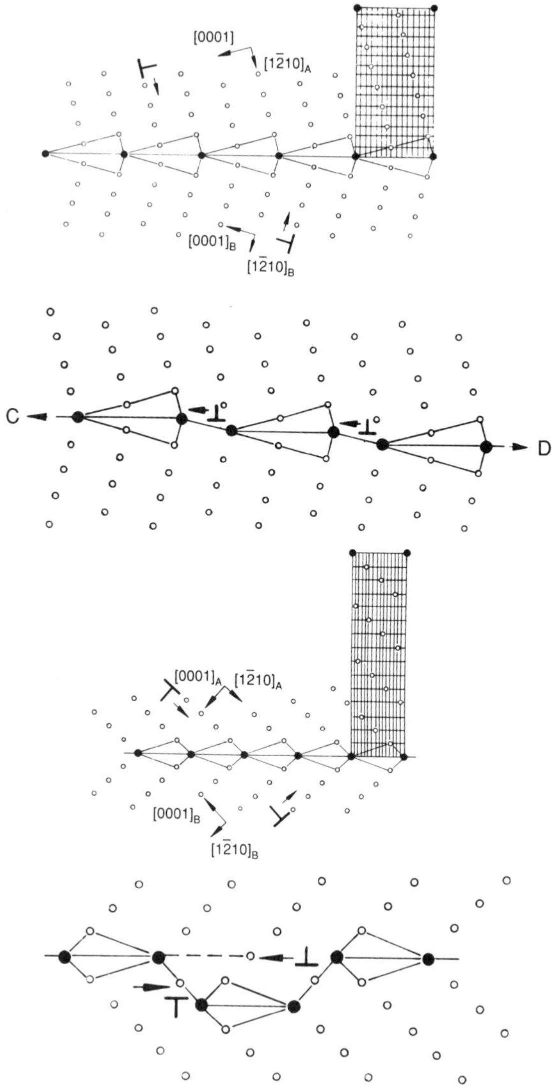

a

b

c

d

Fig. 9 (a) 29.93° $\langle 10\bar{1}0 \rangle$ $\Sigma 15$ $\{1\bar{2}11\}$ symmetric-tilt boundary in zinc. (b) The same boundary after lattice dislocations that run into the boundary have dissociated into grain-boundary dislocations. Glissile GB dislocations move in one direction. Sessile GB dislocations are not shown. (c) 86.16° $\langle 10\bar{1}0 \rangle$ $\Sigma 15$ $\{1\bar{2}14\}$ symmetric-tilt boundary in zinc. (d) The same boundary after lattice dislocations that run into the boundary have dissociated into grain-boundary dislocations. Glissile GB dislocations move in opposite directions. The original position of the boundary is shown by a dashed line. Large black circles represent coincidence sites

7(a) and 9(b). Further, boundary migration becomes oriented and must proceed in the direction of grain B, a process that was actually observed in this experiment (Fig. 8a). Because boundary migration and GBS take place through the motion of the same dislocations, it is important to find out how their values are related. For the 30° $\langle 10\bar{1}0 \rangle$ boundary, the h/b ratio is 1.856 units where h is the height of the grain-boundary dislocation ledge. In the experiment, the average value of the m/s ratio was 2.0 units, which is very similar to the calculated value.

In the case of type 2 bicrystal deformation, as a result of sliding in the basal planes of grains A and B, dislocations run into the 86° $[10\bar{1}0]$ boundary (Fig. 9c and d) where each of the lattice dislocations dissociates into dislocations $4b$ and b (Fig. 9d). Slipping grain-boundary dislocations, arising from the dissociating lattice dislocations that run into the boundary from different grains, have opposite Burgers vectors and ledge signs. Their motion along grain boundaries, promoted by shear stresses, must lead to the mutual annihilation of both the dislocation and the ledges. Perhaps this is the reason that during the deformation of these bicrystals no intensive boundary migration is observed.

Thus, the regular coupling of GBS with migration observed in zinc bicrystals may be regarded from the standpoint of the dislocation nature of sliding, an important argument for the reasonableness of this concept.

Conclusion

The experiments performed prove that the different types of GBS observed in zinc bicrystals are clearly distinguished by their phenomenological and structural parameters. This leads to the assumption that there are differences in their physical nature and atomic mechanisms. The differences observed are clear primarily in the peculiarities of the relationship between intragrain dislocations and a sliding boundary. The problems of the interaction of lattice dislocations with grain boundaries have already been thoroughly investigated (Ref 18-20). The simple diagrams illustrating the behavior of the bicrystal boundaries under different types of GBS (Fig. 10) are based on these studies.

Experiments indicate that under pure sliding there is no nucleation or absorption of lattice dislocations in grain boundaries. In this case, sliding is probably associated with the nucleation of grain-boundary dislocations in the boundary itself (Fig. 10a).

When stimulated GBS develops, lattice dislocations are actively absorbed in the boundary, and, as a result of their dissociation, mobile grain-boundary dislocations are produced (Fig. 10b). The stimulation effect in this case is

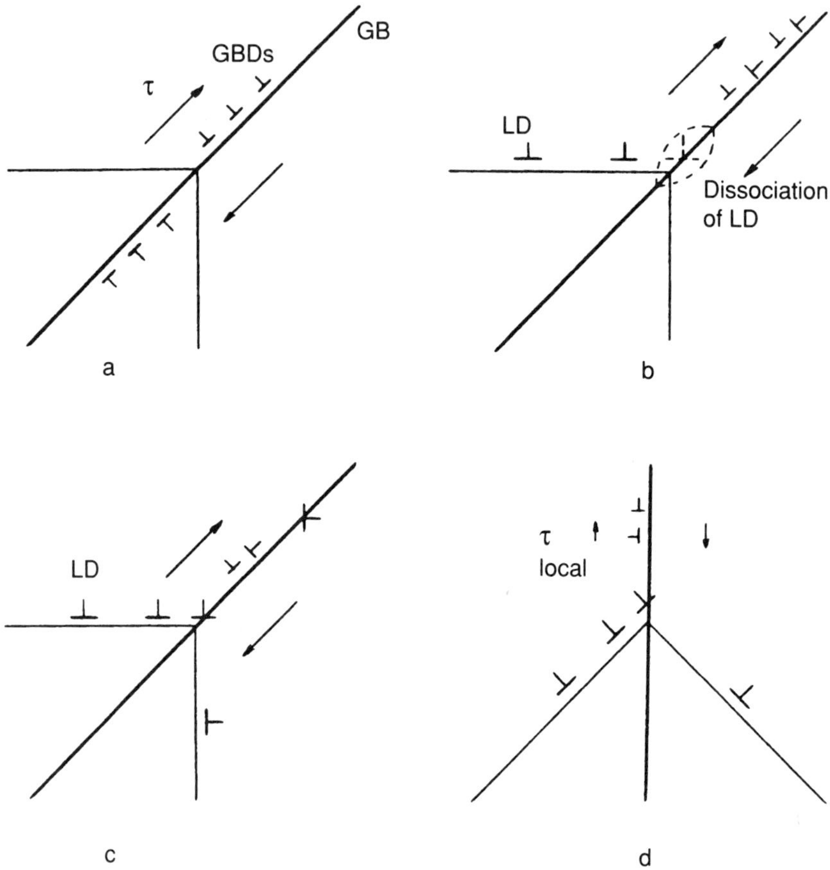

Fig. 10 Diagrams of the interaction of dislocations with boundaries during the development of (a) pure sliding, (b) stimulated sliding, (c) GBS along the hardened boundary, (d) induced sliding

probably due to both the growing density of mobile grain-boundary dislocations and the increasing rate of their motion (Ref 7).

When GBS develops along a hardened boundary (Fig. 10c), only some of the lattice dislocations are absorbed, and they are accumulated in the boundary and boundary regions, thereby hardening the boundary and changing the regularities of sliding.

Under induced GBS, dislocations pile up intensively in the boundary and boundary regions (Fig. 10d). The resulting concentrators of stress provide for the rearrangement of dislocations near the boundary, which in turn induces GBS.

The diagrams in Fig. 10 are also based on the concept of the dislocation nature of GBS, which is supported by a number of experimental results. From the perspective of the present-day concept of intercrystalline boundary structure, random high-angle boundaries can be as sets of structural elements of two different types, their length and sequence being determined by the misorientation parameters of adjacent crystals (Ref 20, 21). It is not quite clear at this time how dislocations move along random boundaries. However, the behavior is similar to that of dislocations in amorphous materials. Experimental and theoretical data (Ref 22, 23) have proved that in an amorphous structure that can be presented as a polycluster, Somigliana dislocations can move with alternating Burgers vectors.

It should be noted that basic regularities of pure and stimulated GBS can be explained, at least qualitatively, from the standpoint of dislocations (Ref 7).

Finally, it must be emphasized that although in this chapter zinc bicrystals are used as model objects, there is reason to believe that the concept of the four basic types of GBS, as well as their dislocation structures now being developed, is applicable to a great variety of metallic materials.

References

1. O.A. Kaibyshev, R.Z. Valiev, and V.G. Khairullin, *Phys. Met. Metallogr. (USSR)*, Vol 56, 1983, p 577
2. R.Z. Valiev, V.G. Khairullin, and O.A. Kaibyshev, *Phys. Met. Metallogr. (USSR)*, Vol 63, 1987, p 396
3. R.Z. Valiev and V.G. Khairullin, *Phys. Met. Metallogr. (USSR)*, Vol 64, 1987, p 1224
4. R.Z. Valiev and V.G. Khairullin, *Phys. Met. Metallogr. (USSR)*, Vol 68, 1989, p 1190
5. R.Z. Valiev and V.G. Khairullin, *Col. Phys.*, C1, Supplement 1, Vol 51, 1990, p C1-685
6. A.D. Sheikh-Ali and R.Z. Valiev, *Phys. Status Solidi (a)*, Vol 117, 1990, p 429
7. O.A. Kaibyshev and R.Z. Valiev, *Grain Boundaries and Properties of Metals*, Metallurgiya Publishing, 1987 (in Russian)
8. O.A. Kaibyshev, R.Z. Valiev, V.V. Astanin, and V.G. Khairullin, *Phys. Met. Metallogr. (USSR)*, Vol 51, 1981, p 193
9. P.B. Ghale, *Phys. Rev.*, Vol 131, 1963, p 174
10. E.S. Wajda, *Acta Metall.*, Vol 2, 1954, p 184
11. V.N. Perevezentsev, V.V. Rybin, and A.N. Orlov, *Poverkhn. Fiz., Khimia*, No. 6, 1982, p 134 (in Russian)

12. V.N. Perevezentsev, in *Theoretical Problems of Defects in Crystals*, Nauka Publishing, 1987 (in Russian)
13. T. Watanabe, M. Yamada, S. Shima, and S. Karashima, *Philos. Mag. A*, Vol 40, 1979, p 667
14. F.R. Chen and A.H. King, *Philos. Mag. A*, Vol 57, 1988, p 431
15. T. Takahashi and R. Horiuchi, *Trans. Jpn. Inst. Met.*, Vol 26, 1985, p 779
16. B. Mikhailovski, *Fiz. Tverd. Tela*, Vol 14, 1972, p 2860 (in Russian)
17. W.A.T. Clark and D.A. Smith, *J. Mater. Sci.*, Vol 14, 1979, p 776
18. H. Gleiter, *Mater. Sci. Eng.*, Vol 52, 1982, p 91
19. M.W. Grabski, *J. Phys.*, Vol 46, Supplement 4, 1985, p 567
20. R.Z. Valiev, V.Y. Gertsman, and O.A. Kaibyshev, *Phys. Status Solidi (a)*, Vol 97, 1986, p 11
21. A.P. Sutton and V. Vitek, *Philos. Trans. R. Soc. (London) A*, Vol 309, 1983, p 1
22. A.S. Bakai, *Z. Physik. Chem. Neue Folge*, Vol 158, 1988, p 201
23. B. Escaig, in *Dislocations in Solids*, H. Suzuki *et al.*, Ed., VSP Netherlands, 1985

Section V:

Migration and Transformation of Interfaces

15

Diffusion-Induced Grain-Boundary Migration in Ceramics

M.D. Vaudin, J.E. Blendell, and C.A. Handwerker
Material Science and Engineering Laboratory
National Institute of Standards and Technology
Gaithersburg, MD 20899

High stresses at grain boundaries and surfaces are produced by the diffusion of solutes into polycrystalline materials where the lattice parameters of the solution vary with the solute concentration. When D_{gb} is much greater than $D_{lattice}$, boundaries may migrate in response to these stresses. Recent observations by the authors of diffusion-induced grain-boundary migration in the Al_2O_3-Cr_2O_3 system using optical, scanning electron microscopy (SEM) and transmission electron microscopy (TEM) techniques are discussed here in terms of the coherency strain theory, in which boundary migration is driven by the elastic energy produced by solute diffusion.

Diffusion-induced grain-boundary migration (DIGM) is one of many phenomena that may occur during the diffusion of a solute in a polycrystalline material. In any diffusion process in which there is substantial solubility of the diffusing species and in which the lattice parameters of the solid solution vary with composition, high stresses can develop wherever diffusion is significant. One possible response to these stresses is dislocation generation and plastic flow, which can even lead to recrystallization. However, when lattice diffusion is orders of magnitude slower than grain-boundary diffusion, the highly stressed regions in the material are confined to thin slabs at the grain boundaries; in this case grain-boundary migration may occur. When neighboring segments of the same boundary migrate in opposite directions, characteristic "wavy" boundary morphologies are frequently produced. Diffusion-induced grain-boundary migration has been observed in many metal systems and a smaller number of ceramic and semiconductor systems. Liquid film migration (LFM) is a similar phenomenon in which a film of liquid is between two grains and the film migrates as a result of diffusion and stress generation in the adjacent grains. The subjects of DIGM and LFM have been reviewed extensively by Handwerker (Ref 1) and King (Ref 2).

In the coherency strain model of DIGM, migration is induced by differences in strain energy across the boundary. The strain arises from the change in lattice parameter with composition (q). In cubic materials, the strain (δ) in the grain-boundary plane is given by:

$$\delta = \frac{[a(q_b) - a(q_o)]}{a(q_o)} = \eta(q_b - q_o) \qquad \text{(Eq 1)}$$

where $a(q_b)$ is the lattice parameter at the composition q_b in the grain-boundary plane and q_o is the composition of the original material. The right-hand side of Eq 1 assumes that da/dq is constant (Vegard's Law) such that:

$$\eta = a/1(q_o)\, da/dq \qquad \text{(Eq 2)}$$

Typically this is a valid assumption for small concentration changes. If the solid solution is constrained to remain coherent with the bulk material, the grain-boundary layer increases its energy and the coherency strain energy per unit volume for cubic materials is given by:

$$f_{coherent} = Y(n)\delta^2 \qquad \text{(Eq 3)}$$

where n is the grain-boundary normal and $Y(n)$ is an orientation-dependent elastic modulus with extrema at $\langle 100 \rangle$ and $\langle 111 \rangle$. When two grains are separated by an asymmetric grain boundary, n and $Y(n)$ are different, leading to a driving force for the migration of the boundary. For noncubic materials, η is a tensor property, and thus the elastic energy density is a complex function of material properties and grain misorientation. However, in any system that is initially stress-free and in which there is a solute-induced lattice parameter change, solute diffusion leads to an energy increase, and the difference in energy increase across the boundary between the grains (E_{diff}) produces a driving force for the migration of the boundary. The compositional and lattice parameter changes caused by the coupled diffusion and migration in a boundary region produce a stressed, alloyed layer that grows in thickness and eventually loses coherence with one or both of its parent grains when the boundary migrates beyond a critical distance (Ref 3).

When the migrating boundary has curvature (K) that is nonzero, E_{diff} contains an additional term, which, for migration away from the center of curvature, is $-2\gamma K$ where γ is the grain-boundary energy. When migration forms wavy boundaries, the magnitude of the curvature term increases with increasing migration distance; when the curvature and coherency strain terms cancel, $E_{diff} = 0$ and migration ceases. Song, Ahn, and Yoon (Ref 4), in LFM experiments in the W-Ni-Fe system, measured the curvature at which the migration direction reverses, i.e. the maximum curvature, and found it to be the right order of magnitude for canceling the effect of the coherency stress.

In this chapter observations of DIGM in the alumina-chromia system using TEM and SEM techniques are described and discussed in light of the coherency strain theory. This system was examined previously by the authors at lower temperatures where no migration was observed, and its reexamination was prompted by a private communication from D.N. Yoon, who observed DIGM in the alumina-chromia oxide scale formed on stainless steel. In the authors' work, the migration of boundaries up to 10 μm has been observed in specimens of alumina (with MgO added as a sintering aid) exposed to chromia vapor for times of the order of 1 h. This migration is a near-surface phenomenon, occurring in the top 5 to 10 μm of the specimen. The temperature range of the experiments (1300 to 1550 °C) is in the complete solid solution region of the phase diagram, well away from both the solidus, $T_m(Al_2O_3) = 2045$ °C, and the miscibility gap, which exists below 950 °C. Alumina and chromia are isostructural with space group $R\bar{3}c$.

Experimental Procedure

Slices of alumina 6 mm square were cut from a bar and polished to a 1 μm diamond finish. For each DIGM experiment, one polished slice was suspended on spacers of the same alumina above a bed of chromia powder in an alumina boat, and the system was isolated with another alumina boat and an alumina crucible (Fig. 1). Twelve DIGM experiments were carried out at temperatures ranging from 1300 to 1550 °C; the times at these temperatures varied from 30 to 120 min. One feature of the experimental setup that proved to be useful when observing the DIGM was that part of each specimen was shaded from direct, unimpeded exposure to the chromia vapor by being

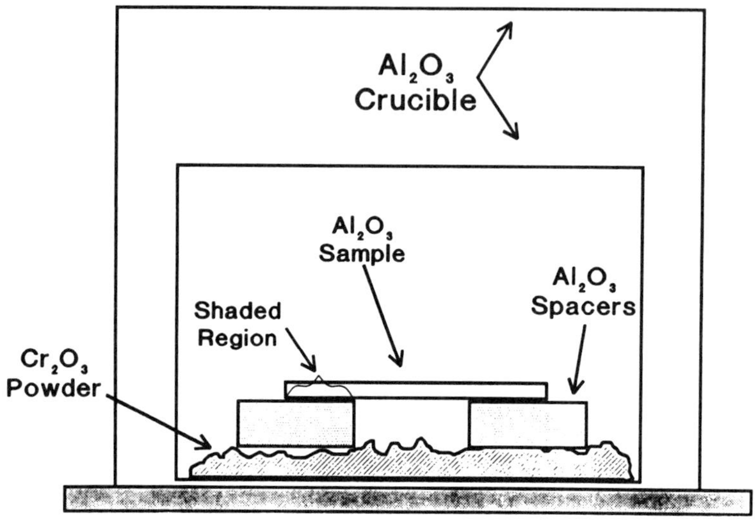

Fig. 1 Experimental setup showing specimen suspended above bed of chromia powder

placed over the alumina spacer (Fig. 1) and developed a surface morphology different from that of the unshaded parts of the specimen, as described below.

After heat treatment, the specimens were examined using optical microscopy, SEM, and energy-dispersive x-ray analysis (EDS). Parts of each specimen were then polished using 0.25 μm diamond paste and 0.05 μm alumina powder to remove the layer of chromia that had been deposited on the surfaces of most of the specimens. The specimens were then reexamined optically and in the scanning electron microscope (SEM). The crystallographic orientation of some surface features of the specimens was determined with an electron backscatter pattern (EBSP) detector and analyzer in the SEM (Ref 5). One specimen (1550 °C/0.5 h) was thinned to electron transparency by dimpling and ion milling from the back side and then examined in the transmission electron microscope (TEM).

Observations

Grain Surfaces. In all 12 DIGM experiments, the morphology of the surface of the specimens changed as the chromia vapor condensed on the alumina grain surfaces, and interdiffusion occurred. It was clear that the grain surface morphologies depended on the duration and temperature of the experiment

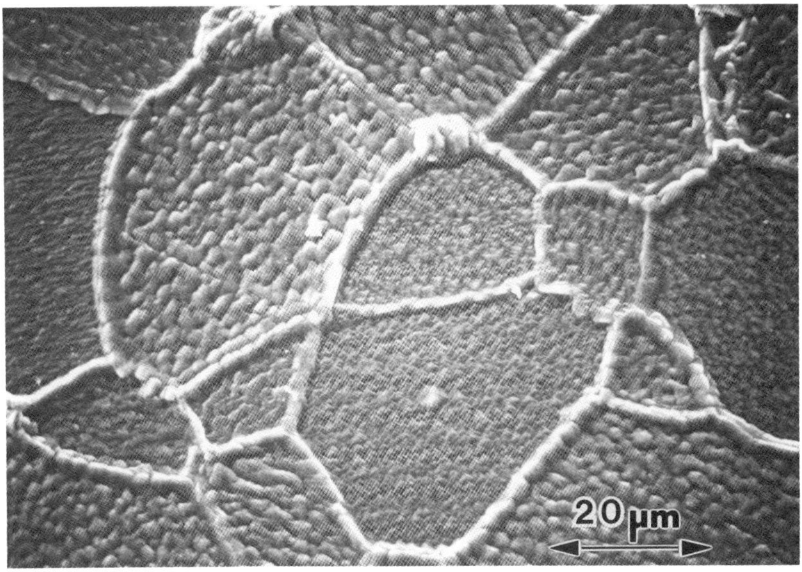

Fig. 2 Secondary electron micrograph of unshaded region of 1500 °C/1-h specimen showing variation of grain surface morphologies

and also on whether the grains were in a shaded or unshaded part of the specimen. In addition, since in some specimens adjacent grains have markedly different appearances, it was inferred that grain orientation affected surface morphology. Figure 2, an SEM micrograph from the unshaded region of the 1500 °C/1-h specimen, demonstrates significant variation in grain surface morphology. Faceting is not observed, but the grain surfaces are "rumpled," rather than planar, as before. The wavelength and amplitude of the morphological instability varies from grain to grain; the morphologies in some grains appear to be crystallographically aligned.

For short times and lower temperatures, SEM and EDS observations showed that a chromia-rich deposit formed on the surface in the form of hemispheric islands that for some grain orientations remained isolated and for others are joined to form a complete layer. From EBSP observations of grains in which the chromia coverage was incomplete, it was determined that the chromia-rich islands were epitaxial with the alumina substrate. The curvature (K) of the (0001) lattice planes in the islands was about 20 μm^{-1}, implying a dislocation density (K/b) of the order of 10^{10} cm^{-2}. In extreme cases, particularly in the shaded region of the specimen, many grains remained flat, and EDS

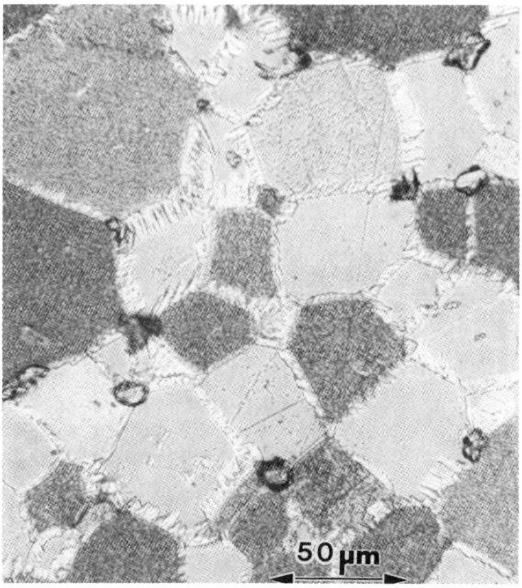

a

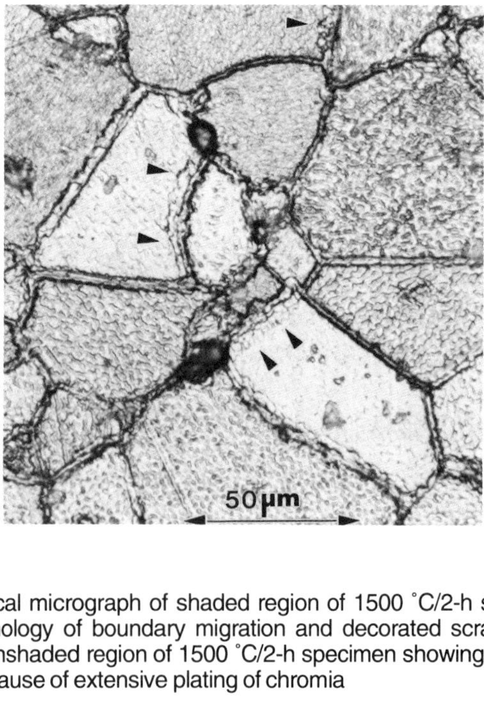

b

Fig. 3 (a) Optical micrograph of shaded region of 1500 °C/2-h specimen showing fingerlike morphology of boundary migration and decorated scratches. (b) Optical micrograph of unshaded region of 1500 °C/2-h specimen showing a different surface morphology because of extensive plating of chromia

Table 1 Surface contrast for different surface normals

Angle between specimen normal and [0001], degrees	Contrast
1300 °C for 1 h	
35.5	Light
39.9	Light
17.5	Light
11.9	Light
29.7	Light
38.2	Dark
57.61	Dark
85.9	Dark
87.6	Dark
67.7	Dark
43.3	Dark
1300 °C for 2 h	
20.2	Light
10.1	Light
12.3	Light
79.1	Dark

detected little or no Cr_2O_3 at the surface. In addition, these grains had a lighter contrast in secondary electron SEM images. Chromia is at least ten orders of magnitude higher in electrical conductivity than is alumina, and the conductivity of alumina is increased by the addition of chromia; low-conductivity areas of a specimen charge more and therefore have lighter contrast in secondary electron images than do higher-conductivity areas. Thus the lighter the contrast in secondary electron images, the lower the chromia concentration. For the 1300 °C/1-h specimen, the misorientation between the specimen normal and the *c*-axis for several grains was measured using EBSPs and correlated with the appearance of the grain (Table 1). It can be seen that the closer the grain surface was to basal orientation, the more likely it was to be of light contrast. The optical micrograph in Fig. 3(a), which is from a shaded region of the 1500 °C/2-h specimen, shows that in the grains with flat surfaces, residual scratches from polishing have become decorated with chromia.

Grain Boundaries. Clear evidence of DIGM is seen in the as-treated specimens heated to 1400 °C or higher, particularly in the shaded regions of the specimens. In Fig. 3 (a), some of the grain surfaces have lighter contrast and have remained flat; significant migration ($\leq$10 μm) has occurred, particularly into the flat grains; grain boundaries between rumpled grains have migrated

Table 2 Migration distances as a function of time and temperature

Temperature, °C	Distance, μm In 0.5 h	Distance, μm In 2 h
1500	3.4 ± 1.3	4.9 ± 2.1
1400	1.9 ± 0.4	3.0 ± 0.9

less, suggesting that in the surface layer, the solution of chromia and/or the development of the rumpled morphology tend to suppress DIGM. The migrated boundaries are in most cases extremely faceted, with a fingerlike or lathlike appearance. From EDS spectra, it is seen that the migrated grain-boundary regions have chromia concentrations of 10 to 15%, and the flat grains have essentially 0% Cr_2O_3. An unshaded area from the same specimen (Fig. 3b) has a completely different appearance. All the grains have rumpled surfaces, and the wavelengths of the surface morphologies are longer than the rumpled grains in Fig. 3(a). Evidence of DIGM is less clear, and where it can be seen (at boundaries indicated by arrows), chromia appears to have plated over the surface after the migration stopped. The grain-boundary regions in this specimen have greater than 60% Cr_2O_3, and the grains have about 30%. It should be borne in mind that the generation volume for x-rays at this voltage (20 KeV) extends to at least 1 μm in depth, and therefore an EDS spectrum represents a weighted average over this depth. The profiles of chromia concentration in these various areas are the result of the simultaneous deposition and diffusion of chromia, accompanied, in the grain-boundary regions, by grain-boundary migration; thus the profiles are expected to vary with depth.

After polishing the specimens to remove the surface layer of nearly pure chromia, optical and scanning microscopy showed that the boundary migration penetrated below the surface layer, producing the same faceted interface morphology. Figure 4(a) is an optical micrograph of the 1550 °C/1-h specimen; in the lower portion of the micrograph, the alloyed regions produced by boundary migration are lighter than the surrounding grains, and EDS indicates that the chromia layer on the grain surfaces has been completely removed. This area should be compared with the regions in which the grain surfaces were found to be chromia rich and are lighter than the grain boundaries in the micrograph. Evidently, the reflectivity of a chromia-alumina solid solution is proportional to chromia content; this correlates with the higher electrical conductivity of chromia compared to that of alumina. In the SEM, the contrast is reversed, and an SEM image of the polished surface of the 1550 °C/0.5-h specimen (Fig. 4b) shows the migration regions to be darker than the

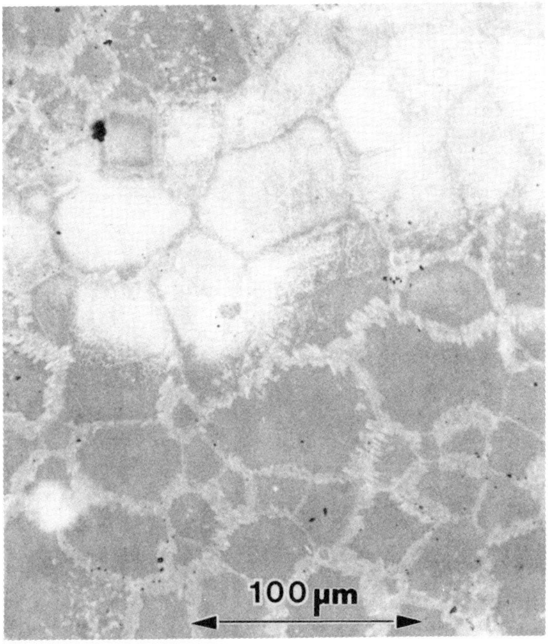

a

b

Fig. 4 (a) Optical micrograph of 1500 °C/1-h specimen after polishing. (b) Secondary electron micrograph of 1550 °C/0.5-h specimen

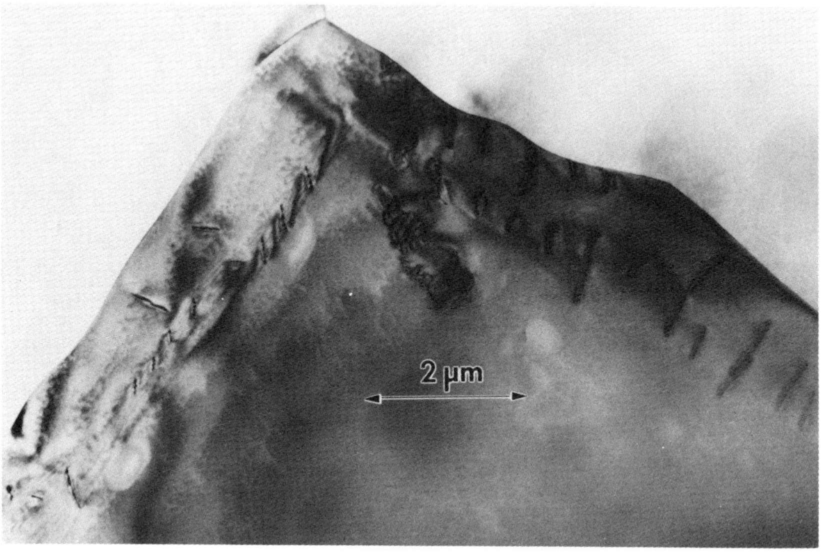

Fig. 5 Bright-field TEM micrograph of 1550 °C/0.5-h specimen

grains. The micrographs in Fig. 3 and 4 indicate that the migration distance varies considerably from boundary to boundary. Migration distances were measured for four specimens: 1400 °C/0.5-h, 1400 °C/2-h, 1500 °C/0.5-h, and 1500 °C/2-h specimens. The boundaries fell into two categories, those in which the migration distance was relatively uniform along the length of the boundary, and those in which there was significant variation in distance (by factors of two or three in some cases). The data for the boundaries with uniform migration are presented in Table 2.

TEM observations of a thin foil taken from the 1550 °C/0.5-h specimen revealed many of the classic features of DIGM: wavy boundaries, alloyed regions, and misfit dislocations. In most cases, an array of dislocations approximately parallel to each grain boundary was observed, with no detectable misorientation across it. For each boundary region, the chromia concentration measured by EDS between the current boundary position and the dislocation array was typically 1 to 3%, and the concentration dropped sharply on either side to an undetectable level. Thus, it is deduced that very close to the original position of most of the grain boundaries exists an array of misfit dislocations accommodating the change in alumina lattice parameters caused by the solute.

Table 3 Misfit dislocation spacing

Spacing (h), nm	Cr_2O_3 concentration (q), %	Burgers vector (b), nm
480	0.027	0.58
460	0.018	0.38
620	0.020	0.55
440	0.030	0.59
560	0.030	0.38

Figure 5 is a bright-field image of the corner of a grain in which the grain boundaries have migrated a distance of 0.8 to 1.2 μm from the center of the strongly diffracting grain. The interface between the pure and alloyed alumina is clear not only from the misfit dislocations, which run from the top to bottom surfaces of the foil (and are close to end on in the micrograph), but also from the complex distribution of fringes, which are particularly visible in the boundary region on the left of the micrograph. The misfit dislocations in this array fall into two different sets, distinguished by their line directions. One set (the majority set) is regularly arranged at the interface, although only one of the minority set has incorporated into the interface structure, where it is associated with a step. The chromia concentration in the alloyed region of the left-hand boundary is 1.8%, and the average dislocation spacing is 0.45 μm. Table 3 summarizes the average dislocation spacing and chromia concentration for five boundaries.

Discussion

These observations of DIGM in the alumina-chromia system are similar to observations in metal systems such as Au-Ag. The most obvious differences lie in the fingerlike nature of the grain-boundary morphologies, created by DIGM, and in the surface rumpling that occurred in all experiments. The observations of surface rumpling that vary as a function of temperature, orientation, and solute supply (shaded or unshaded) all point to a nucleation and growth mechanism of epitaxy, with the density of nucleation sites being a function of surface orientation. Surface diffusion and vapor transport in this system are very rapid at these temperatures such that a radical of CrO^+ or CrO_2^- condensing on the alumina surface rapidly diffuses to a site of low energy such as a surface step, scratch, preexisting epitaxial island, grain boundary, or dislocation.

The tendency of both free surfaces and grain boundaries in ceramics to be faceted has been noted (Ref 6, 7). The surface morphologies developed here are not equilibrium shapes, since the surface layer is thickening and changing its stress state throughout the experiment. Hence the regular surface faceting observed in, for example, equilibrated NiO-MgO solid solutions (Ref 6) is not expected here. In contrast to the grain-boundary morphologies seen at and near the surface, the TEM observations presented here do not show faceting of the grain boundaries, probably because the TEM foil came from several microns below the original specimen surface, although the exact depth is not known.

The variation of average migration distance with time and temperature has the expected trends: increasing the temperature or time increases the migration distance. For each experiment some migration may have occurred as the specimen was heated to the experimental temperature, but allowing for this effect, the data in Table 2 show that migration at the surface is not linear with time. The coherency strain theory predicts a constant migration rate only if there is no change in grain-boundary length or curvature. When the curvature and grain-boundary length increase, the rate should decrease with time.

As discussed above, the driving force for DIGM for the noncubic crystal classes is a complex issue. An algorithm for calculating the driving force as a function of boundary normal, solute concentration, and distortion tensor for any crystal class has been developed and will be published soon. The elastic energy density, $f_{coherent}$, can be expressed as $A(q_b - q_o)^2$, where A is dependent on boundary normal. Symmetry arguments show that for alumina-chromia, which is rhombohedral, A is the same for all prismatic planes since the distortion tensor is isotropic for boundary normals in the basal plane. For prismatic planes at room temperature, A = 1.10 GJm^3, and for the basal plane at room temperature, A = 1.01 GJm^{-3}; the difference, 900 MJm^{-3}, indicates that considerable driving force for boundary migration can be developed.

In simple terms, the solute-induced elastic strain is approximately ηq, where η is the average of the distortion tensor components, η_{11}, η_{22}, and η_{33}. For alumina-chromia, $\eta_{11} = \eta_{22} = 0.041$, $\eta_{33} = 0.049$, and $\eta = 0.044$. A solute concentration of q in the basal plane produces a stress of Aq/η, which for $q = 0.03$ gives a stress of 680 MPa. When the thickness of the diffusion zone is greater than a certain critical value, these high stresses lead to the formation of misfit dislocations (Ref 4). Because the boundary planes in the TEM foil studied are approximately perpendicular to the plane of the foil, and the dislocation spacings are small compared to the foil thickness, only a limited portion of each misfit dislocation network is visible. However, correlations have been made between the Table 3 data on misfit dislocation spacing (h) and solute concentration (q) in the alloyed region. Two assumptions can be made:

first, that all of this elastic strain is relieved by dislocations and, second, that the Burgers vectors of these dislocations are parallel to the grain-boundary plane. Based on these assumptions, the magnitude of the dislocation Burgers vector (b) is given by b = $h\eta q$. Calculated values of b are listed in Table 3 and should be compared with the two smallest Burgers vectors in alumina, $1/3\langle 2\bar{1}\bar{1}0\rangle$ and $1/3\langle 1\bar{1}00\rangle$, with magnitudes 0.476 nm and 0.512 nm, respectively. If the relief of elastic strain is incomplete, the calculated b will be greater than these values; if the Burgers vectors are not parallel to the interface, the calculated b will be less than these values. The degree of agreement between calculated and allowed values of b is within 20%, which is good. The misfit dislocations in the left-hand boundary of Fig. 5 are arranged in facets and steps. The analysis of the relationship between the grain-boundary plane and the faceting of misfit dislocation arrays is the subject of future work.

Acknowledgment

The authors gratefully acknowledge the useful discussions held with Dr. W. Craig Carter and partial financial support from the U.S. Office of Naval Research.

References

1. C.A. Handwerker, in *Diffusion Phenomena in Thin Films and Microelectronic Materials*, D. Gupta and P. Ho, Ed., Noyes Publication, 1988, p 245-322
2. A.H. King, *Int. Mater. Rev.*, Vol 32, 1987, p 173-189
3. W.A. Jesser and D. Kuhlmann-Wilsdorf, *Phys. Status Solidi*, Vol 19, 1985, p 95-105
4. Y.D. Song, S.T. Ahn, and D.N. Yoon, *Acta Metall.*, Vol 33, 1985, p 1907-1910
5. D.J. Dingley, *Scanning Electron Microscopy (USA)*, Part 2, Scanning Electron Microscopy, Inc., 1984, p 569-575
6. C.A. Handwerker, M.D. Vaudin, and J.E. Blendell, *J. Phys.*, Vol 49 (C5, No. 10), 1988, p 367-373
7. M.D. Vaudin, M. Rühle, and S.L. Sass, *Acta Metall.*, Vol 31, 1983, p 1109-1116

Section VI:

Electrical Properties of Interfaces

16

Electrical Properties of Grain Boundaries

D.A. Smith
IBM, T.J. Watson Research Center
P.O. Box 218
Yorktown Heights, NY 10598
C.S. Nichols
Department of Materials Science and Engineering, Bard Hall,
Cornell University, Ithaca, NY 14853

The electrical properties of grain boundaries have been shown to exhibit a range of dependencies on the crystallography and chemical composition of the grain boundary. The normal resistivity of aluminum depends systematically on the dislocation density, as does the critical current in superconducting oxide, $YBa_2Cu_3O_{7-\delta}$. The diversity of dielectric breakdown fields in zinc oxide-based varistors can be related to the intergranular phases present. Remarkably, there does not appear to be a grain misorientation dependence of barrier height in bicrystals of silicon. In contrast there is a relation between structure and minority carrier lifetime as measured by electron beam induced conductivity (EBIC) on grain boundaries in silicon.

Grain boundaries and grain-boundary-dependent phenomena are of great importance in a wide variety of electronic applications. The underlying principle is that grain boundaries offer a potential barrier to majority carrier transport and that this barrier can be tailored, by control of the structure and composition of the boundary, to give desirable properties (Ref 1). Examples are zinc oxide-based varistors, where the breakdown voltage depends on the grain size and the grain-boundary composition (Ref 2), titanium oxide- or tin oxide-based sensors for gas, strain, or photon detection (Ref 3), where the conductivity depends on the grain-boundary barrier properties modulated by some external stimulus; and strontium titanate capacitors, where the capacitance is governed by the width of the depletion region at the grain boundary (Ref 4). In other applications the issue is to minimize rather than exploit the special properties of grain boundaries. Examples are the control of texture, analyzed in Ref 5, to optimize the critical current in high-T_c superconductors and the degenerate doping of polysilicon for gate electrodes in field effect transistors. A proper understanding of such behaviors and their further development and more refined application can be achieved through the elucidation of a correlation between the structure and properties of, first, individual interfaces and, second, polycrystalline materials.

It is well recognized that the existence of five macroscopic degrees of freedom renders impracticable the characterization of the whole spectrum of boundary structures even for a particular material. Instead the direction of research efforts has been to attempt to clarify general rules through the study of particular boundaries. This approach has enjoyed only limited success. For low-angle grain boundaries, the principles are established. Even though it is not always obvious what dislocations will occur, it is indisputable that a dislocation network in which the overall Burgers vector is normal to the rotation axis will be the essence of the grain-boundary structure (Ref 6). Furthermore, it is relatively straightforward to deduce the structure of a mixed boundary from those of the corresponding tilt and twist boundaries. This is done by projecting the dislocation structures of the simple boundaries onto the desired plane and then reconstructing unstable nodes (Ref 6-8). No such procedure can be developed for high-angle grain boundaries. This is because, in contrast to the low-angle case in which crystal coordination is preserved between dislocations, the variation of the boundary structure between the dislocations in a high-angle boundary as the orientation of the boundary plane changes is not known. High-resolution transmission electron microscopy and computer simulation are both limited to relatively simple structures so that the nature of the arbitrary boundary remains inaccessible to structural characterization. However, irrespective of the particular material, the notion of some kind of ordered structure in which defects may be embedded is prevalent for interfaces (Ref 9-11).

The conclusion of both modeling and experimental studies (Ref 12-20) is that the grain-boundary core structure is as crystal-like as is possible; thus in metals, low-energy relaxed structures are associated with maximal coordination, which is equivalent to the minimization of the grain-boundary excess volume (Ref 21); in semiconducting materials, reconstruction at grain boundaries minimizes the density of dangling bonds and antisite defects (Ref 22, 23); in ionic materials, the guiding principle is the preservation of charge neutrality concomitant with a coordination shell of unlike species (Ref 24). The properties of selected individual interfaces and polycrystalline materials have also been measured. In some instances both the structure and the properties of particular interfaces are known. Even in these cases the chemical composition of the interfacial region is much less well characterized. For polycrystalline materials, only recently has attention been focused on the varying nature of the individual boundaries in a polycrystalline ensemble.

Although the challenge of establishing a structure-property relationship for interfaces is daunting, it is fortunate that both the properties and the atomic structure of a number of grain boundaries are known, and the bulk of what follows here is focused on this subset of all possible interfaces.

Primary Dislocation Model of Interfaces

Numerous observations have been made of interfacial defects, together with a taxonomy of the admissible kinds of defect (Ref 25). However, for the majority of the boundaries where measurements of electrical behavior have been made, the low-angle boundary model of Read and Shockley, extended to what at first might seem rather large rotation angles, gives an appropriate description of structure.

The Read-Shockley dislocation model for low-angle grain boundaries (Ref 6) establishes a correlation between dislocation density (d) and misorientation as follows:

$$S = (u \times V)\theta \qquad \text{(Eq 1)}$$

where S is the Burgers vector (b) traversed by a probe vector V lying in a grain boundary with a rotation angle θ about a rotation axis u which for small angles reduces to:

$$d = \theta / b \qquad \text{(Eq 2)}$$

According to modeling and experimental studies, these equations seem to apply to all [001] symmetrical-tilt boundaries and to [110] symmetrical-tilt boundaries with a misorientation angle less than 26.5°.

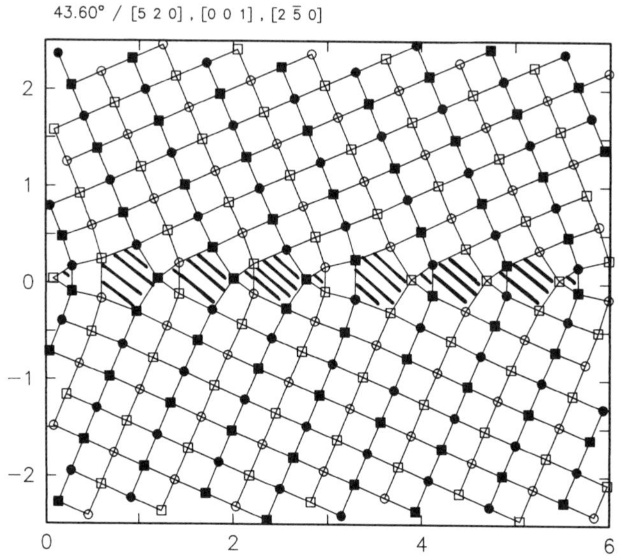

Fig. 1 An (001) projection of the calculated relaxed structure for the Σ = 29, 43.60°/[001] (520) symmetrical-tilt boundary. The four types of symbols represent the ...abcda... stacking sequence of the (004) atom planes. The hatched areas indicate the dislocation cores. Courtesy of A.A. Levi

In a cubic material, for example, silicon, even when the disorientation is as large as possible for an [001] axis, i.e., 45°, the coordination polyhedra corresponding to the cores of lattice dislocations are readily discernible. This can be seen in Fig. 1, which is an (001) projection of the calculated relaxed structure for the Σ = 29, 43.60°/[001] (520) symmetrical-tilt boundary. The four types of symbols represent the abcda stacking sequence of the (004) atom planes. The lines joining the various symbols represent the projections of interatomic bonds; almost invariably these bonds are inclined to the plane of the projection. The numbers on the frame enclosing the figure express the length scale in units of the lattice parameter. The Miller indexes above the figure identify the boundary normal, the rotation axis, and the boundary trace, in that order. The core region of the grain boundary consists almost entirely of dislocation cores (hatched area) with occasional groups of crystal-coordinated atoms. The dislocation core consists (in projection) of a five-sided figure that is joined, apex to apex, to a three-sided figure.

Relaxed structures of the kind described above may be used as a first approximation (essentially neglecting the energy of antisite defects) to the corresponding grain boundaries in GaAs by simply replacing alternate silicon atoms with gallium and arsenic. If in addition the number of antisite defects is minimized, plausible structures result. High-resolution transmission electron microscopy observations (Ref 12, 13) and x-ray scattering experiments (Ref 26) reveal features that can be interpreted as an almost but not quite perfect periodic array of lattice dislocations, giving a Burgers vector density that accommodates the total disorientation. These observations are consistent with the results of computer calculations. Similar conclusions can be drawn from studies of ⟨001⟩ and ⟨110⟩ tilt grain boundaries in metals and ionic materials. The same principles apply to ceramic high-T_c superconductors but may break down at smaller misorientations because of the extension of the dislocation core by dissociation.

Properties

The task of developing a synthesis relating structure to properties of individual interfaces is simplified by recognizing that invariably the behavior in any particular case involves a core process, possibly in series with an interaction with a long-range field of the boundary. The electrical properties of grain boundaries belong to the class of phenomena in which the interface is static and, as will be explained for the simple interfaces discussed here, the behavior of the boundary is the sum of the contributions of the behaviors of the individual defects. There seems to be a general trend in grain-boundary phenomena in which essentially two classes of behavior are exhibited. The two classes are associated with low-angle and special boundaries on the one hand and general boundaries on the other. These generalizations are shown in Fig. 2(a) and 2(b) for a low-angle boundary and the vicinity of a special boundary. There are two kinds of dependency of property on structure. There is always an extremum in behavior for special and low-angle boundaries (for the low-angle case, the singularity is at a misorientation of 0°, i.e., the perfect crystal). The extremum may involve either a local minimum or a local maximum. Energy, electrical resistance, and segregation are minimized at both special and low-angle boundaries (Ref 27-29). Critical current and dielectric breakdown field are greatest for special boundaries (Ref 30-32). Key dependencies that provide a link between many of these properties are the grain-boundary excess energy and the association between energy and excess volume. For the grain boundaries relevant to this discussion, the link is even more explicit and is the dislocation density.

More quantitative expressions can be developed to relate the behavior of defects in interfaces to the properties of the interface as a whole. Such developments depend on a relationship of the form:

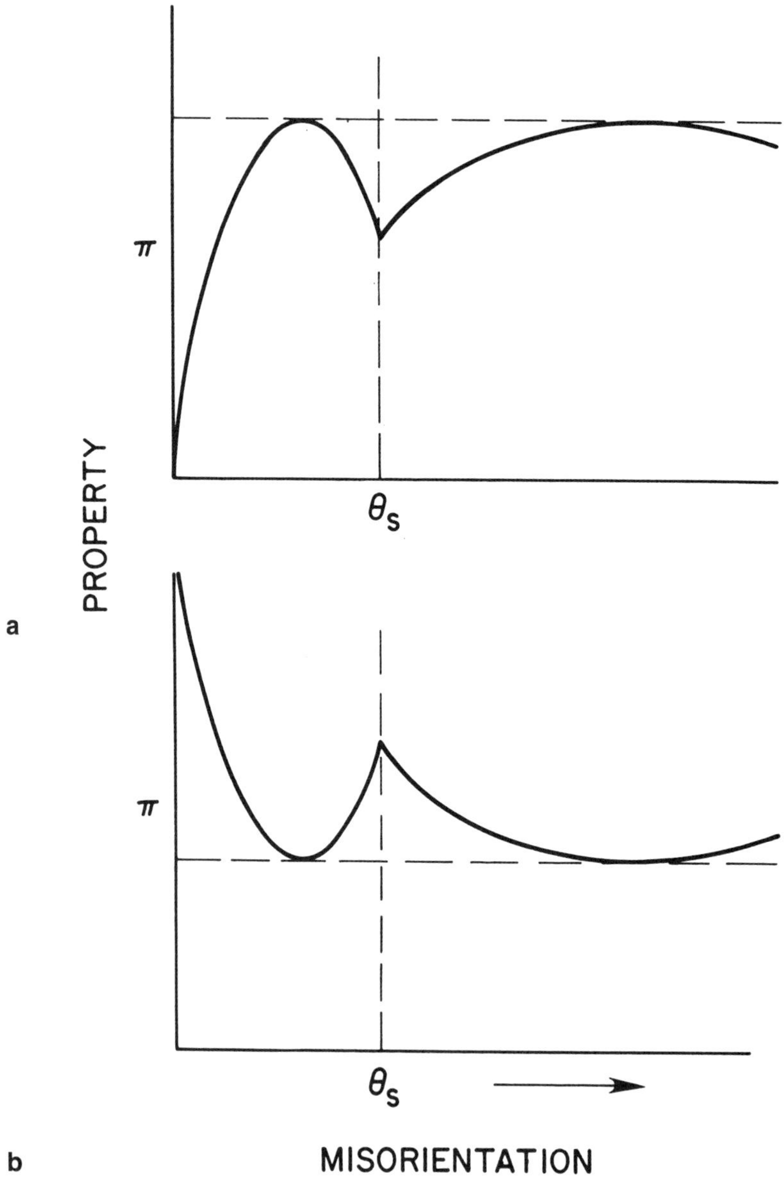

Fig. 2 The schematics in (a) and (b) show the behavior of properties such as energy, electrical resistance, and segregation, which are minimized at both special and low-angle boundaries, and critical current and dielectric breakdown field, which are greatest for special boundaries. The behavior of the general boundary is indicated by the dashed line

Table 1 Specific resistivities in units of $10^{-25}\Omega\,m^{-3}$ for edge and screw dislocations for misorientation $\theta < 22^\circ$ and $\theta > 22^\circ$

Angle	Screw	Edge
$\theta < 22^\circ$	1.2	2.6
$\theta > 22^\circ$	0.91	1.0

$$\Pi = \pi_d \rho \quad \text{(Eq 3)}$$

where Π is the property of the interface containing a density of defects ρ each with properties π_d. The reader will appreciate that Eq 3 has the same form as the Read-Shockley expression for the elastic energy of a low-angle grain boundary, where, neglecting the core term, π_d is $Gb^2/4\pi(1-\upsilon)\ln\theta$ and ρ is θ/b.

In 1990, Nakamichi reported the measurement of the resistivity of individual boundaries in {001} textured zone refined aluminum (Ref 28); the crystallographic character of each boundary was measured by back reflection Laue photography and trace analysis. It was found that an expression of the same form as Eq 3 could account for all the observations. The explicit values of the coefficients π_d are given in Table 1. These are specific resistances of edge and screw dislocations. It should be noted that the coefficients change discretely according to the rotation angle. The change occurs at a misorientation that may relate to the disappearance of good crystal between the dislocation cores.

An alternate expression, applicable to cases in which the material between the defects is not inactive, is a law of mixtures in which the interface property Π is written as:

$$\Pi = \pi_d X + (1 - X)\pi_i \quad \text{(Eq 4)}$$

where X is the effective area fraction of defects and π_i represents the behavior of the interface between the defects.

The misorientation dependence of J_c in high-temperature superconductors (Ref 28) is relatively well described by Eq 4 and Fig. 2(b), at least for certain simple boundary configurations. Some insight into the extent of the perturbation offered by dislocations is revealed by the high-resolution electron micrograph shown in Fig. 3. The result of the climb dissociation of the edge dislocations with Burgers vector magnitude c, 1.17 nm, into three c/3 partial dislocations (Ref 33) is to offer a perturbation to the superconducting pairs of electrons that extends for about 5 nm in the boundary plane. Accordingly, each dislocation is associated with a slab of material that is likely to have a low critical current, and these slabs become continuous, at a misorientation

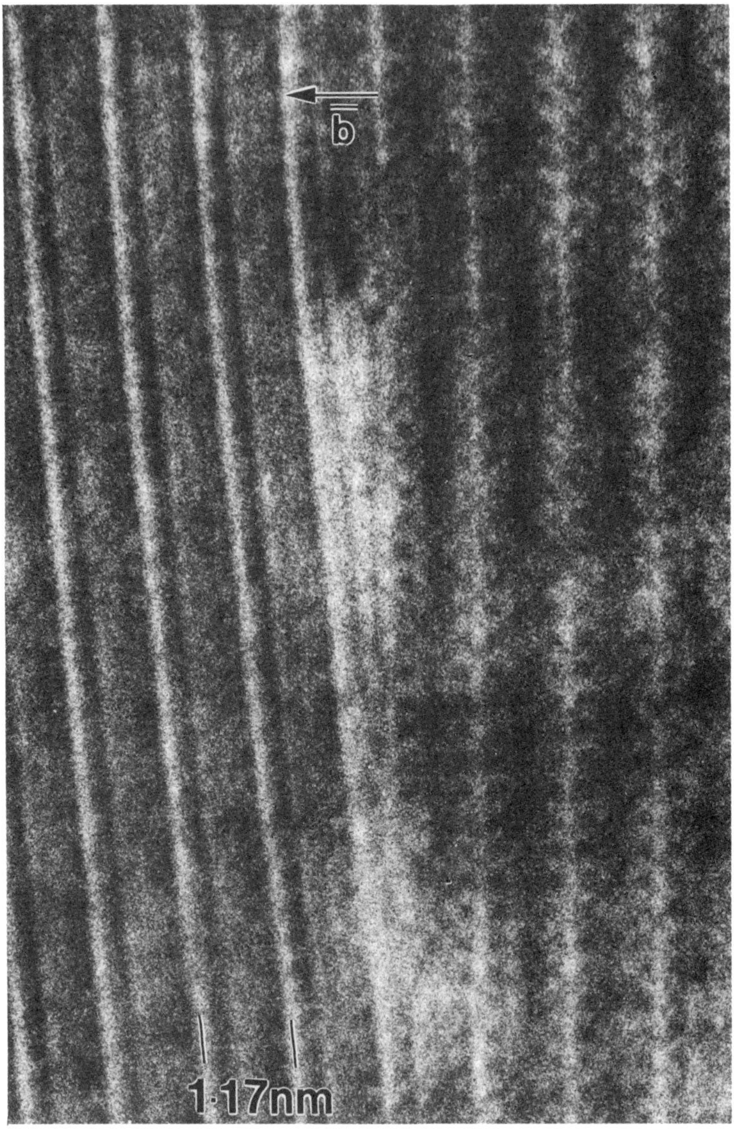

Fig. 3 Extended core of a c dislocation in the yttrium barium cuprite high-temperature superconductor. Courtesy of Dr. M.F. Chisholm and *Philosophical Magazine*

of about 12°(applying Eq 2), which is of the same order as the observed saturation of the decrease in critical current as a function of rotation angle (Ref 30).

The properties of grain boundaries in polysilicon, a representative semiconductor, are surprisingly little understood and evidently very sensitive to dopants, thermal processing, and impurities (Ref 34-36). Grain boundaries in semiconductors are frequently found to offer large resistances to the passage of majority carriers. It is generally accepted that localized defect states in the forbidden gap trap carriers give rise to a potential barrier resembling two Schottky barriers back-to-back. It has been suggested that such deep defect states are of extrinsic origin, meaning that they are associated with some physical or chemical process occurring preferentially at the boundary, segregation of impurities, or gross precipitation, as opposed to arising intrinsically from dangling or distorted bonds (Ref 37). Subsequent work has treated the limits of low dopant concentration in which the transport is diffusion limited and the high dopant concentration limit is barrier limited (Ref 38). By analogy with other grain-boundary properties, the barrier height was expected to be structure sensitive. However thus far three separate investigations (Ref 36, 39, 40) have suggested that *clean* grain boundaries are *not* electrically active in the sense of offering a barrier to the transport of majority carriers. The example discussed below illustrates the principles and findings common to each investigation.

Grain boundaries in boron-doped silicon on insulator thin films produced by zone melt recrystallization (ZMR) were characterized by measuring current-voltage (I-V) characteristics. The ZMR process has the effect of purifying the sample in the manner of zone refining and producing a range of boundaries in a configuration convenient for electrical characterization. The material is formed as a ribbon and is comprised of ribbonlike grains. The films have [001] texture and consist of grains that are elongated in the growth direction and separated by tilt boundaries.

The doping level was varied in the range 10^{15} to 10^{18} cm^{-3}; no electrical activity was detected at any boundary (the tilt angles were in the range of 0 to 25°). The resistance of all specimens agreed with that calculated from the bulk resistivity, and current was linearly dependent on voltage, while for a given current the voltage varied linearly along the specimen. Figure 4 shows a logarithmic plot of current density versus voltage measured on a sample of bulk resistivity of 0.4 Ω cm. The circles are the measured data (the error is estimated to be considerably smaller than these symbols), while the two solid curves were calculated assuming the presence of a single active grain bound-

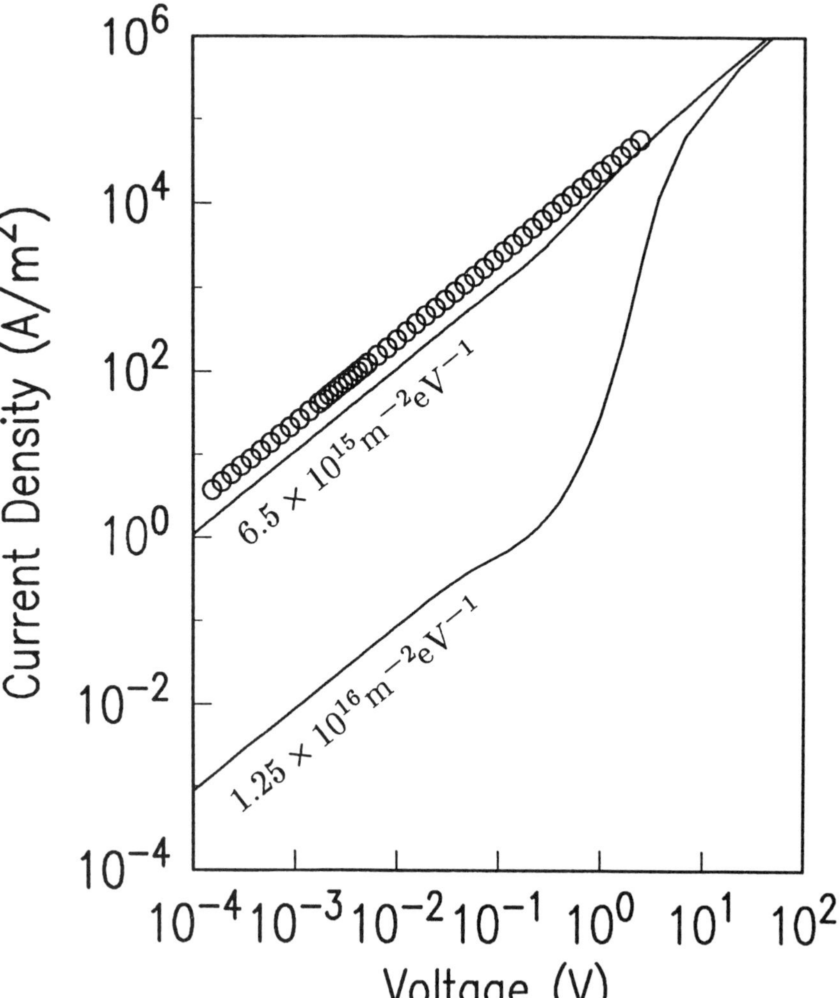

Fig. 4 Current voltage characteristics for specimen doped at 10^{16} cm^{-3}. Circles are experimental data; curves were calculated assuming a single active boundary with defect state densities 6.5×10^{15} and 1.25×10^{12} cm^{-2}eV^{-1}. Courtesy of Dr. P.V. Evans

ary in the specimen with a uniform density of defect states of the numbers shown. The calculated curves show a low-voltage ohmic region, followed by a rise in current as the barrier collapses and then a flat band condition. From

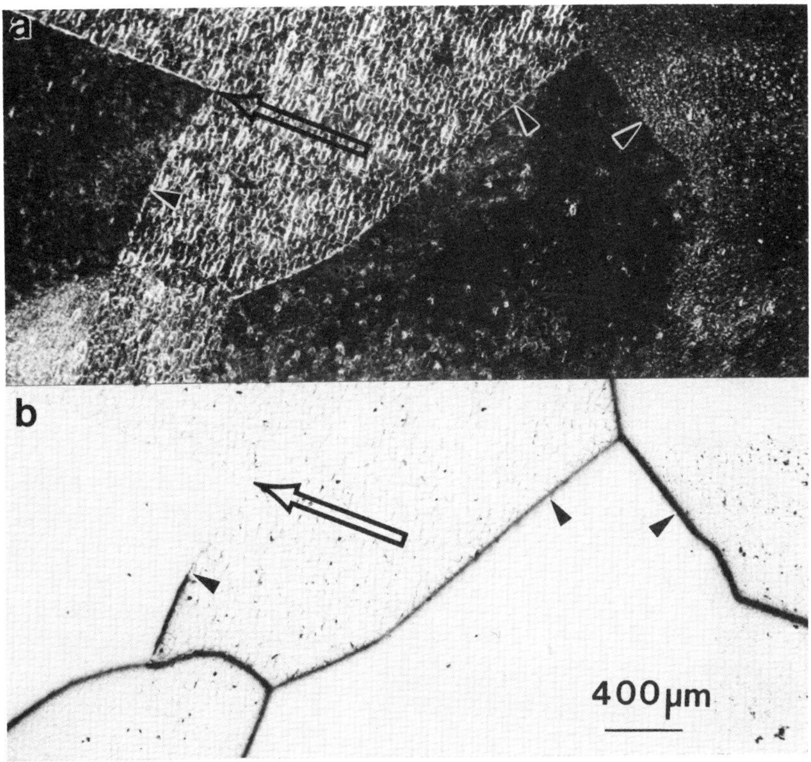

Fig. 5 A complementary pair of scanning electron microscope images of polycrystalline silicon recorded (a) in the secondary electron mode and (b) in the EBIC mode. Note the variations of the EBIC contrast within and between boundaries. Courtesy of Dr. S.F. Nelson

examination of the derivatives of the current voltage curves, an estimated upper limit of the density of defect states at this typical grain boundary is 10^{11} cm^{-2}. Typical reported-literature values are in the range from 10^{12} to 10^{13} cm^{-2} (Ref 37).

It is not clear at present whether the disappearance of electrical activity that accompanies the ZMR process is the consequence of a change in the crystallographic character of the grain boundaries or a zone-refining phenomenon or both.

The results of EBIC measurements provide a counterexample in which a property is structure dependent but at the same time the particular structural feature may be an extraneous part of the structure. As Fig. 5 shows, the EBIC signal varies not only from boundary to boundary but also within a particular boundary where the structure would be expected to be similar throughout. It is likely that a grain boundary contains extraneous dislocations and/or local compositional fluctuations. The contributions of extraneous defects that are not in local equilibrium and of compositions that fluctuate are generic issues and represent a recurrent complication in the establishment of structure-property relationships for interfaces. The difficulty is that nonequilibrium defects depend on the processing history of the material rather than on the crystallographic nature of the interfaces involved.

Collective Behavior

Although the properties of single interfaces are a key to such fields as bandgap engineering (Ref 41, 42) and the supermodulus effect (Ref 43), for many materials the main issue is understanding the behavior of a polycrystal containing a multiplicity of different interfaces, not a bicrystal. This involves, in principle, consideration of the coupling of grain boundaries to other grain boundaries, of grain to grain, and of grain-boundary properties to those of grain interiors.

Previous work on polycrystalline materials has sought to establish a connection between a single measure of the grain size (d), and the physical properties of the polycrystal. Fuchs (Ref 44) derived an expression in which the resistivity is proportional to the inverse of the grain size. The boundary-to-boundary variations of behavior were neglected in developing the model. The description of grain-boundary properties in terms of a single-length scale gives a good description of the relationship between grain size and the normal resistivity of metals (Ref 45). Nevertheless a full synthesis requires the incorporation of boundary-to-boundary variations in properties and the coupling of processes in the grain boundaries and the grain interiors. For manganese zinc ferrites doped with calcia, the contribution of the grain boundaries to the electrical resistivity is of the kind sketched in Fig. 1a in Ref 31. The resistivity of a polycrystalline sample containing two kinds of boundaries can be modeled using the insights of percolation. For example, if the proportion of special boundaries exceeds 0.13 in a three-dimensional random polycrystal, these special boundaries form a spanning cluster and thus offer a low resistance path. For the two-dimensional case, a spanning cluster exists when the proportion of special boundaries exceeds 0.35. A similar approach applies to

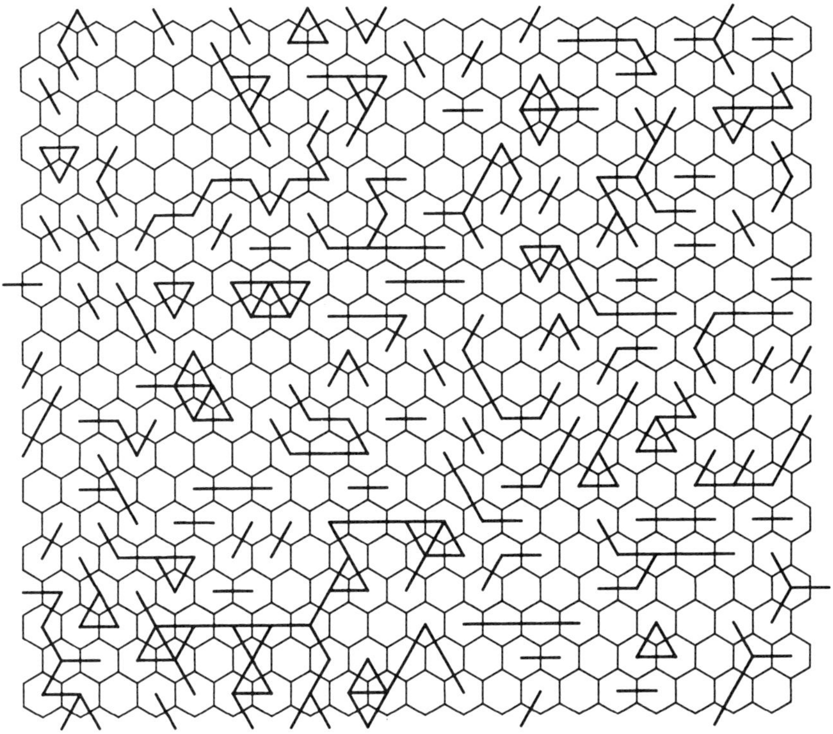

Fig. 6 A sample microstructure showing the hexagonal grains and the clusters of low-angle boundaries. Grains in the same cluster are connected by a line through the centers of the grains. A smaller sample than is generally used in the simulations is shown so that the bonds are visible

the problem of the varistor behavior of polycrystalline materials, where the population of grain boundaries can be differentiated by dividing into a number of classes. For the case of a zinc oxide-based varistor with 10 mol% additives (oxides of bismuth, antimony, and cobalt) four classes of boundaries were identified (Ref 2). These are characterized by barrier heights that range between 0.4 and 3.6V. The barrier heights were shown to correlate with the nature of the phases present at the boundaries. The concept of differentiation of boundaries into a small number of classes lends itself to modeling and has been successfully applied to the determination of the critical currents of textured polycrystalline superconductors (Ref 5).

An initial structure consisting of randomly oriented, equiaxed hexagonal grains is generated by the computer. The grain boundaries are then placed in two distinct categories depending on their misorientation angle, which for present purposes are low angle and high angle. The cutoff angle (θ_c) between these two classes is a computational variable, the value of which can be selected according to an appropriate criterion, which could be, for example, energy or excess volume. The result of the classification of grain boundaries is a new feature, the cluster of grains bounded by a particular kind of interface, and new length scales such as the cluster size and its radius of gyration.

An example of a microstructure and the clusters that comprise it is shown in Fig. 6, where the centers of grains with a common low-angle boundary are connected. This particular structure was generated for $\theta_c = 15°$ in an $\langle 001 \rangle$ fiber-textured, cubic polycrystal. It is notable for the ramified nature of the clusters and their interpenetration. For this particular value of θ_c, the microstructure is close to (but below) the percolation threshold. Clearly in the vicinity of the threshold the dependence of properties on grain size deviates from the simple power laws that apply far from the threshold.

Discussion

The preceding description of some key aspects of the electrical properties of interfaces in materials provokes some general observations and discussion. It is impractical to expect that there will ever be a comprehensive body of data from which information concerning the characteristics of particular grain boundaries in specific materials may be obtained in the same manner as from, for example, tables of physical constants. There are too many variables for the collection of such a body of knowledge to be feasible, quite apart from the issue of how such a compendium might be used in practice. The research strategy has been to study particular, relatively simple interfaces such as tilt boundaries with low index rotation axes with the goal of identifying generic characteristics of boundaries. However, this approach places too much emphasis on boundaries that may be described in terms of primary dislocations, and also it is unable to address the question of the nature of the "general" boundary. Modeling supports the view that tilt boundaries are atypical and suggests that as the crystallographic parameters change, less symmetrical configurations exhibit less variety of behavior than do tilt boundaries (Ref 20). The unifying concept seems to be that the excess volume is minimized at grain boundaries in metals and that this quantity is not strongly dependent on rotation angle or axis except for symmetrical-tilt boundaries. The implication of this insight and its extension to nonmetals is that there exists the potential for a considerable advance in the quantitative description of the properties of a polycrystalline material through the simplification resulting from classifying boundaries as twins and low-angle boundaries, boundaries comprised of periodic arrays of primary dislocations, or all others. A somewhat similar proposition, which has been made elsewhere (Ref 35), leads directly to the possibility of engineering the properties of materials (Ref 46).

It can be argued that segregation effects have been neglected at the expense of structural studies. This is substantially a result of the later development of high-spatial-resolution techniques for chemical characterization, but impressive progress is currently being achieved by so-called quantitative chemical lattice imaging (Ref 47). On the same note, it has been difficult to determine the structure of nonperiodic objects by conventional techniques. However, scanning tunneling microscopy and Fresnel imaging (Ref 48) in the transmission electron microscope offer the potential for addressing these issues, which are essential to a broader understanding of grain-boundary mediated behavior.

Conclusions

The electrical properties of grain boundaries exhibit a range of dependencies on grain-boundary crystallography and chemical composition. For the particular case of boundaries that can be described as arrays of primary dislocations, the properties scale with the dislocation density except for the barrier height in silicon. Grain boundaries and grain-boundary-dependent phenomena are of great importance in a wide variety of electronic applications and can either be exploited or minimized according to the particular application by controlling the chemical composition of the boundary, the boundary misorientation, and the crystallographic texture.

References

1. F. Greuter and G. Blatter, *Semicond. Sci. Technol.*, Vol 5, 1990, p 111
2. E. Olsson and G.L. Dunlop, *J. Appl. Phys.*, Vol 66, 1989, p 3666
3. L.L. Kasmerski, Ed., *Polycrystalline and Amorphous Thin Films and Devices*, Academic Press, 1980
4. N. Yamoaka, *Ceram. Bull.*, Vol 65, 1986, p 698
5. C.S. Nichols and D.R. Clarke, *Acta Metall.*, submitted for publication
6. W.T. Read and W. Shockley, *Phys. Rev.*, Vol 78, 1950, p 275
7. W. Bollmann, *Crystal Defects and Crystalline Interfaces*, Springer-Verlag, 1970
8. J.N. Calvert, P.M. Hazzledine, and D.A. Smith, *Scr. Metall.*, Vol 11, 1977, p 513
9. W.A.T. Clark and D.A. Smith, *Philos. Mag.*, Vol 38, 1978, p 367
10. R.W. Balluffi, Y. Komem, and T. Schober, *Surf. Sci.*, Vol 32, 1972, p 68
11. C.P. Sun and R.W. Balluffi, *Philos. Mag. A*, Vol 46, 1982, p 49

12. O. Krivanek, S. Isoda, and K. Kobayashi, *Philos. Mag.*, Vol 36, 1977, p 931
13. W. Krakow and D.A. Smith, *Ultramicroscopy*, Vol 22, 1987, p 47
14. K.L. Merkle, *J. Phys.*, Vol 51, 1990, p C1-251
15. J.L. Rouviere and A. Bourret, *J. Phys.*, Vol 51, 1990, p C1-329
16. H. Ichinose and Y. Ishida, *Philos. Mag.*, Vol 43, 1981, p 1253
17. H. Ichinose and Y. Ishida, *J. Phys. Coll.*, Vol 46, 1985, p C4-39
18. D.A. Smith, V. Vitek, and R.C. Pond, *Acta Metall.*, Vol 25, 1977, p 475
19. P.D. Bristowe and A.G. Crocker, *Acta Metall.*, Vol 25, 1977, p 1363
20. R.C. Pond, V. Vitek, and D.A. Smith, *Acta Crystallogr. A*, Vol 35, 1979, p 689
21. D. Wolf, *J. Mater. Res.*, Vol 5, 1990, p 1708
22. J. Hornstra, *Physica*, Vol 25, 1960, p 409
23. J.T. Wetzel, A.A. Levi, and D.A. Smith, *J. Appl. Phys.*, Vol. 69, 1991, p 2048
24. P.W. Tasker and D.M. Duffy, *Philos. Mag. A*, Vol 47, 1983, p 817
25. R.C. Pond, *Dislocations and the Properties of Real Materials*, F.R.N. Nabarro, Ed., North-Holland, 1989, p 8
26. M.R. Fitzsimmons and S.L. Sass, *Acta Metall.*, Vol 36, 1988, p 3103
27. M.C. Inman and H.R. Tipler, *Metall. Rev.*, Vol 8, 1963, p 105
28. I. Nakamichi, *J. Sci.*, Hiroshima Univ., Vol A, 45, 1990, p 49
29. A. Fraczkiewicz and M. Biscondi, *J. Phys. Coll.*, Vol 46, 1985, p C4-497
30. D. Dimos, P. Chaudhari, and J. Mannhart, *Phys. Rev. B*, Vol 41, 1990, p 4038
31. J.Y. Laval and M.H. Berger-Pinet, *J. Phys.*, Vol 47, 1986, p C1-329
32. M.H. Berger and J.Y. Laval, *J. Phys.*, Vol 51, 1990, p C1-965
33. M.F. Chisholm and D.A. Smith, *Philos. Mag. A*, Vol 59, 1989, p 181
34. C.Y. Wong, C.R.M. Grovenor, P.E. Batson, and D.A. Smith, *J. Appl. Phys.*, Vol 57, 1985, p 438
35. A. Fedotov, B. Evtody, L. Fionova, J. Iliashuk, E. Katz, and L. Poljak, *J. Cryst. Growth*, Vol 104, 1990, p 186
36. P.V. Evans, C.V. Thompson, and D.A. Smith, *Appl Phys. Lett.*, submitted for publication
37. J.H. Werner, *Inst. Phys. Conf.*, Vol 104, 1989, p 63
38. S.F. Nelson and P.V. Evans, *J. Appl. Phys.*, in press
39. S.F. Nelson, Ph.D. thesis, Cornell University, 1991
40. T. Benabbas, C. Canabel, J.Y. Laval, J.L. Pastol, and N. Dinh, *J. Phys. Coll.*, Vol 51, 1990, p C1-439
41. L. Esaki and R. Tsu, *IBM J. Res. Dev.*, Vol 14, 1970, p 61
42. R. Dingle, H.L. Stormer, A.C. Gossard, and W. Weigmann, *Appl. Phys. Lett.*, Vol 33, 1978, p 665
43. W.M.C. Yang, T. Tsakalakos, and J.E. Hilliard, *J. Appl. Phys.*, Vol 48, 1977, p 876
44. K. Fuchs, *Proc. Camb. Philos. Soc.*, Vol 34, 1938, p 100
45. P.V. Andrews, M.B. West, and C.R. Robertson, *Philos. Mag.*, Vol 19, 1969, p 887
46. T. Watanabe, *Mater. Forum*, Vol 11, 1988, p 284
47. A. Ourmazd, F.H. Baumann, M. Bode, and Y. Kim, *Ultramicroscopy*, Vol 34, 1990, p 237
48. W.M. Stobbs and F.M. Ross, *Philos. Mag.*, Vol A63, 1991, p 1, 37

17

SEM/EBIC Study of the Recombination Activity of Grain Boundaries in Germanium

Nouar Tabet
Solid State Physics Laboratory
Institute of Physics, University of Constantine
25000, Constantine, Algeria

The electron beam induced current (EBIC) mode of the scanning electron microscope (SEM) has been used extensively to study the recombination activity of various extended defects in semiconductors (Ref 1). Different theories have been proposed to analyze the dark EBIC contrast at defects such as dislocations (Ref 2-4) and grain boundaries (GBs) (Ref 5, 6). However, most of the experimental results that have been published to date deal with silicon. The purpose of this chapter is to review the main results related to the electrical activity of GBs in germanium. GBs with high misorientations have been found to be inactive in the as-grown specimen (no EBIC contrast) while most of the subgrain boundaries (SGBs) were strongly active (dark EBIC contrast). After annealing, a bright EBIC contrast appeared at defects that were inactive before the thermal treatment. The origin of such contrast has been related to the passivation of the recombination centers at GBs due to impurity segregation during cooling of the Ge ingots from the melt. The temperature effect in both

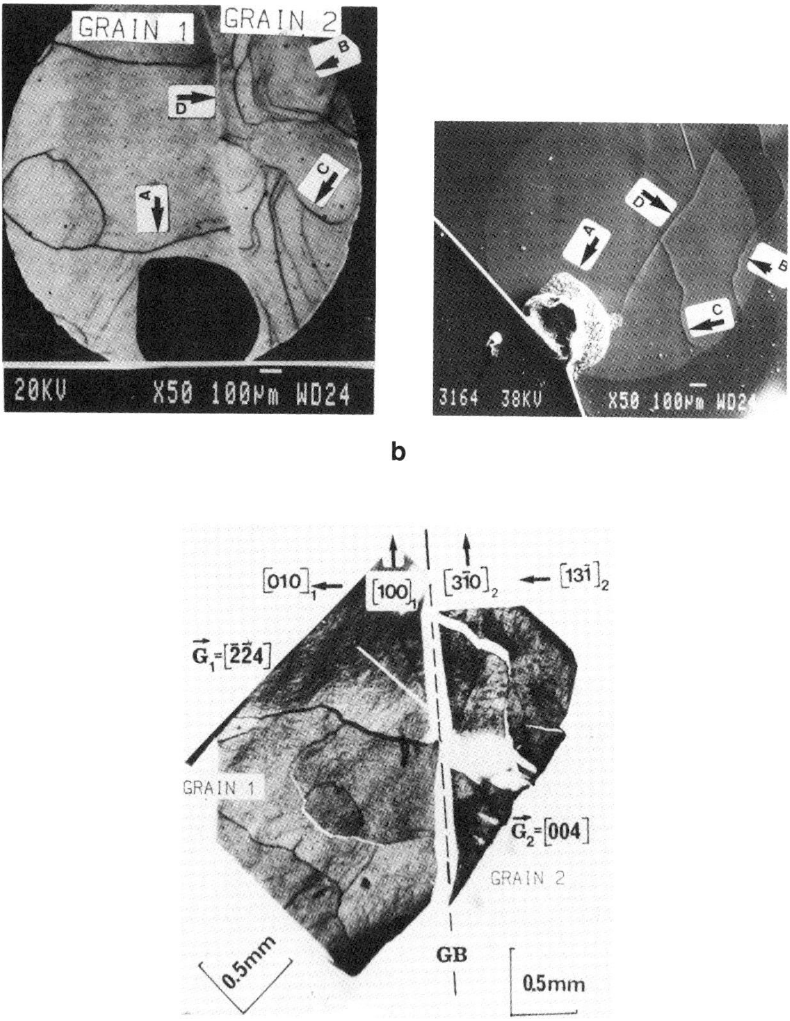

a b

c

Fig. 1 (a) EBIC and (b) SEM images of the same area. D is a grain boundary with large misorientation, as can be seen on x-ray images (c) performed separately on the adjacent grains. G1 and G2 are the diffraction vectors. The divergence of the x-ray beam is 24. The incident planes are $(110)_1$ and $(312)_2$

dark and bright EBIC contrasts has been analyzed. It has been found that dark contrast at SGBs decreases as the temperature increases. The analysis of the bright contrast variation with the specimen temperature has shown the existence of a hole trapping center that acts as a recombination center on a near-valence band level as the temperature decreases and/or the injection level increases.

Experiment

The germanium polycrystals were obtained by slow cooling from the melt. They were Sb doped with 0.4 Ohm-cm resistivity. The EBIC images were obtained using Au/Ge Schottky contacts, a JEOL-SEM 840, and a Keithley 427 amplifier.

Thermal treatments were performed at $T = 812$ °C during 2 h, followed by quenching in air. The Ge specimens were in silica ampules filled with argon.

EBIC Observations

As-Grown Material. Figure 1 shows EBIC and SEM images (Fig. 1a and b) and x-ray images (Fig. 1c) obtained on the same area. The defect marked D is a highly misoriented GB, while A, B, and C are subgrain boundaries, as can be seen on the x-ray images.

It can be noted that D is inactive (no EBIC contrast). This result was unexpected since it is well known that highly misoriented GBs present a high recombination efficiency (except for coincident orientations).

The SGBs present different electrical behaviors: A is strongly active (high and uniformly dark contrast), B is inactive (no contrast), and C is locally active (nonuniform contrast). Since the germanium ingots were obtained by slow cooling from the melt, it seems possible to relate the absence of any recombination activity at some GBs and SGBs to a passivation effect due to impurity segregation.

The various electrical behaviors observed at SGBs can be related to the influence of the structure of these defects (type of dislocations at the SGB plane, electric charge, etc.) on the impurity segregation phenomenon. The observations made after annealing seem to be in good agreement with the passivation hypothesis, as will be shown in the following section.

Effect of Annealing. Figure 2 shows an EBIC image obtained after annealing at $T = 812$ °C for 2 h. The same area was previously observed in the as-grown specimen (Fig. 1a). (The Schottky contact has been removed and performed again after annealing.)

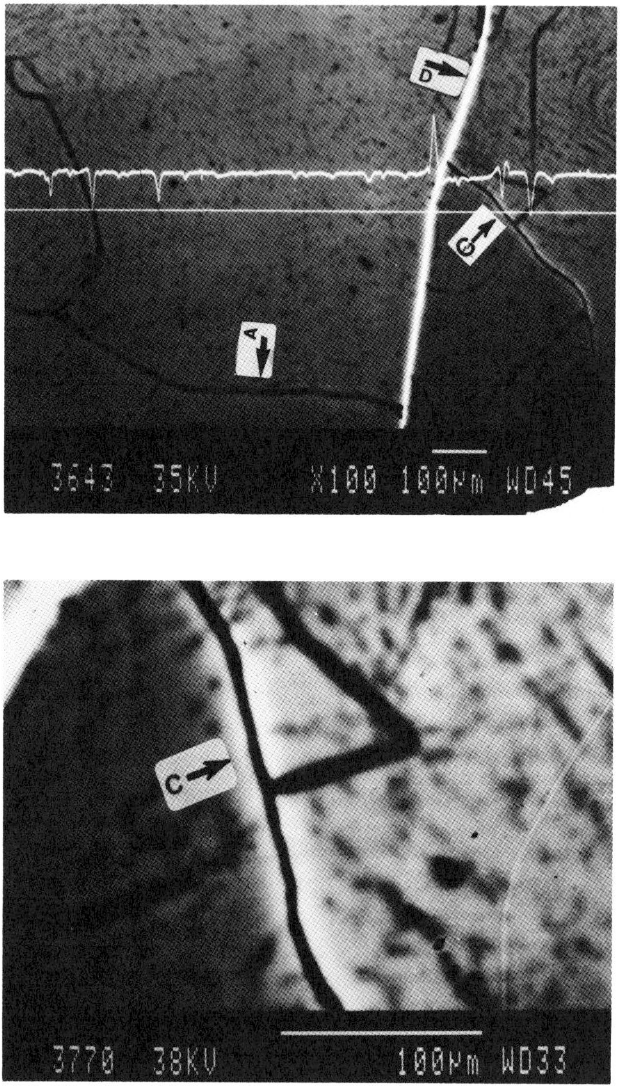

a

b

Fig. 2 (a) EBIC image obtained after annealing, on the same area observed in Fig. 1. Note the appearance of bright contrast at the grain boundary (GB) marked D and near the subgrain boundary (SGB) marked C. (b) Detail of defect C at higher magnification

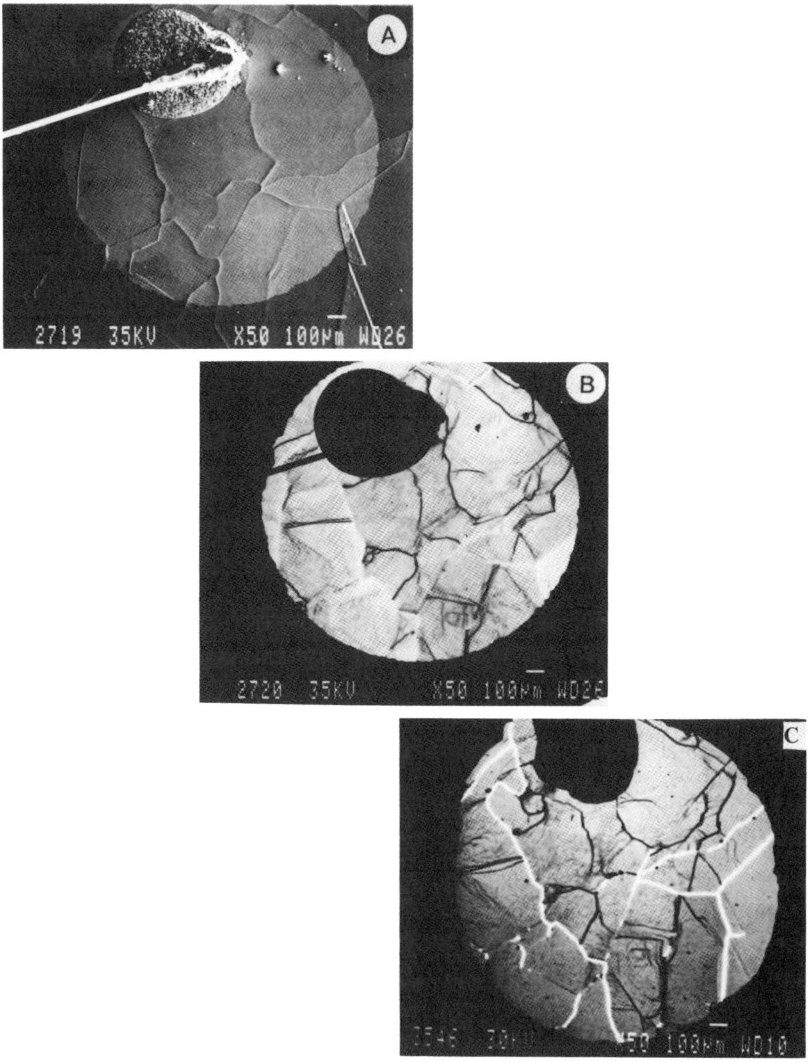

Fig. 3 (a) SEM and (b) EBIC images of the same area of an as-grown specimen. (c) An EBIC image of the same area after annealing at $T = 812\,^{\circ}C$ for 2 h. Most of active defects (with a dark contrast) are subgrain boundaries. Inactive defects in the as-grown specimen present bright EBIC contrast after annealing

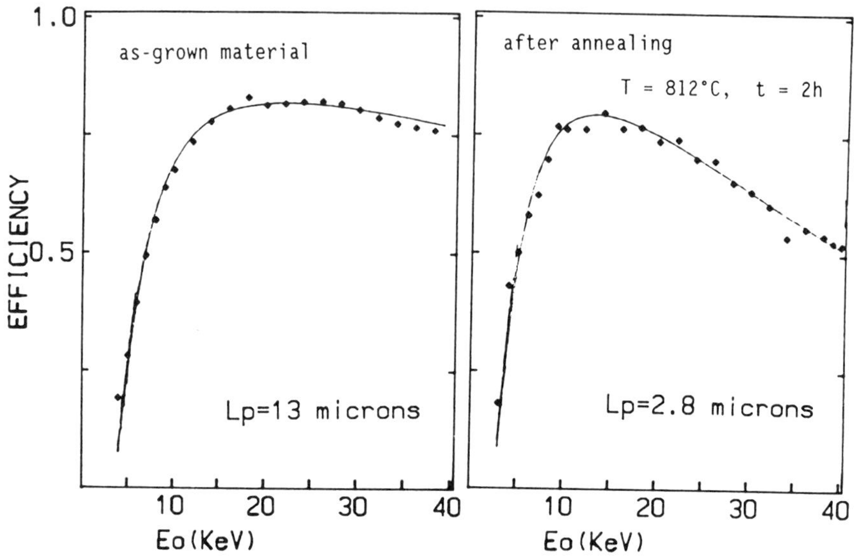

Fig. 4 Collection efficiency curves obtained before and after annealing. The experimental curves have been fitted using an original model (Ref 12). The diffusion length values (L_p) deduced from these results showed a decrease of L_p from 13 to 2.8 μm after annealing

Comparing Fig. 1(a) and Fig. 2(a), it can be seen that the most conspicuous effect of the annealing is the appearance of a bright EBIC contrast at the GB marked D (Fig. 2a). Subgrain boundary B, which is not visible on Fig. 2(a), also presents a bright contrast. Subgrain boundary C shows a bright contrast in the adjacent zones, but a dark contrast remains at the SGB plane, as can be seen in Fig. 2(b). These observations confirm that impurity segregation occurred at GBs that were inactive (D and B) or locally active (C). The recombination centers at these GBs are passivated, and the adjacent areas are purified. Figure 3 shows similar results obtained on another specimen. Most SGBs are active before and after annealing, while GBs present no contrast after annealing. Collection efficiency measurements showed that the minority carrier diffusion length (L_p) within the grains decreased from 13 to 2.8 μm after annealing (Fig. 4). Calculations of bright EBIC contrast were made assuming that L_p takes a higher value at GB areas ($L_p > 13$ μm) inside the grains ($L_p = 3$ μm). A good agreement with the experimental results was obtained (Ref 7). Dark EBIC contrast at subgrain boundary A was measured for different values of the incident beam energy (E_o). Figure 5 shows that the contrast in the annealed

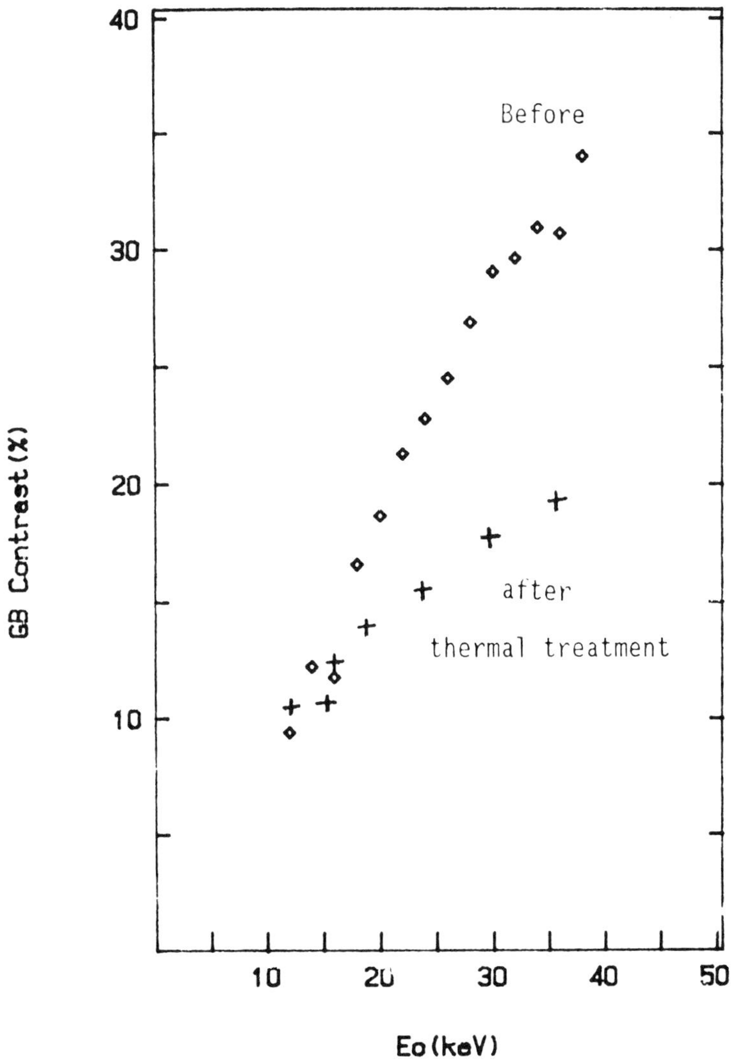

Fig. 5 Variation of grain-boundary maximum contrast versus beam energy before and after annealing at $T = 812$ °C for 2 h

specimen is lower and its variation with E_0 is smoother. This result is in good agreement with Marek's theory (Ref 5), which predicts that the dark EBIC contrast decreases as the diffusion length of the minority carriers decreases.

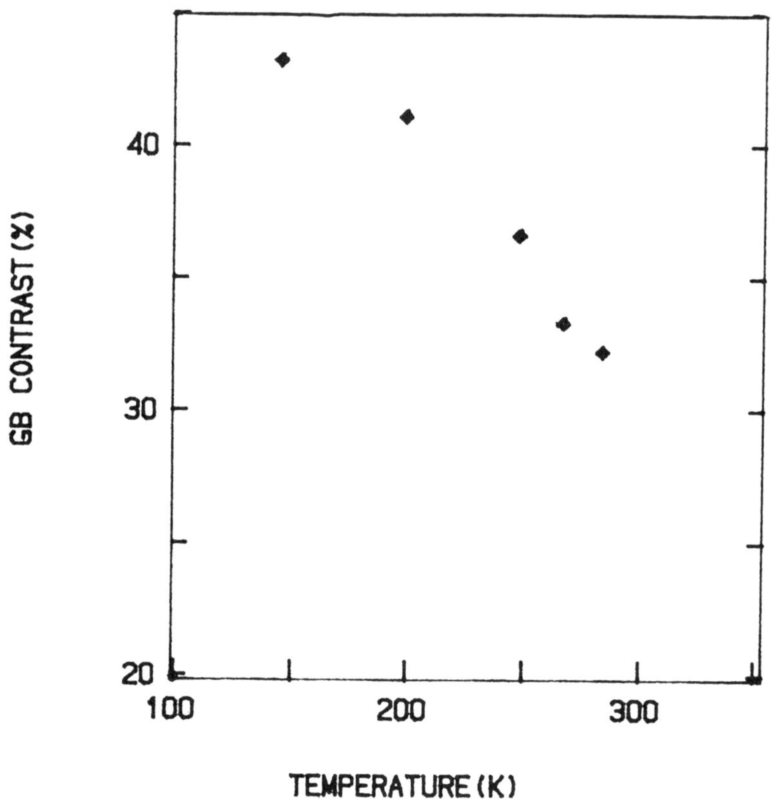

Fig. 6 Variation of GB dark contrast on annealed specimen

Effect of Thermal Treatment

Temperature Effect on Dark EBIC Contrast. Figure 6 shows the dark EBIC contrast variation with the temperature specimen in the temperature range of 140 to 300 K. The measurements were performed at the SGB marked A in Fig. 1. The maximum EBIC contrast is defined as:

$$C_{max} = (I_B - I_{GB})/I_B$$

where I_{GB} and I_B are, respectively, the collected currents on the boundary and far from it, inside the adjacent grains. The results show that the contrast decreases as the temperature increases.

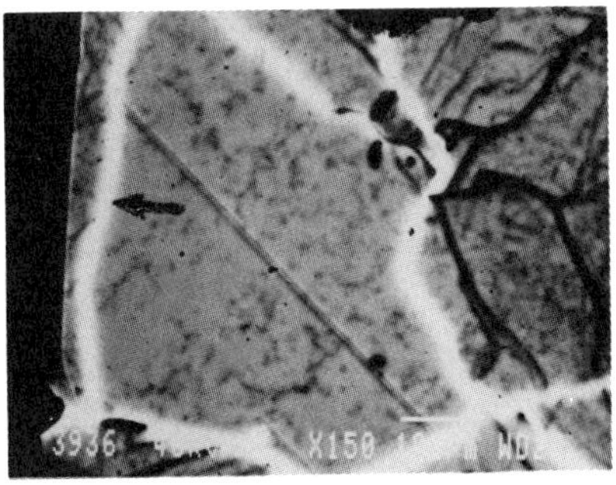

a

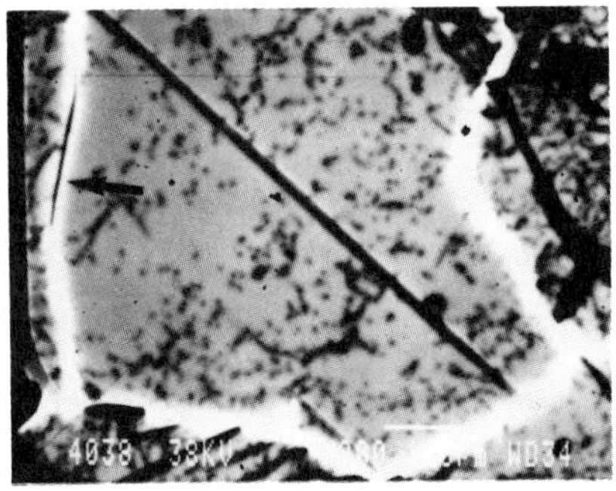

b

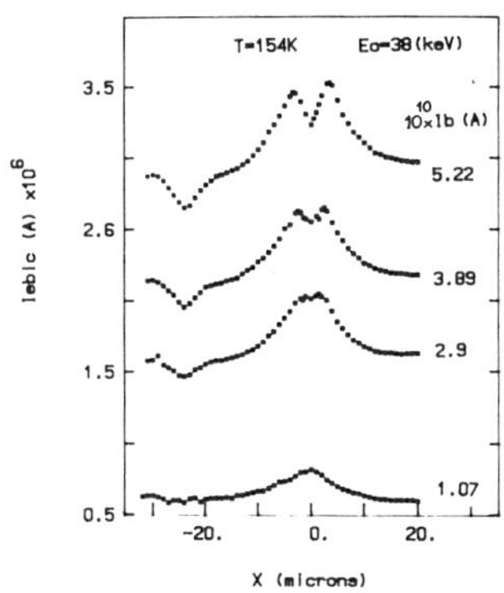

c

Fig. 7 EBIC images. (a) At room temperature ($T = 300$ K). (b) At $T = 154$ K. Note the appearance of dark contrast at low temperature. (c) EBIC profiles across the GB indicated by the arrow (in a and b), recorded for different values of the incident intensity (I_b)

Following Marek's and Donolato's theories, the GB EBIC contrast depends mainly on the recombination velocity (V_s) and the diffusion length (L_p) within the adjacent grains. Considering the first order approximation (Marek's theory), the maximum contrast can be expressed in the form:

$$C_{max} = V_s \cdot f(L_p)$$

where f is a function that depends only on L_p.

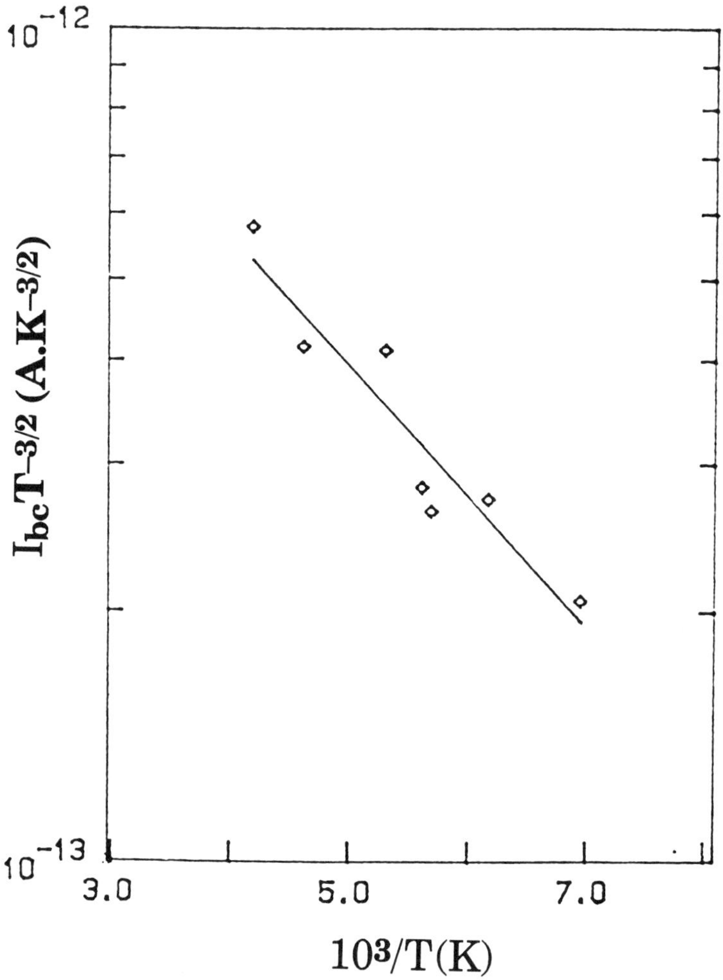

Fig. 8 The variation of the critical current (I_{bc}) as a function of temperature. The trap level (E_t) is given by $E_t = -d\,[\,\ln(I_{bc}T^{-3/2})\,/\,d(1/kT)\,]$.

The measurements of EBIC signal inside the grains showed a small decrease (a few percent) of the collected current as the temperature decreased. Thus, we can assume that the diffusion L_p either remains quasiconstant or decreases slightly as the temperature decreases. Consequently, the variation of the contrast can be related mainly to the variation of the recombination velocity (V_s).

Following Seager's theory (Ref 8), V_s is proportional to the capture section (σ). The variation of the Fermi level with the temperature specimen affects the occupancy of the band gap levels, and hence the σ varies. However, a reliable analysis of the temperature dependence of σ requires a good knowledge of the recombination mechanism at the grain boundary being considered.

Finally, it is worth pointing out that Ourmazd *et al.* (Ref 9) have reported a decrease of EBIC contrast at dislocations in Si as the temperature decreases.

Temperature Effect on Bright EBIC Contrast. The evolution of the bright EBIC contrast as the specimen temperature varies has been observed. Figure 7 shows two EBIC images of the same GB obtained for two temperatures, T = 300 K (Fig. 7a) and T = 154 K (Fig. 7b).

It has been observed that a dark contrast appears as the specimen temperature decreases. Moreover, EBIC profiles have been recorded for different injection levels (obtained by using different primary-beam intensities, I_b). Figure 7c shows that a transition from one maximum profile to two maximum profiles occurs for a critical value, I_{bc}, of the primary-beam intensity. It has been observed that I_{bc} depends on the temperature: the lower the temperature, the smaller the I_{bc}. These observations suggest that trapping centers exist at the GB plane. These traps act as recombination centers when the temperature decreases or the injection level increases. Such a transition occurs when the capture rate of electrons from the conduction band (C_n) becomes equal to or greater than e_p, the emission rate of holes from the trap level to the valence band ($C_n / e_p > 1$). Following Ryvkin (Ref 10), the amplification factor $K_p = C_n / e_p$ can be expressed as:

$$K_p = \frac{\gamma_n \, n}{\gamma_p N_v \exp - E_t/kT} \qquad \text{(Eq 1)}$$

where γ_n and γ_p are, respectively, the capture coefficient for electrons and for holes; N_v is the density of states in the valence band; and E_t is the trapping level. The valence band energy is taken as a reference for energies (eV = 0), and n is the concentration of electrons. It can be seen from the above relation that the critical value giving $K_p = 1$ can be expressed as a function of the trapping level (E_t):

$$I_{bc} \propto n_c = (\gamma_p/\gamma_n)\, N_v \exp(-E_t/kT) \qquad \text{(Eq 2)}$$

A detailed analysis has been carried out assuming the coexistence of a trapping level (E_t) near the valence band and a single recombination center in the gap (Ref 11). The results of calculations showed that I_{bc} can be expressed in a form quite similar to that of the above relation. A value of E_t equal to 0.03 eV has been deduced from I_{bc} measurements for different temperatures (Fig. 8).

Conclusion

SEM/EBIC technique has been successfully used to investigate the electrical activity of GBs in germanium. Random GBs with large misorientations have been found to be inactive in the as-grown specimen and presented a bright EBIC contrast after annealing. Most subgrain boundaries were strongly active. Impurity segregation during the cooling of the germanium ingots from the melt seems to be responsible for the passivation effect. Annealing induced a degradation of the diffusion length inside the grains, and it has been shown that the appearance of bright EBIC contrast can be related to this degradation. EBIC measurement at low temperatures showed that trapping centers exist at some GBs that present bright contrast. These traps act as recombination centers as the temperature decreases and/or the injection level increases.

Acknowledgment

This work has been performed in Laboratoire de Physique des Materiaux, Centre National de la Recherche Scientifique de Bellevue, France. The author wishes to thank Dr. C. Monty and Dr. Y. Marfaing for numerous and fruitful discussions.

References

1. H.J. Leamy, *J. Appl. Phys.*, Vol 53, 1982, p 51
2. L. Pasemann, *Phys. Status Solidi (a)*, Vol 84, 1984, p 133
3. C. Donolato, *Phys. Status Solidi (a)*, Vol 65, 1981, p 649
4. J.L. Farvacque and B. Sieber, *Rev. Phys. Appl.*, Vol 25, 1990, p 353
5. J. Marek, *J. Appl. Phys.*, Vol 53, 1982, p 1454
6. C. Donolato, *J. Appl. Phys.*, Vol 54 (No. 3), 1983, p 1314
7. N. Tabet, C. Monty, and Y. Marfaing, in *Polycrystalline Semiconductors*, Vol 35, J.H. Werner, H.J. Moller, H.P. Strunk, Ed., Springer-Verlag, 1989, p 89
8. C.H. Seager, *J. Appl. Phys.*, Vol 52, No 6, 1981, p 3960
9. A. Ourmazd and G.R. Booker, *Phys. Status Solidi (a)*, Vol 55, 1979, p 771
10. S.M.R. Ryvkin, in *Photoelectric Effects in Semiconductors*, Consultants Bureau, 1964
11. N. Tabet, C. Monty, and Y. Marfaing, *Rev. Phys. Appl.*, Vol. 24, 1989, p 871
12. N. Tabet and R.T. Tarento, *Philos. Mag. B*, Vol 59, No 2, 1989, p 243

18

Adhesion and Short-Range Forces between Crystalline Surfaces Effects of Surface Lattice Mismatch

Patricia M. McGuiggan
3M Company, St. Paul, MN 55144
Jacob N. Israelachvili
Materials Department and Department of Chemical and Nuclear Engineering, University of California, Santa Barbara, CA 93106

Using the Surface Forces Apparatus, the authors have measured the adhesion forces and interaction potentials between two mica surfaces as a function of the orientation, i.e. twist angle (θ), of the surface lattices. The forces were measured in distilled water and an aqueous KCl solution. In water, apart from a relatively angle-independent "baseline" adhesion, a sharp adhesion peak (energy minima) occurred at the twist angle $\theta = 0^\circ$, where there was maximum alignment of the surface lattices. The energy decreased by 50% at $\pm 1^\circ$ from this peak. In aqueous KCl solution, because of potassium ion adsorption, an oscillatory structural force was measured where the adhesive minima occurred at discrete separations corresponding to an integral number of water layers. The adhesion energies corresponding to the first three potential minima were angle dependent near $\theta = 0°$, again decreasing by 50% at ±1° from $\theta = 0°$. The repulsive maxima were also affected near $\theta = 0$, showing that the whole interaction potential between two surfaces in liquids depends on the orientation of the surface lattices and that these effects can extend at least four molecular layers (about 1.0 nm film). Such ultrathin liquid films actually "freeze" and become solid or solidlike and may be considered quasi epitaxially ordered next to the surfaces.

The structure-property relationship of interfaces has been an important area of interest in recent years (Ref 1-4). Mismatched interfaces give rise to grain boundaries with physical properties that are different from those of the bulk material and often dictate the strength of the composite material. For example, the diffusion coefficient is usually strongly enhanced at the boundary, impurities tend to segregate towards the boundary, and the strength of the interface is generally weaker than the surface free energy of the crystal (Ref 4). Control of the properties of grain boundaries is often essential to obtaining the desired property of the overall material.

The interaction between two solid surfaces across a liquid film depends on the structure of the liquid molecules and the interaction with the surfaces (Ref 1, 5). At separations below about ten molecular layers, liquid molecules tend to pack, giving rise to a force that is usually an oscillatory function of the distance. These structural or oscillatory forces reflect the forced ordering of the liquid molecules into discrete layers when constrained between two surfaces. A single, solid surface can also impose a structure on another solid adsorbate or liquid in contact with it, but when confined between *two* solid surfaces, the solvent structure is much enhanced and liquid films thinner than three to four layers can be made to undergo a liquid-to-solid phase transition in such molecularly thin films (Ref 6).

The molecular structure of the liquid is known to affect the packing and thus the interaction between two solid surfaces across the liquid, especially within the last few molecular solvent diameters (Ref 1). The crystal structure and orientation of the surface lattices is also expected to affect the net interaction.

In vacuum, the surface energy of a crystalline solid has been shown experimentally through measurement to depend on the specific exposed plane (Ref 7, 8). In air, the studies include only indirect measurements including thermal grooving experiments and experiments in which smoke particles are deposited on single-crystal planes (Ref 8). These studies show that there are peaks in the adhesion energy at certain lattice "coincidence" angles, in agreement with the standard theory of Read and Shockley (Ref 9). These lattice coincidence angles refer to the mutual alignment of the atoms from two opposing surfaces, as in the original lattice crystal. All these studies have been limited to metal and metal oxide surfaces. Direct measurements and theoretical studies of the surface structure effects of two crystalline surfaces interacting across a *liquid* medium or separated by an intercalated layer remain unexplored, even though a wide variety of interactions occur across such condensed phases.

The authors now report direct measurements of the adhesion forces between mica surfaces in distilled water and an aqueous KCl solution. Such forces reflect the effect of the surface lattice orientation on the surface-induced liquid structure and give insight into the energetics of grain boundaries.

Experimental

Adhesion, or "pull-off," forces were directly measured using the Surface Forces Apparatus designed by Israelachvili and shown in Fig. 1. This apparatus also measures surface forces as a function of separation between two solid surfaces (Ref 10, 11). In these studies, the surfaces used were muscovite mica that had been cleaved to produce molecularly smooth surfaces free from steps. The mica surfaces were glued to cylindrical silica support disks using Epon 1004 epoxy adhesive. The surfaces were positioned face to face in cross-cylinder geometry. The interaction between two crossed cylinders is geometrically similar to the interaction between a sphere and a flat surface. The two cleaved mica surfaces could be mounted so that contact occurred in the same orientation as exists in the perfect crystal. This corresponds to $\theta = 0°$.

The back sides of the mica surfaces were silvered. As white light was passed normally through the surfaces, an optical interference pattern was formed. Observation of the fringes of equal chromatic order allowed the separation between the mica surfaces to be measured to ±1 Å and also gave information regarding the shape of the surfaces, such as the contact area and radii of curvature of the two cylinders (Ref 11).

The lower silica disk was mounted on the end of the spring. When the remote end of the spring was displaced, the adhesive force between the surfaces held them together. However, when the resulting force on the surface exceeded the adhesive, or pull-off, force, the surface jumped apart.

The final surface separation (jump distance) (D_j), is measured by the optical technique. The pull-off force (F_o) is calculated by multiplying the spring constant (K) by the jump distance (D_j):

$$F_o = KD_j \qquad \text{(Eq 1)}$$

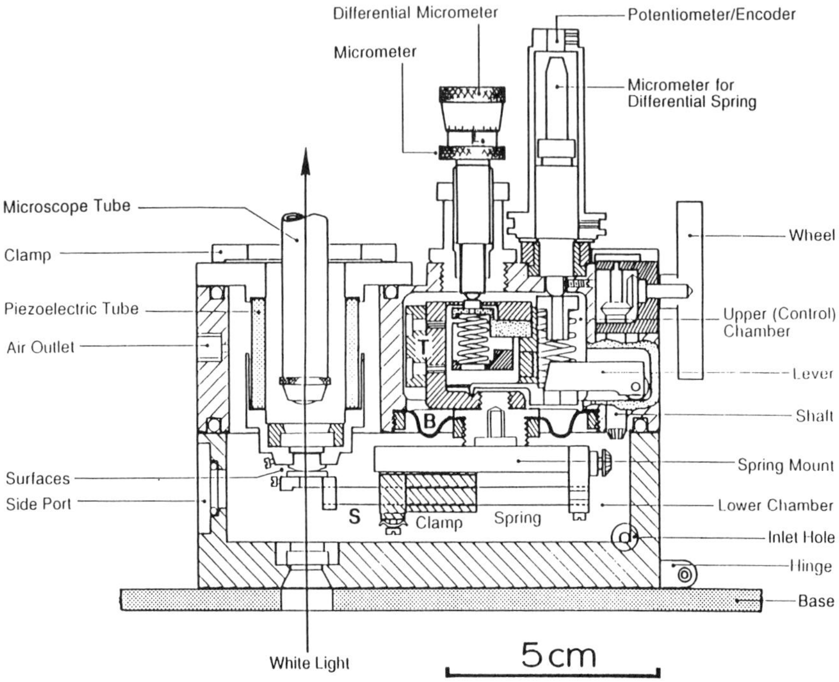

Fig. 1 The Surface Forces Apparatus (SFA Mk III) for measuring forces between surfaces. The micrometers and the piezoelectric tube accurately control the distance. The lower half of the instrument is exposed to the solvent. This is separated from the mechanical parts by the bellows (B). The force-measuring spring (S) can be clamped at varying points to change the spring constant

For these experiments, a ring was fitted around the top surface mount to allow measurement of the twist angle. After the surfaces were aligned, the adhesion force was measured. The upper mount was then rotated slightly (with the surfaces apart), and the adhesion force was again measured.

Theory

The adhesion energy (E_o) between two *flat* surfaces can be calculated from the adhesion force (F_o) between two cylindrically curved surfaces of radii R_1 and R_2 according to the Derjaguin approximation (Ref 1):

$$E_o = F_o/2\pi\,(R_1R_2)^{1/2} = \mathrm{K}D_j/2\pi(R_1R_2)^{1/2} \quad \text{(Eq 2)}$$

By definition, the surface energy is twice the solid/liquid interfacial energy (γ_{sl}). Thus:

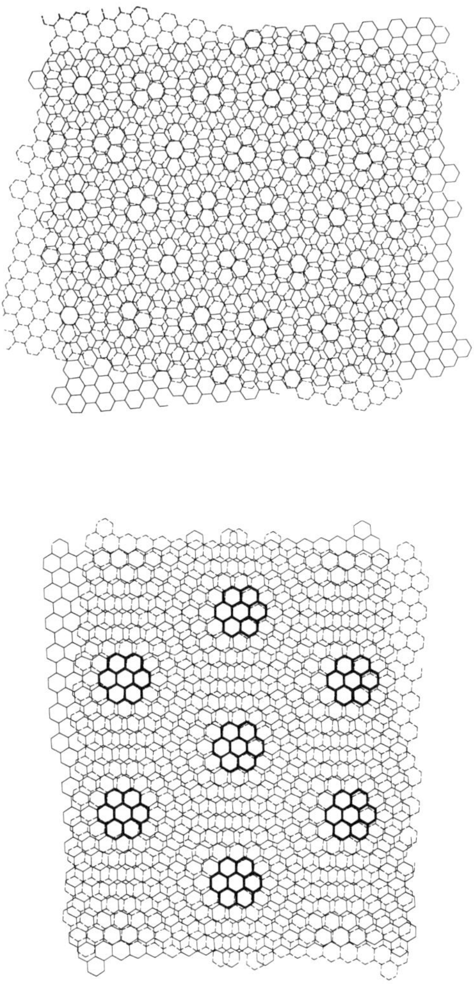

a

b

Fig. 2 Schematic diagrams of two contacting hexagonal lattice surfaces, similar to the oxygen atoms on the mica basal plane. (a) 15° rotation of one surface lattice showing the misalignment of the surface lattices. (b) Two hexagonal lattices near maximum alignment. The darker seven-membered rings correspond to possible alignment of the atoms when they are near coincidence

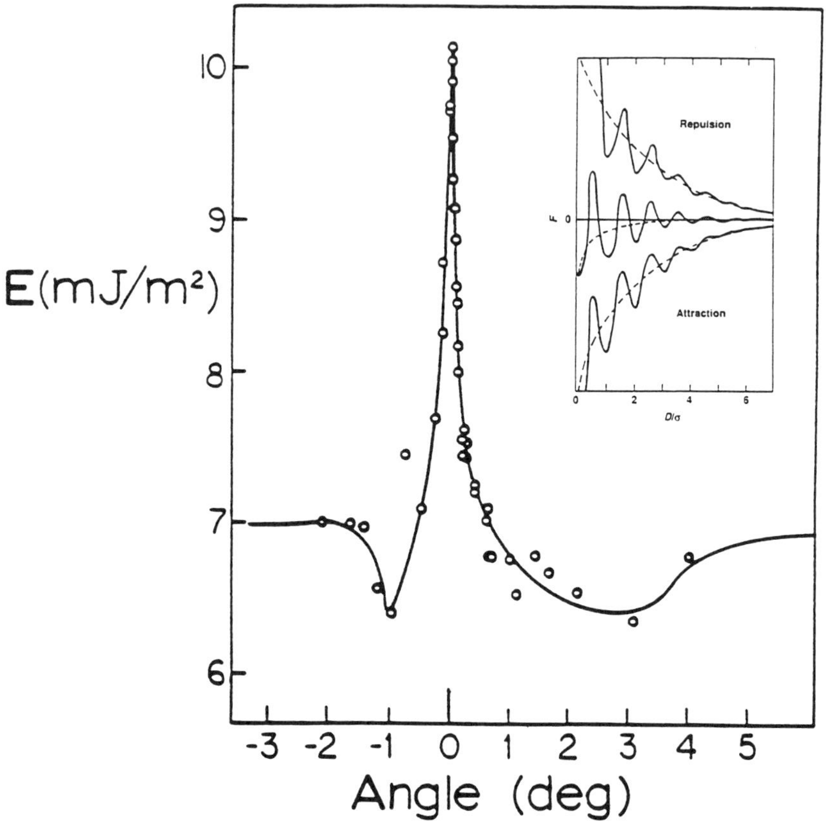

Fig. 3 Measured adhesion energy (E_0) versus twist angle (θ) in distilled water. Inset: Possible short-range interaction forces versus distance for mica surfaces interacting across aqueous solution. The periodicity of the structural force corresponds to the diameter of the solvent molecule (σ). The interaction forces can be altered by adsorption, changing the surfaces from attractive and hydrophobic to repulsive and hydrophilic. This additional monotonic force is superimposed on the oscillatory or structural force

$$E_0 = 2\gamma_{sl} \quad \text{(Eq 3)}$$

If the adhesion force between the surfaces is due entirely to van der Waals forces, Lifshitz theory predicts the force between two crossed cylinders to be given by:

$$F_0 = A(R_1R_2)^{1/2}/6D_0 2 \quad \text{(Eq 4)}$$

where A is the Hamaker constant and D_o is the cut-off distance between the two surfaces when they are in adhesive contact. The value of D_o is typically close to the 0.16 to 0.2 nm (Ref 1). The Hamaker constant for mica interacting across water is $A = 2.2 \times 10^{-20}$J. Combining Eq 2 and 4 gives:

$$E_o = 2\gamma_{sl} = F_o/2\pi(R_1R_2)^{1/2} \quad \text{(Eq 5)}$$

$$= -A/12\pi D_o^2 \quad \text{(Eq 6)}$$

Lifshitz theory is a continuum theory and assumes no discrete structure of the surfaces or of the intervening liquid medium. However, as previous experiments have shown, there is a preferential orientation of two interacting surface lattices. This implies that the surface lattice structure and interlattice interactions are important. The surface oxygen atoms of the mica basal plane form a hexagonal network. If another hexagonal lattice were to contact it, the interface would most likely be out of register or crystal alignment, as shown in Fig. 2(a). This increases the surface energy (decreases the adhesion energy) since an atom located at the interface is strained with respect to the surface lattices. Minimum surface energy (maximum adhesion energy) should correspond to maximum lattice coincidence, which for mica surfaces occurs at $\theta = 0°, \pm 60°, \pm 120°$, and 180°. If the atoms that are within a certain critical distance are allowed to relax into the equilibrium unstrained positions, discrete areas of maximum coincidence are formed surrounded by grain boundaries, as shown in Fig. 2(b).

The excess surface energy of a low-angle dense-grained boundary (D) in a vacuum has been proposed by Read and Shockley to vary with θ according to:

$$D = D_o\theta(C - \ln\theta)\,\text{J/m}^2 \quad \text{(Eq 7)}$$

where C and D_o are constants that depend on the character of the boundary. The first term, $D_o\theta C$, represents the contribution of each dislocation to the total grain-boundary energy. This is proportional to the dislocation density. The second term, $D_o\theta\ln\theta$, represents the interaction between these dislocations.

Results

The adhesion energy between two mica surfaces in distilled water as a function of θ of the mica lattices is shown in Fig. 3. The narrow peak centered around $\theta = 0°$ is superimposed upon a baseline adhesion energy of 6.5 ± 1.0 mJ/m^2. The half-width of the peak, $D_{\theta/2}$, was calculated from $\pm(E_{max} - E_{baseline})/2$ and was measured at $0.6 \pm 0.2°$. Similar results between mica surfaces in distilled water were previously reported at other maximum coincidence angles, including 60°, 120°, and 180°. These peaks are much sharper than previously reported with metals and metal oxides and indicate a strong dependence of the interface on intralayer energy on the interlayer bond strength.

In dilute electrolyte solutions, the mica surfaces are negatively charged because of the dissolution of the surface potassium ions. As the surfaces are brought together, an electrostatic repulsion is measured. With increased electrolyte ion concentration, hydrated potassium ions bind to the surface and give rise to a short-range monotonically repulsive hydration force. This hydration force is dependent on the solution conditions, such as pH, type of cation, and temperature (Ref 1). In addition, as the surfaces are forced together, the solvent molecules between them begin to pack. This packing constraint produces an oscillatory force that is superimposed on the monotonic force. The interaction between mica surfaces in aqueous electrolyte solutions is shown in the inset in Fig. 3. In dilute electrolyte solutions (curve A), attractive van der Waals forces overcome both the oscillatory force and the repulsive electrostatic force, and the surfaces jump into contact. In more concentrated solutions, the oscillatory-hydration force is measured, as shown in curve C. In curve B, the maxima are repulsive, and the minima are attractive. The minima occur at some discrete separation corresponding to an integral number of water layers with the adhesion energy quantized for that minimum.

Figure 4 gives the adhesion energy of the "primary", second, third, and fourth minima as a function of θ. These minima correspond to the surface separations of 0.3 nm, 0.6 nm, 0.9 nm, and 1.2 nm, respectively. The forces were measured in 10^{-3} M KCl solution. The equilibrium force distance profile for this concentration corresponds to the Fig. 3 inset, curve B. The primary, second, and third minima show an orientation dependence similar to the orientation dependence observed in distilled water, with a baseline adhesion 30 to 50% smaller than the maximum adhesion. The half-width of the peaks was $D_{\theta/2} = 1.0°$. The fourth minimum was very weak (adhesion energies ≃0.2 ± mJ/m^2). Any deviations from this baseline value would have been beyond the authors' measuring limits.

The maxima of the oscillatory force curve were also found to be orientation dependent. For $\theta < 1°$, the force maxima was reduced, or, in other words, less force was required to push the surfaces together. The primary minimum was only accessible when the surfaces were within $\theta = 0.5°$. At larger orientations, the primary minimum could not be obtained under accessible loads, showing that the primary maximum for misaligned surfaces is higher than for aligned

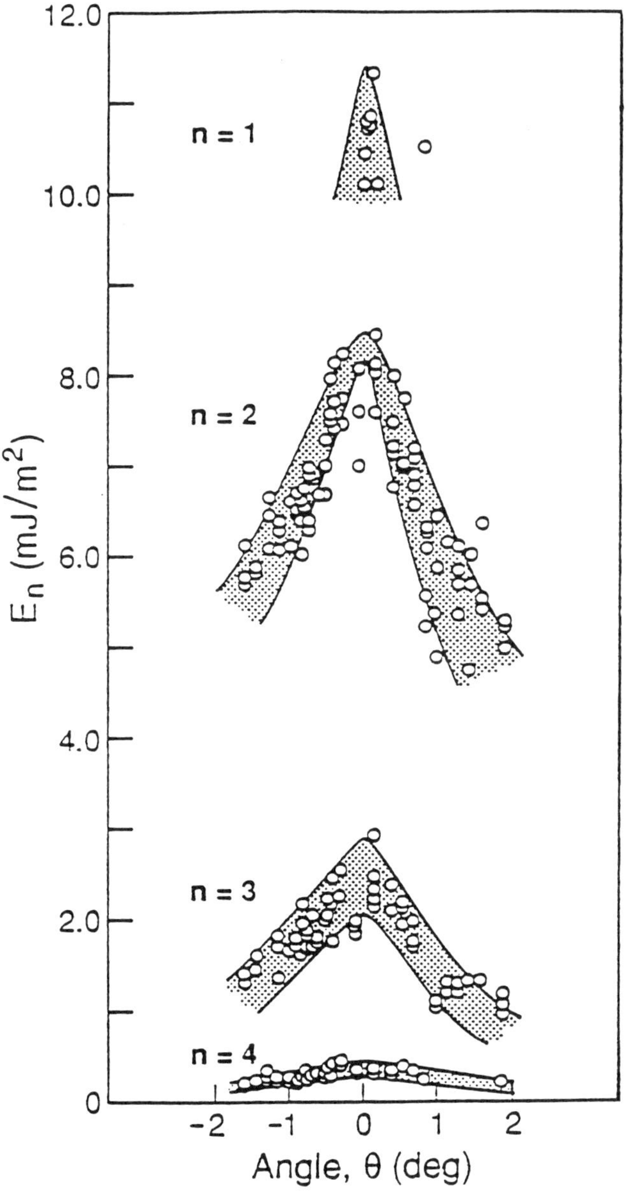

Fig. 4 Measured adhesion energies (E_n) versus twist angle (θ) in 1 nM KCl. The *n* represents the number of integral layers of solvent molecules separating the mica surfaces. The mean periodicity in the oscillations was 0.27 ± 0.05 nm

surfaces. This indicates that the whole interaction potential between surfaces is affected near $\theta = 0°$.

Discussion

The results show that the whole interaction potential between two solid crystalline surfaces in aqueous solutions is dependent on the small angle twist boundary of the two lattices. The peak in the measured adhesion values corresponds to maximum coincidence of the surface lattice atoms, and the surface alignment effects extend at least four molecular solvent layers.

The peak width is much smaller than those found for metals and metal oxide surfaces. In general, the half-width of peaks such as NiO is about 10° at $\theta = 0°$, with other smaller but equally broad peaks measuring at less dense coincidence angles (Ref 12). The main factor determining the shape of the adhesion peak must be the relationship between the strengths of the interlayer and of the intralayer bond forces . If the interlayer forces are comparable to the intralayer forces, as is the case for metallic interfaces, the surface atoms are easily pulled out of their equilibrium positions. However, in the case of mica, the interlayer van der Waals forces are at least two orders of magnitude weaker than the intralayer covalent bonds in the siloxane network of the mica lattice. Thus, the rigid mica lattice will not allow relaxation of its surface atoms until the two surfaces are in near perfect alignment.

The peak width was independent of the area of contact. In a distilled water and KCl solution, the area of contact varied between 0.5 and 8×10^{-5} cm^2, yet the peak width remained unchanged. Simple geometric arguments show that a rotation of the surface lattice by 1° gives lattice matching about 30 nm radially from the center. In all cases, the radius of contact was at least two orders of magnitude larger than this. The results are thus due to an intrinsic property of the system and are not limited by the finite size of the contact area.

Smaller, secondary peaks at less dense coincidence angles were not measured with mica. This indicates that the exposed surface oxygen atoms play the major role in determining the mismatch energies. The underlying atoms, such as silicon, contribute to the overall baseline energy. The absence of any secondary peaks at other angles further indicates that the highly rigid mica lattice will not allow isolated atoms to relax out of their positions by the weak van der Waals forces for the formation of small patches of aligned atoms.

The adhesion force at maximum alignment is still an order of magnitude smaller than the force required to cleave mica from its pristine state. If only van der Waals forces are considered, the lower adhesion values measured can be accounted for by assuming the mica surfaces are still separated by one layer of water molecules (0.3 nm). If one could also force this last layer of molecules out from between the surfaces, it is likely that the adhesion peak would be much stronger. The results reported are confined to mica surfaces interacting in an aqueous KCl solution. The effect that a different lattice surface structure or different solvent molecules that do not pack as readily into the hexagonal lattice of the mica surfaces might have on the shape of the adhesion peak remains to be seen. An investigation of different surfaces such as metals would also give insight into the relationship between the intralattice and the interlattice forces in determining the nature of the grain boundary.

Lifshitz theory describes the interaction between surfaces in a liquid. However, the theory breaks down at small distances because it is a continuum theory and neglects solvent structure effects. Similarly, there is currently no coincident site lattice theory for two surfaces interacting across liquids. However, Eq 7 can be fitted to describe the experimental data in distilled water near maximum coincidence once the measured energy is subtracted from the baseline energy. Other mathematical treatments, such as a linear or Gaussian distribution, can also fit the data. This suggests that only the first term in Eq 7 may be important, implying that the interactions between dislocations are small in the rigid mica system. This would be consistent with the small angular width of the peaks, which indicates a low density of dislocations.

Theoretical considerations including Monte Carlo and the molecular dynamic simulations of two surfaces interacting in a liquid have concentrated on the structure of the liquid but have largely ignored the surface structure (Ref 13, 14). Recently, the frictional properties of a thin film were modeled (Ref 15). These studies show that the stick-slip motion observed when two molecularly smooth surfaces slide past each other is caused by a thermodynamic transition of the molecules in the thin film. This is dependent upon the surface crystal orientation and the direction of sliding. Direct measurements of the wear of a rutile single crystal showed a change as much as 700% different from the sliding direction and crystal plane changes (Ref 16). More experimental and theoretical studies are needed on the surface structure effects on the static and dynamic properties of thin films.

Besides friction and adhesion, the interaction of surfaces across thin films is important in material-processing operations such as ceramic processing, densification at high temperatures, and grain-boundary modification. An understanding of all the contributing factors at the molecular level is crucial to an understanding of the physical properties of materials.

References

1. J.N. Israelachvili, *Intermolecular and Surface Forces,* Academic Press, 1985
2. J.A. Pask and A.G. Evans, Ed., *Ceramic Microstructures '86: The Role of Interfaces,* Vol 21, *Materials Science Research,* Plenum Press, 1987
3. J.B. Wachtman, Jr., Ed., *Structural Ceramics,* Academic Press, 1989
4. M. Rühle, *J. Phys. C, Solid State Phys.,* Vol 46, 1985, p C4-281
5. J.N. Israelachvili and P.M. McGuiggan, *Science,* Vol 241, 1988, p 795
6. J.N. Israelachvili and S.J. Kott, *J. Colloid Interface Sci.,* Vol 129, 1989, p 461
7. A.G. Guy, *Essentials of Materials Science,* McGraw-Hill, 1976
8. J. Ferrante and J.R. Smith, *Phys. Rev.,* Vol 78, 1981, p 1400; G. Dhalenne, M. Dechamps, and A. Revcolevschi, *J. Am. Ceram. Soc.,* Vol 65, 1979, p 267
9. W.T. Read and W. Shockley, *Phys Rev.,* Vol 78, 1950, p 275
10. J.N. Israelachvili and P.M. McGuiggan, *J. Mater. Res.,* Vol 5, 1990, p 2232
11. J.N. Israelachvili and G.E. Adams, *Faraday Trans. 1,* Vol 74, 1978, p 975
12. J. Ferrante and J.R. Smith, *Phys. Rev.,* Vol 78, 1981, p 1400
13. W. van Megan and I. Snook, *Faraday Trans. 2,* Vol 75, 1979, p 1095; R. Kjellander and S. Marcelja, *Chem. Scr.,* Vol 25, 1985, p 73
14. D.L. Henderson and M. Lozada-Cassou, *J. Colloid Interface Sci.,* Vol 114, 1986, p 180
15. P.A. Thompson and M.O. Robbins, *Science,* Vol 250, 1990, p 792
16. D.H. Buckley and K. Miyoshi, in *Structural Ceramics,* Vol 29, J.B. Wachtman, Jr., Ed., Academic Press, 1989, p 300

19

Electromigration in Thin Films

C.L. Bauer
Department of Metallurgical Engineering and Materials Science
Carnegie Mellon University
Pittsburgh, PA 15213

The phenomenon of electromigration involves diffusion-controlled mass transport induced by current flow in a conducting element, either through coulombic force exerted on individual lattice ions or through momentum transfer between electrons and lattice ions (electron wind) (Ref 1-3). Since the magnitude of each effect is proportional to current density, deleterious behavior associated with electromigration increases as film dimensions decrease (for constant current), resulting in the degradation of desirable properties and eventual failure of the conducting element. Mass transport is often facilitated, especially at moderate temperatures, by short-circuit diffusion associated with lattice defects, such as dislocations, grain boundaries (Ref 4-6), interphase boundaries, and/or external surfaces. Therefore, the extent of electromigration is affected by concomitant microstructure and phase morphology.

Electromigration is usually expressed quantitatively by assigning an effective charge to the (positive) host ions qz^*, where q and z^* denote, respectively, the nominal coulombic (ionic) charge and an empirical (dimensionless) parameter, which accounts for differences between the effective strength of the coulombic field and the efficiency of momentum transfer between electrons and positive ions. Accordingly, the electromigration force (F_e) exerted on charge qz^* is given by:

$$F_e = -qz^*E = -qz^*\rho j \qquad \text{(Eq 1)}$$

where E denotes the electric field, ρ is electrical resistivity, and j is current density. Likewise, the net mass flux (J), under usual drift conditions, is given by the (Nernst-Einstein) expression:

$$J = NMF = N(D/kT)F_e \qquad \text{(Eq 2)}$$

where N is the number of atoms per unit volume (reciprocal atomic volume), D is the diffusion coefficient of the rate-controlling species, k is the Boltzmann constant, T is absolute temperature, and M is the atomic mobility (D/kT) of the rate-controlling species. Therefore, mass flux (J) is directly related to current density (j).

The purpose of this chapter is to consider (mass) flux divergence at microstructural inhomogeneities (grain-boundary triple junctions) in thin films and thereafter to develop relationships between the corresponding microstructure and physical characteristics, such as the growth and decay of holes and hillocks and the ultimate time to failure. These effects are especially important in thin films because of the large current density associated with the small cross-sectional area available for conduction, the small grain size (of the order of film thickness), and the mass transport along grain boundaries. The combined effect of these phenomena often promotes the premature failure of electronic components. Relationships between electromigration and microstructure are developed by considering cumulative flux divergence at grain-boundary triple junctions during the early stages of electromigration. First, the growth of individual holes and hillocks resulting from cumulative flux divergence at isolated grain-boundary triple junctions is considered and related to fractional change of electrical resistance ($\Delta R/R$); then, analysis is extended to distributions of grain-boundary triple junctions defining global microstructure; and, last, the generation and subsequent healing of electromigration damage is interpreted in terms of the growth and decay of holes and hillocks at grain-boundary triple junctions and related to $\Delta R/R$ and ultimate time to failure.

The remainder of this chapter is divided into several sections: first, various fundamental mechanisms for electromigration are reviewed in the section below. Then, in the sections following, electromigration associated with isolated grain-boundary triple junctions, electromigration associated with distributions of grain-boundary triple junctions, and the healing of corresponding damage are considered.

Fundamental Mechanisms for Electromigration

Four examples of electromigration at microstructural inhomogeneities in thin films are shown in Fig. 1 in which an imbalance of (mass) flux produces grain-boundary migration (Fig. 1a), hole placement (Fig. 1b), saddle movement on an underlying rail (Fig. 1c), and hole (or hillock) growth at a grain-boundary triple junction (Fig. 1d). In each example, it is assumed that mass transport is facilitated by short-circuit interfacial diffusion only and that grain boundaries, interfacial boundaries, and free boundaries (surfaces) are either perpendicular or parallel to the film surface. Both assumptions are well satisfied in thin films at moderate temperatures, in which grain size equals or exceeds film thickness and mass transport by bulk diffusion is negligible. In the first example (Fig. 1a), grain-boundary velocity (v_{gb}) is computed from the condition for mass continuity between grains I and II and is given by:

$$v_{gb} = (D_{gb}/kT)F_e \qquad \text{(Eq 3)}$$

where D_{gb} denotes an appropriate grain-boundary diffusion coefficient. Therefore, diffusion-controlled mass transport induced by electromigration produces grain-boundary migration. This effect is related to the stability of bamboo structures in narrow conducting films (Ref 7, 8).

In the second example (Fig. 1b), velocity of a cylindrical hole (v_h), defined by radius (r), is computed from the rate of mass transport along the cylindrical wall and is given by:

$$v_h = (a^4/4r^3d)\,(D_s/kT)F_e \qquad \text{(Eq 4)}$$

where a is nearest-neighbor atomic distance ($1/N^{1/3}$), d is film thickness, and D_s is surface diffusion coefficient. Therefore, diffusion-controlled mass transport induced by electromigration produces hold displacement, which varies with the dimensions of the hole. This effect is related to the stability of fine dispersions in conducting films.

In the third example (Fig. 1c), the velocity of a high-conductivity (thin film) saddle, defined by thickness (d), width (w), and length (l) ($l >> d$), is positioned on a low-conductivity (thin film) rail so that current is deflected from the rail through the saddle (Ref 9, 10). Resultant saddle velocity (v_s) is given by:

$$v_s = (A_i/wd)(D_i/kT)F_e \qquad \text{(Eq 5)}$$

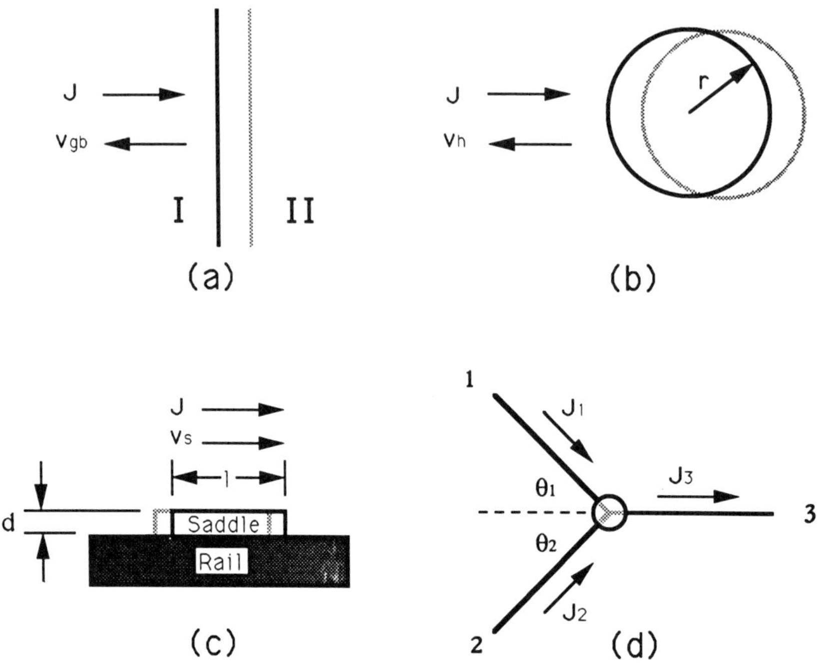

Fig. 1 Four examples of electromigration at microstructural inhomogeneities in thin films, in which an imbalance of (mass) flux (J) produces (a) grain-boundary migration, (b) hole displacement, (c) saddle movement on an underlying rail, and (d) hole (or hillock) growth at a grain-boundary triple junction. Shaded and black images denote, respectively, positions before and after electromigration

where D_i denotes the appropriate diffusion coefficient for mass transport through cross-sectional area A_i. (In principle, v_s is independent of saddle length.) If the predominant mechanism for mass transport is controlled by diffusion through the film, A_i is equal to wd and Eq 5 reduces to the equation for atomic drift velocity ($v = MF$). However, cross-sectional areas for surface and grain-boundary diffusion are given approximately by $2aw$ and $\delta wd/\lambda$, respectively, where δ is effective grain-boundary thickness and λ is average grain size ($\lambda >> \delta$). Therefore, diffusion-controlled mass transport induced by electromigration produces saddle movement, which varies with the dimensions of the saddle and the diffusion mechanism. This effect is related to the stability of multilevel interconnections in thin film circuits (Ref 11, 12).

In the fourth example (Fig. 1d), cumulative flux divergence (ΔJ) at the intersection of three grain boundaries (grain-boundary triple junction) is given by:

$$\Delta J = (NF_e/kT)\,\Sigma\,(D_i \cos\theta_i) \qquad \text{(Eq 6)}$$

where D_i is the diffusion coefficient of, and θ_i is the angle subtended by, grain boundary i with respect to the applied electric field. Therefore, diffusion-controlled mass transport induced by electromigration produces cumulative flux divergence at grain-boundary triple junctions, resulting in depletion (holes) or accumulation (hillocks) of matter. This effect is related to the stability of thin polycrystalline conducting films.

It may be concluded that diffusion-controlled mass transport is induced by current flow in the vicinity of a variety of microstructural inhomogeneities. A particularly deleterious phenomenon is the formation of holes and hillocks at grain-boundary triple junctions. Holes increase concomitant electrical resistance, in turn increasing both current density and local temperature (through increased Joule heating), resulting in open circuits. On the other hand, hillocks often produce bridges between neighboring conducting elements, resulting in short circuits. In either case, the generation of electromigration damage eventually leads to the (catastrophic) failure of a conducting element. The remainder of this chapter analyzes the growth and decay of holes and hillocks and evaluates their effect on the electrical resistance and ultimate failure of conducting thin films.

Electromigration Associated with Isolated Grain-Boundary Triple Junctions

With reference to the coordinate system shown in Fig. 2(a), mass flux along an idealized conduit i is given, in two dimensions, by reexpressing Eq 1 and 2 as:

$$J_i = (N\langle D\rangle/kT)\, z^* q\rho j Y_i \qquad \text{(Eq 7)}$$

where

$$Y_i = (D_i/\langle D\rangle) \cos \phi_i \qquad \text{(Eq 8)}$$

In this elementary example, flux depletion at one end of the conduit equals flux accumulation at the other end. (Obviously, flux divergence is zero when the current i is perpendicular to the conduit and maximum when i is parallel to the conduit.) In general, Y_i may be summed over any number of intersecting conduits to evaluate the net cumulative flux divergence, as shown in Fig. 2(b). Grain boundaries in thin films may be represented in an analogous manner since they are usually inclined normal to the film and extend from one surface to the other. In this case, Y_i provides a measure of flux divergence due to both variation of atomic mobility (D_i) and driving force ($\cos \phi_i$) in grain boundary i. For grain-boundary triple junctions, $D_i/\langle D\rangle$ may be replaced by exp

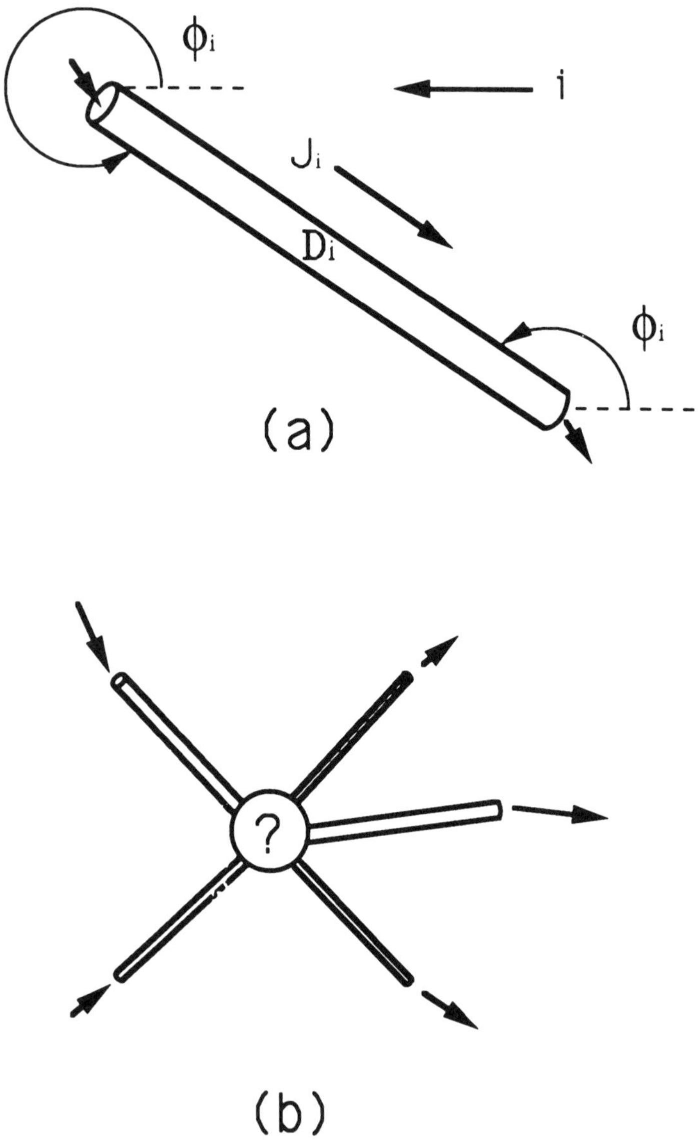

Fig. 2 (a) Flux divergence at extremities of a single conduit, characterized by conductivity (diffusion coefficient) D_i and inclination ϕ_i with respect to the direction of applied current (i). (b) Flux divergence at the intersection of several conduits, characterized by different conductivities and inclinations

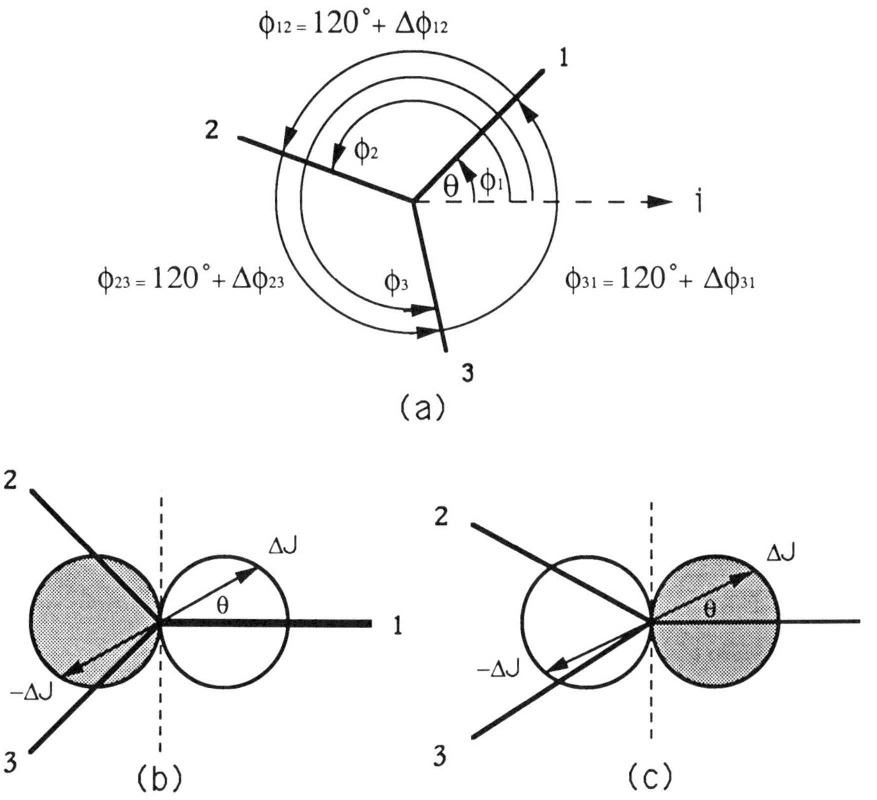

Fig. 3 Origins of flux divergence at grain-boundary triple junctions: (a) definition of grain-boundary inclination ϕ_i and angle subtended by adjacent grain boundaries ϕ_{ij} ($\Sigma\phi_{ij}$ = 360°) with respect to the direction of applied current (i). (b) Flux divergence at a grain-boundary triple junction, in which D_i varies from boundary to boundary (denoted by a variable cross section of the boundaries) but fi remains constant. (c) Flux divergence at a grain-boundary triple junction, in which ϕ_i varies from boundary to boundary (denoted by variable angle subtended by adjacent boundaries) but D_i remains constant. The relative magnitude of flux divergence (ΔJ) as a function of θ (= θ_1) is denoted by the corresponding circles in (b) and (c), in which shaded circles denote positive flux divergence (formation of hillocks) and open circles denote negative flux divergence (formation of holes). In all cases, flux divergence integrated over all θ vanishes (area of shaded circles equals area of open circles), ensuring conservation of mass for a random distribution of grain-boundary triple junctions, regardless of the direction of applied current. Also, a unique direction (dashed lines) is defined where the flux divergence always vanishes

($-\Delta Q_i/kT$), where ΔQ_i denotes the deviation of activation energy for diffusion in grain boundary i from the mean value $<Q>$. Accordingly, Eq 8 reduces to (Ref 13):

$$\Delta Y = \Sigma Y_i = A(\Delta Q_i, \Delta\phi_{ij}) \cos\theta + B(\Delta Q_i, \Delta\phi_{ij}) \sin\theta \qquad \text{(Eq 9)}$$

where θ denotes the angle subtended by the electric current and a reference grain boundary (Fig. 3):

$$\Delta Q_i = Q_i - <Q> \qquad \text{(Eq 10)}$$

$$\Delta\phi_{ij} = \phi_{ij} - 120° \qquad \text{(Eq 11)}$$

and where ϕ_{ij} denotes the angle subtended by adjacent grain boundaries, and both $A(\Delta Q_i, \Delta\phi_{ij})$ and $B(\Delta Q_i, \Delta\phi_{ij})$ are constants for a given grain-boundary triple junction, obtained by an appropriate summation of ΔQ_i and $\Delta\phi_{ij}$ over all three boundaries. If angles subtended by adjacent grain boundaries are equal ($\Delta\phi_{ij} = 0$) and diffusion coefficients of all grain boundaries are equal ($\Delta Q_i = 0$), ΔY vanishes. In this special case, flux divergence at grain-boundary triple junctions cannot occur, regardless of the direction of applied current. Therefore, electromigration is produced by distributions of grain-boundary inclinations and/or distributions of grain-boundary diffusion coefficients. Accordingly, ΔY provides a useful measure of susceptibility of an individual triple junction to degradation (hole or hillock growth) by electromigration. In general, flux divergence vanishes at the intersection of any number of grain boundaries (n-tuple junction) when $\Delta\phi_i = D_i = 0$, regardless of the direction of applied current.

The effect of distributions of grain-boundary diffusion coefficients and/or grain-boundary inclinations is shown more clearly in Fig. 3, in which the origins of flux divergence at grain-boundary triple junctions are specified. With reference to grain-boundary inclination (ϕ_i) and the angle between adjacent grain boundaries (ϕ_{ij}) ($\Sigma\,\phi_{ij} = 360°$), defined relative to the direction of applied electrical current i in Fig. 3(a), two special cases are considered. In the first case, D_i varies from boundary to boundary (denoted by a variable cross section of grain boundaries), but ϕ_i is constant, as shown in Fig. 3(b). In the second case, ϕ_i varies from boundary to boundary (denoted by a variable angle subtended by adjacent grain boundaries), and D_i is constant (Fig. 3c). The relative magnitude of flux divergence (ΔJ) as a function of θ ($= \theta_1$) is represented by the corresponding circles in Fig. 3(b) and (c), where shaded circles denote positive flux divergence (formation of hillocks) and open circles represent negative flux divergence (formation of holes). In all cases, flux divergence

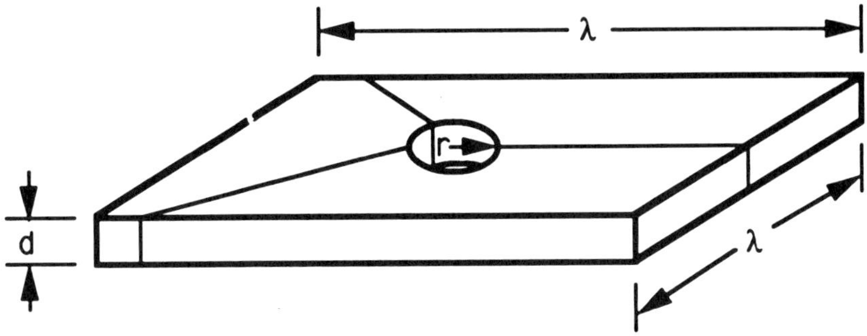

Fig. 4 Cylindrical hole of radius (r), centered on a grain-boundary triple junction in a thin film characterized by thickness d, length λ, and width λ

integrated over all θ vanishes (area of shaded circles equals area of open circles), ensuring the global conservation of mass for a random distribution of grain-boundary triple junctions, regardless of the direction of the applied current. The maximum diameter of a circle (maximum ΔJ) is given by:

$$\Delta Y_{max} = (A^2 + B^2)^{1/2} \quad \text{(Eq 12)}$$

Also, a unique direction is defined where flux divergence always vanishes (dashed lines). The corresponding angle α subtended by the common circle axis and reference grain boundary 1 (direction of maximum ΔJ) and angle β subtended by the axis separating adjacent circles (dashed lines) and reference grain boundary 1 (direction of zero flux divergence) are given by:

$$\tan \alpha = B/A = \cos \beta \quad \text{(Eq 13)}$$

Accordingly, flux divergence at a given grain-boundary triple junction is maximized when current is applied at angle α, and the flux divergence vanishes when current is applied at angle β ($\alpha = \beta \pm 90°$). (For both examples in Fig. 3, $\alpha = 0 \pm 180°$ and $\beta = \pm 90°$). In general, when any two angles ϕ_{ij} are equal in magnitude and sign, maximum flux divergence occurs in a direction parallel to the common grain boundary, whereas when any two angles ϕ_{ij} are equal in magnitude but opposite in sign, zero flux divergence occurs in a direction parallel to the common grain boundary. Further details concerning flux divergence at grain-boundary triple junctions are being published elsewhere (Ref 14).

Finally, it is useful to relate flux divergence at a given grain-boundary triple junction to a measurable quantity. An especially useful quantity is electrical resistance (R), since R can be measured easily and continuously during the application of an electrical field. The fractional change of electrical resistance ($\Delta R/R$) may be estimated with reference to Fig. 4, which shows a cylindrical hole, defined by radius r, centered on a grain-boundary triple junction in a thin film that is characterized by thickness d, length λ, and width λ. For the early stages of flux divergence, $\Delta R/R$ is given by (Ref 15):

$$\Delta R/R = \pi r^2/\lambda^2 = \Delta Y\,(t/t_0) \qquad \text{(Eq 14)}$$

where t is time and

$$t_0 = \lambda^2 kT/\langle D\rangle \delta z^* q\rho j \qquad \text{(Eq 15)}$$

and δ denotes (constant) grain-boundary thickness. The last term in Eq 14 is obtained by equating total volume associated with flux divergence ($\Delta J d\delta t$) to the volume of the cylindrical hole ($\pi r^2 d$). Although Eq 15 is strictly correct for a cylindrical hole during the very early stages of electromigration ($r \ll \lambda$), it may be correct for other shapes, current crowding, etc. (Ref 16). The important point is that r provides a useful link between microstructure (flux divergence at a grain-boundary triple junction) and a measurable quantity (electrical resistance). Also, it is assumed that corresponding hillocks do not affect $\Delta R/R$ since current cannot be sufficiently deflected through the small volume. Therefore, the fractional change of electrical resistance ($\Delta R/R$) associated with an individual grain-boundary triple junction should vary linearly with time (t) during the early stages of electromigration and be characterized as slope ($1/t_0$) that varies exponentially with reciprocal temperature. Moreover, ΔY provides a useful measure of the susceptibility of grain-boundary triple junctions in the section below.

Electromigration Associated with Distributions of Grain-Boundary Triple Junctions

The above analysis of current-induced flux divergence is based on the growth of isolated holes and hillocks at grain-boundary triple junctions. Actual thin-film conducting elements, however, are characterized by a polycrystalline microstructure containing a complex distribution of triple junctions. Three different (two-dimensional) distributions of grains in thin films are shown in Fig. 5: the distribution of regular hexagonal grains (Fig. 5a), the distribution of irregular hexagonal grains (Fig. 5b), and the distribution of (five- to seven-sided) irregular polygonal grains (Fig. 5c). For the array of

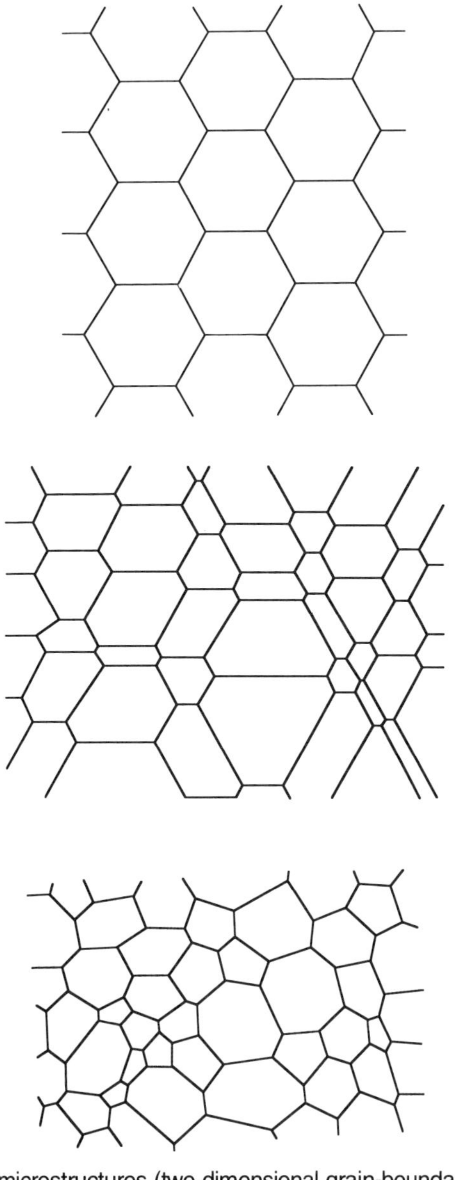

a

b

c

Fig. 5 Possible microstructures (two-dimensional grain-boundary configurations) in thin films. (a) Distribution of regular hexagonal grains. (b) Distribution of irregular hexagonal grains. (c) Distribution of (five- to seven-sided) irregular polygonal grains. Cumulative flux divergence at grain-boundary triple junctions can occur only for the distribution of irregular polygonal grains, regardless of the direction of applied current regular hexagonal grains (Fig. 5a), all grain-boundary triple junctions are characterized by angles subtended by adjacent grain-boundary inclinations of 120° ($\Delta\phi_{ij} = 0$), implying that corresponding grain-boundary surface tensions are identical. Since grain-boundary surface tension and (short-circuit) grain-boundary diffusion are related to grain-boundary structure, it is conceivable that all grain boundaries are characterized by an identical diffusion coefficient. Under these conditions, net flux divergence at all grain-boundary triple junctions vanishes, rendering this microstructure immune to degradation by electromigration, regardless of the direction of applied current. For the distribution of irregular hexagonal grains (Fig. 5b), all grain-boundary triple junctions are again characterized by angles subtended by adjacent grain-boundary inclinations of 120° ($\Delta\phi_{ij} = 0$). Therefore, net flux divergence at all grain-boundary triple junctions vanishes, rendering this microstructure also immune to degradation by electromigration, regardless of the direction of applied current. For the distribution of (five- to seven-sided) irregular polygonal grains (Fig. 5c), however, angles subtended by adjacent grain-boundary inclinations and corresponding grain-boundary diffusion coefficients are not equal ($\Delta\phi_{ij} \neq 0$), resulting in local cumulative flux divergence at grain-boundary triple junctions. This (usual) distribution of grains (microstructure) is susceptible to degradation by electromigration, causing a distribution of holes and hillocks to form at grain-boundary triple junctions during electromigration, according to the magnitude and sign of ΔY_i. Only when $\Delta DQ_i = \Delta\phi_i = 0$ is the corresponding microstructure immune to degradation by electromigration.

The idealized microstructures shown in Fig. 5 may be compared with damage by electromigration (formation of holes and hillocks) in actual conducting elements subjected to large current densities (≥1 mA/cm^2). Typical examples are presented in Fig. 6, in which scanning electron micrographs of gold on a substrate of semi-insulating gallium arsenide (Fig. 6a) and Al-1 wt% Si on a substrate of (oxidized) silicon (Fig. 6b) are pictured after severe degradation by electromigration. Generally, damage (holes and hillocks) occurs at grain-boundary triple junctions. In the first example, several features may be noted: holes and hillocks (at least for the latter stages of electromigration) are highly irregular in shape, the volume of holes does not necessarily equal the volume of hillocks, and the density of grain-boundary triple junctions (~10^8/cm^2) greatly exceeds the combined density of holes and hillocks (~10^6/cm^2). Therefore only a small fraction of all grain-boundary triple junctions is susceptible to severe degradation by electromigration. In the second example, several features also may be noted: the hole/hillock pair is separated

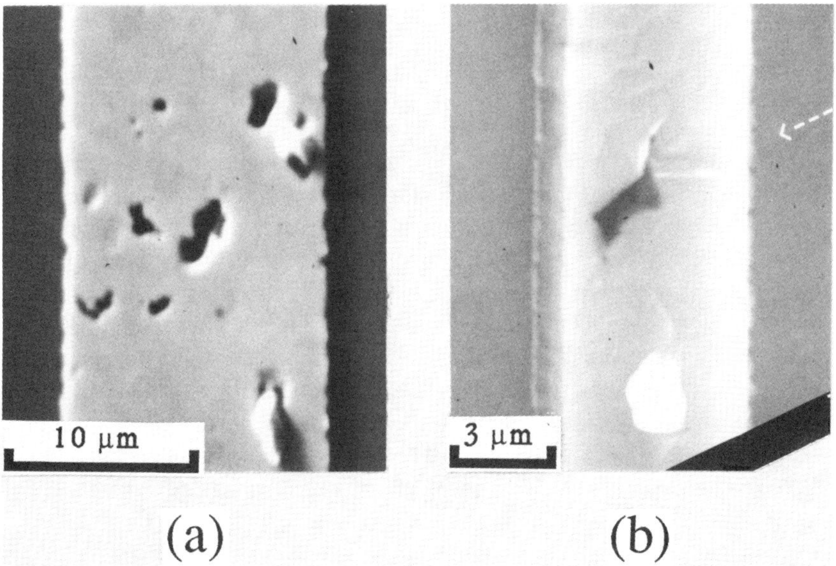

Fig. 6 Scanning electron micrographs of typical damage produced by electromigration at grain-boundary triple junctions. (a) 30 nm gold on a substrate of semi-insulating gallium arsenide subjected to a current density of 3 mA/cm^2 for 80 h at 220 °C. (b) 1 μm Al-1 wt% Si on a substrate of (oxidized) silicon subjected to a current density of 3 mA/cm^2 for 48 h at (nominal) room temperature

by about 5 μm; since hillocks are produced by electron flow downstream from holes, the sign of z^* (Eq 1) is positive, implying that, in this case, the dominant driving force for electromigration results from momentum transfer between electrons and positive lattice ions; and, again, the density of grain-boundary triple junctions greatly exceeds the combined density of holes and hillocks. From these examples and many others (Ref 17), it is well established that the details of microstructure play a key role in the formation and growth of holes and hillocks that result in the ultimate failure of conducting thin films.

A principal challenge for the complete understanding of electromigration in thin films is the formulation of self-consistent procedures to combine the contributions of individual grain-boundary triple junctions properly in order to obtain a global description of macroscopic properties. This challenge involves three central issues: the self-consistent combination of grain-boundary triple junctions to produce a realistic microstructure; the proper summation of characteristics of individual grain-boundary triple junctions defined by this microstructure during the early stages of electromigration; and the extension to the latter stages of degradation by electromigration, in which localized heating, complex hole and hillock shapes, the consumption of neighboring grain-boundary triple junctions by growing holes, etc., complicate analysis enormously. In fact, a correct procedure for the summation of the characteristics of individual grain-boundary triple junctions must include, at least , proper interconnection of all grain boundaries, continuity of mass transport between grain-boundary triple junctions, compatibility of grain-boundary surface tensions defining angles ϕ_{ij}, dependence of grain-boundary diffusion coefficients on both grain misorientation and grain-boundary inclination, and self-consistency of grain-boundary surface tensions with the crystallographic nature of adjacent grains. In addition, a two-dimensional grain structure, in which mass transport is confined to the corresponding grain boundaries, is usually assumed. These formidable constraints have thus far prevented the development of an exact description of damage produced by electromigration and the computation of corresponding effects on physical properties.

In spite of the complexities associated with the extension of the behavior of a single grain-boundary triple junction to a distribution of triple junctions, several approximate procedures have been devised (Ref 4, 5, 18). One recent approach involves the selection of random orientations of grain-boundary triple junctions, i.e., random θ and random distributions (within arbitrary limits) of ΔQ_i and $\Delta\phi_{ij}$, disregarding interconnectability of individual triple junctions (Ref 18). This approach is justified when the number of grain-boundary triple junctions is sufficiently large that average effects can be determined. A distribution of holes and hillocks is generated by a Monte Carlo method, hillocks are subsequently disregarded, and individual values of $(\Delta R/R)_i$ are summed over all holes to obtain macroscopic values for $\Delta R/R$ as a function of time, temperature, and current density. In effect, however, the largest holes contribute disproportionately to the degradation process. Therefore, a good approximation is achieved by considering the hole produced by the largest (maximum) combination of ΔD_i and $\Delta\phi_{ij}$, as specified by Eq 12. Although maximization of Eq 12 is somewhat arbitrary, this expression reduces to a simple combination of $\Delta\phi_{max}$ and $\Delta Q_{max}/kT$ for small deviations from the mean values (120° and $<Q>/kT$). Usually, deviations of several percent from the mean values provide satisfactory agreement with many experimental results. Therefore, the effective behavior of a large number of grain-boundary triple junctions may be replaced by the grain-boundary triple junction characterized by the largest flux divergence, due either to an imbalance in grain-boundary mass conductivities (mobility) and/or to grain-boundary inclination (driving force). Moreover, a random distribution of θ, ΔD_{max}, and $\Delta\phi_{max}$ produces results that mimic actual experimental results for the distribution of times to failures (log-normal distribution) (Ref 18). Other variations, based on random

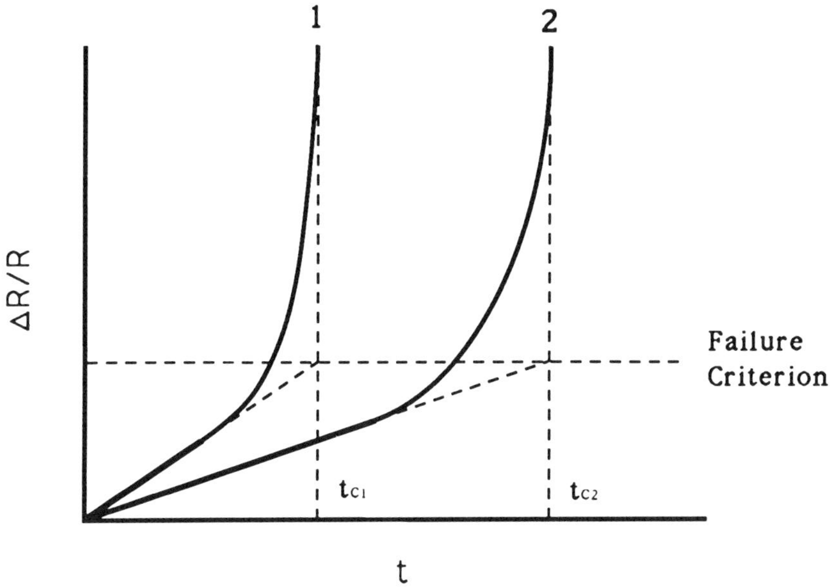

Fig. 7 Fractional change of resistance $\Delta R/R$ as a function of time (t) for two values of initial slope. The time to failure ($\Delta R/R = \infty$) is defined by the initial slope ($\Delta Y/t_0$) and a failure criterion, i.e., $r \rightarrow \lambda$

flux divergence, have been equally successful (Ref 4, 5, 19). On the other hand, methods insisting on more strict adherence to constraints associated with the summation of individual grain-boundary triple junctions are, at present, not sufficiently developed to compare with actual experimental results or predict reliable mean times to failure (Ref 20).

As mentioned previously, the complexities associated with the summation of individual grain-boundary triple junctions may be circumvented by replacing a complex distribution of triple junctions with a single, effective triple junction. The gross approximation involves only two elementary parameters: the susceptibility of an effective grain-boundary triple junction to degradation by electromigration (ΔY or equivalent) and the definition of a failure criterion. These two parameters may be combined by reexpressing Eq 14 in terms of an effective grain-boundary triple junction as:

$$[d(\Delta R/R)/dt]_t = 0 = \Delta Y/t_0 = (\Delta R/R)_c/t_c \quad \text{(Eq 16)}$$

where $[d(\Delta R/R)/dt]_t = 0$ is the initial slope of $\Delta R/R$ $(= \Delta Y/t_0)$, $(\Delta R/R)_c$ is a failure criterion, and t_c is (mean) time to failure. Interrelationships between these parameters are shown in Fig. 7, in which the fractional change in electrical resistance ($\Delta R/R$) is plotted as a function of time (t) for two examples of initial slope (1 and 2). Clearly, the ultimate time to failure (t_{ci}) is equal to the failure criterion divided by the initial slope, as expressed by Eq 16. The failure criterion may be expressed as:

$$(\Delta R/R)_c = \frac{\pi r_c^2}{\lambda^2} \qquad \text{(Eq 17)}$$

by analogy with Eq 14. It is interesting to note that, in many cases, the product $[d(\Delta R/R)/dt]\ t_c\ [= (\Delta R/R)_c]$ is observed to equal a constant (~0.20), thereby yielding the relationship $r_c \approx 0.25\ \lambda$. Therefore, the failure of a complex microstructure due to cumulative flux divergence at grain-boundary triple junctions may be analyzed by the failure of a single, effective grain-boundary triple junction. Failure is observed to occur when the radius of a (cylindrical) hole approaches a sizable fraction of the grain size. Accordingly, the determination of mean time to failure reduces to the determination of $\Delta Y/t_0$ for the effective grain-boundary triple junction. In general, because holes are somewhat elongated rather than cylindrical in shape, an interconnecting network of holes is probably formed before r approaches λ. In any event, the determination of both an appropriate slope for $\Delta R/R$ and the definition of a failure criterion circumvents many complexities associated with the latter stages of electromigration with surprisingly good results.

In the above discussion, it has been assumed that a sufficient number of grain-boundary triple junctions is present to validate an averaging process. When the number of triple junctions in a conducting element is too small to justify the assumption of a random distribution, i.e., when λ approaches a lateral dimension (w) of the conducting element, other procedures must be devised for combining the characteristics of individual grain-boundary triple junctions to obtain a reliable global value (Ref 21). In effect, these procedures involve the appropriate summation of localized electrical resistance in some combination of series and parallel. Generally, ultimate time to failure decreases with decreasing values of w/λ since the growth of a single hole resulting from abnormal flux divergence at a localized grain-boundary triple junction is more likely to result in an open circuit. As w/λ approaches or becomes less than unity, however, the spacing of grain-boundary triple junctions along a conducting element is reduced below $1/\lambda$, thereby decreasing the probability of premature failure. In the limit of $w << \lambda$ (i.e., monocrystalline or bamboo structure), the ultimate time to failure is controlled by some other mechanism, since grain-boundary triple junctions are no longer available to produce localized cumulative flux divergence. Even though the ultimate time to failure may increase when the number of grain-boundary triple junctions is reduced, the corresponding standard deviation usually increases also, because the conducting element is no longer characterized by average values of flux divergence. Therefore, additional complications arise when the number of grain-boundary triple junctions is limited by the lateral dimensions of the conducting element.

In sum, the extension of the characteristics of a single grain-boundary triple junction to a distribution of grain-boundary triple junctions is extremely complicated. However, good agreement with experimental results can be obtained by an appropriate averaging process. In effect, most processes merely involve (implicitly or explicitly) the characteristics of an effective grain-boundary triple junction and a failure criterion. This criterion is equivalent to equating the effective radius of the corresponding hole to a sizable fraction, for example, one-fourth, of the separation between neighboring grain-boundary triple junctions (grain size). When the number of grain-boundary triple junctions is limited by the lateral dimensions of a conducting element, however, random averaging procedures are invalidated, and procedures based on small numbers of grain-boundary triple junctions must be used.

Healing of Damage Caused by Electromigration

According to Eq 1, damage (holes and hillocks) caused by electromigration, as measured by $\Delta R/R$, should be completely reversible when the applied current is reversed, for example, $j \rightarrow -j$ (Ref 22). A typical example is presented in Fig. 8, in which $\Delta R/R$ is plotted as a function of time before and after current reversal in an Al-1 wt% Si conducting element for the temperatures ranging from 150 to 200 °C. It is evident that $\Delta R/R$ increases (approximately) linearly with time until the current is reversed and then decays (heals) rapidly and nonlinearly. Two important features should be noted: first, the rate of decay of $\Delta R/R$ immediately after current reversal ($t = 166$ min) is greater than the rate of growth immediately before current reversal and , second, the full amount of $\Delta R/R$ cannot be reclaimed (incomplete healing). Repeated current reversals produce similar results, with the amount of healing being related to the amount of prior degradation (Ref 14). This phenomenon is related to the electromigration under alternating (pulsed) current (Ref 23).

The rate of increase of $\Delta R/R$ immediately preceding current reversal and the rate of decrease of $\Delta R/R$ immediately succeeding current reversal

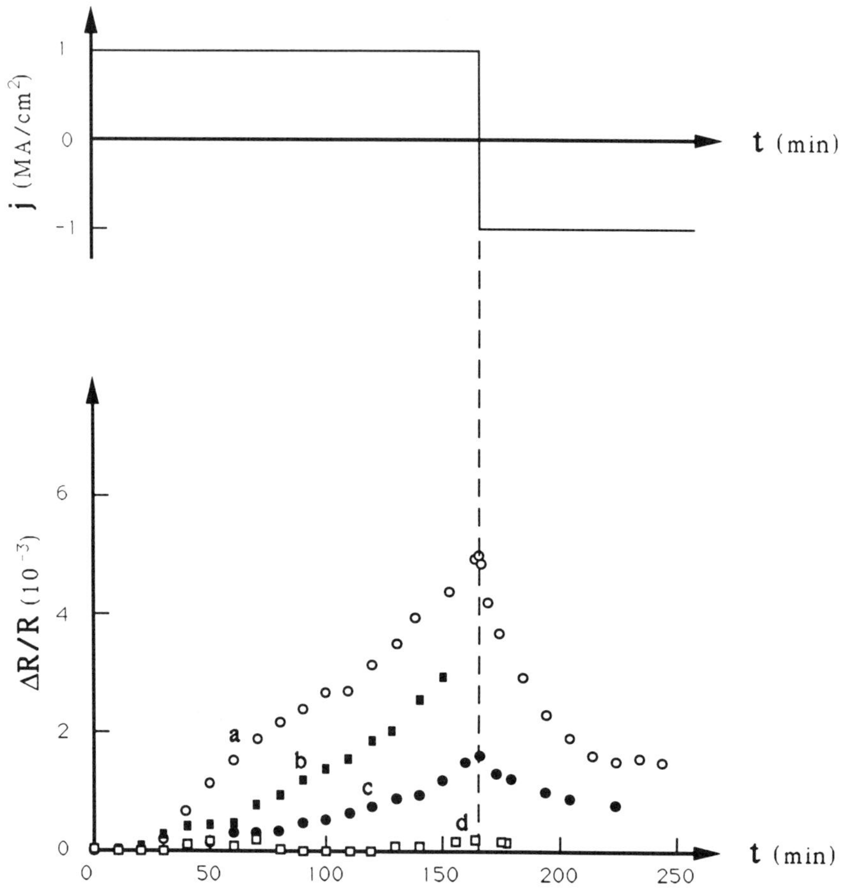

Fig. 8 Variation of fractional change of electrical resistance ($\Delta R/R$) with time (t) at (a) 200 °C, (b) 185 °C, (c) 175 °C, and (d) 150 °C in narrow conducting thin films of Al-1 wt% Si. When current is reversed (t = 166 min), $\Delta R/R$ decreases, resulting in (partial) healing of prior electromigration damage

($j \rightarrow -j$) are especially meaningful quantities, since neither temperature drift nor variations of Joule heating ($j^2\rho$) can produce a singularity in $\Delta R/R$. In general, the rate of change of electrical resistance may be expressed as (see Eq 14):

$$d(\Delta R/R)/dt = (\delta \langle D \rangle \Delta Y/\lambda^2 kt)\, F \qquad \text{(Eq 18)}$$

where F denotes an arbitrary force. Therefore, $(\Delta R/R)/dt$ is proportional to a (driving) force that produces mass transport by grain-boundary diffusion.

Since $d(\Delta R/R)/dt$ immediately preceding current reversal does not equal $d(\Delta R/R)/dt$ immediately succeeding current reversal in either magnitude or functional behavior, a force, in addition to the electromigration force (F_e), must be attained. This force could be due to capillary effects associated with the creation and annihilation of free surfaces (Ref 13) or due to the generation of stress near holes, hillocks, and connecting grain boundaries (Ref 10). Since the first effect should be the largest for the incipient stage of hole/hillock formation or annihilation and the second effect should be the largest when steady state is achieved, the effect of stress is assumed to override the capillary effect.

The effect of localized stress may be introduced by considering the (uniform) stress gradient ($d\sigma/dx$) that resists mass transport along a grain boundary connecting two adjacent triple junctions, as first considered by Blech (Ref 9). The minimum force required to activate hole/hillock growth is obtained by equating the corresponding force $F_B = (1/N)\, d\sigma/dx$ to the electromagnetic force (F_e), or:

$$F_B = (1/N)\, \sigma_0/l_B = z^* q \rho j = F_e \qquad \text{(Eq 19)}$$

where σ_0 is the effective stress barrier and l_B is the minimum (Blech) length necessary to activate cumulative flux divergence. (Typically, $o_0 \approx 1 \times 10^{10}$ dyn/cm^2 and $l_B \cong 15$ μm for j = 1 mA/cm^2 in aluminum.) The net force immediately preceding (–) and succeeding (+) current reversal, therefore, is:

$$F_{\pm} = z^* q \rho j\, (\pm 1 - l_B/l) \qquad \text{(Eq 20)}$$

where l denotes the actual distance between adjacent holes/hillocks. Accordingly, the ratio of $d(\Delta R/R)/dt$ (slope) immediately preceding and immediately succeeding current reversal (s_-/s_+) is given by:

$$s_+/s_- = F_+/F_- = -(1 - l_B/l)/(1 + l_B/l) \qquad \text{(Eq 21)}$$

Results stemming from a large number of experiments on both gold and aluminum conducting elements indicate that $s_+/s_- \approx -1/3$ (see Fig. 8), or $l \approx 2l_B$. In comparison, l_B is computed to be about 5 μm for aluminum (j = 3 mA/cm^2), in satisfactory agreement with many experimental observations (see Fig. 6b). Accordingly, the healing of prior damage produced by electromigration is associated with localized stress in the vicinity of grain boundaries in thin films. In effect, Eq 20 indicates the cumulative flux divergence cannot occur when the driving force is insufficient ($l \le l_B$) or when the diffusion path is too large ($l >> l_B$). Apparently, optimum conditions are achieved when $l \approx 2l_B$.

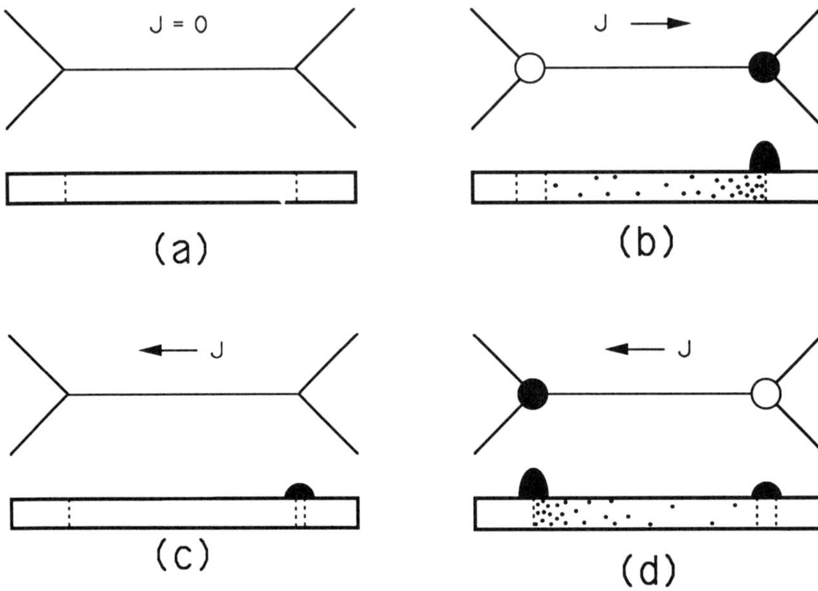

Fig. 9 Damage and subsequent (partial) healing of grain-boundary triple junctions. (a) Initial (top and cross-sectional) configuration of two adjacent grain-boundary triple junctions. (b) Formation of a hole and hillock due to an imbalance of (mass) flux (*J*). (c) Partial healing of prior damage due to a reversed imbalance of *J* (current reversal). (d) Interchange of hole and hillock due to continued imbalance of *J*. Small dots represent a stress gradient in the connecting grain boundary produced by a corresponding concentration gradient

The incomplete healing during current reversal of damage produced by electromigration is explained by Fig. 9, which shows top and cross-sectional views of an idealized pair of triple junctions connected by a common grain boundary. Before the application of current density (j), the entire system is in (metastable) equilibrium (Fig. 9a). When current producing mass flux (J) is applied, matter is displaced, thereby producing cumulative flux divergence and resulting in (for example), a hole/hillock pair, separated by a grain boundary under (compressive) stress (Fig. 9b). When the current is reversed, however, current cannot be easily shunted through the existing hillock; rather, sufficient volume is removed from beneath the hillock to heal (fill in) the hole completely (Fig. 9c). In this case, residual damage remains since the corresponding increase in $\Delta R/R$ produced by the underlying hole is not completely nullified by the residual hillock, even though the net volume change is zero. Continued current reversal converts holes to hillocks and vice versa, but permanent damage increases, since the residual hillock has not been completely absorbed into the underlying hole (Fig. 9d). Thus, the degree of healing is related to prior damage produced by electromigration, as well as mechanical stress produced by localized flux divergence at grain-boundary triple junctions.

Summary

Computational and experimental results, along with corresponding discussion, clearly demonstrate that the degradation and subsequent healing of thin films by electromigration are intimately related to concomitant microstructure. Important relationships between these phenomena based on cumulative flux divergence at grain-boundary triple junctions, have been developed. Important features include:

- Cumulative flux divergence at grain-boundary triple junctions represents an important mechanism for the generation of electromigration damage in thin films. In all cases, effects are due to a delicate balance between current-induced mass transport along grain boundaries, microstructure, and (localized) mechanical stress
- Flux divergence at grain-boundary triple junctions can occur only for distributions of grain-boundary diffusion coefficients and/or distributions of grain-boundary inclinations; otherwise, the corresponding microstructure is immune to degradation by electromigration, regardless of the direction of applied current
- The construction and analysis of a network of interconnected triple junctions is difficult, but often a complex network may be replaced by a single, effective triple junction when the corresponding distribution is sufficient to justify a random averaging process
- When grain size approaches the lateral dimensions of a conducting element, procedures involving a small number of grain-boundary triple junctions must be devised. In the limit of small lateral dimensions, both mean time to failure and the corresponding standard deviation increase
- Both damage and healing caused by electromigration can be interpreted in terms of cumulative flux divergence at grain-boundary triple junctions. Beyond a certain level of damage, however, only incomplete healing can be achieved

In general, consideration of flux divergence at grain-boundary triple junctions provides valuable insight into diffusion-controlled processes associated with the early stages of electromigration in thin films, as well as guidelines for the control of degradation and the prediction of ultimate time to failure in electronic components. Remaining unresolved issues are, first, the experimental investigation of the character of individual grain-boundary triple junctions and second, the development of procedures for combining the characteristics of individual grain-boundary triple junctions to determine macroscopic properties, such as the variation of $\Delta R/R$ with time, temperature, and current density and the ultimate time to failure of the conducting element.

Acknowledgment

Helpful discussions and/or experimental assistance by M. Genut, Z. Li, S. Mahajan, A.G. Milnes, and P.F. Tang and support from the International Business Machines Corporation Center for Thin Film Sciences at Carnegie Mellon University are gratefully acknowledged.

References

1. H.B. Huntington and A.R. Grone, *J. Phys. Chem. Solids*, Vol 20, 1961, p 76
2. H.B. Huntington, *Thin Solid Films*, Vol 25, 1975, p 265
3. F.M. d'Heurle and P.S. Ho, in *Thin Films—Interdiffusion and Reactions*, J.M. Poate, K.N. Tu, and J.W. Mayer, Ed., John Wiley & Sons, 1978, p 243
4. M.J. Attardo and R. Rosenberg, *J. Appl. Phys.*, Vol 41, 1970, p 2381
5. R. Rosenberg and M. Ohring, *J. Appl. Phys.*, Vol 42, 1971, p 5671
6. P.B. Ghate, *Solid State Technol.*, March 1983, p 113
7. S. Vaidya, T.T. Sheng, and A.K. Sinha, *Appl. Phys. Lett.*, Vol 36 (No. 6), 1980, p 464
8. I. Kinsbron, *Appl. Phys. Lett.*, Vol 36 (No. 12), 1980, p 986
9. I.A. Blech and E. Kinsbron, *Thin Solid Films*, Vol 25, 1975, p 327
10. I.A. Blech, *J. Appl. Phys.*, Vol 47, 1976, p 1203
11. C.F. Dunn, F.R. Brotzen, and J.W. McPherson, *J. Elec. Mater.*, Vol 3, 1986, p 273
12. H.H. Hoang and J.M. McDavid, *Solid State Technol.*, Vol 121, 1987
13. P.F. Tang, C.L. Bauer, A.G. Milnes, and S. Mahajan, Paper presented at Fourth International Conference on Quality in Electronic Components, Bordeaux, France, 1989
14. Z. Li, Ph.D. Thesis, Carnegie Mellon University, 1991
15. C.L. Bauer and P.F. Tang, Electromigration in Thin Films, in *Proceedings of the Conference on Metals and Alloys*, DIMETA-88, 1988
16. M. Genut, Z. Li, C.L. Bauer, S. Mahajan, P.F. Tang, and A.G. Milnes, *Appl. Phys. Lett.*, Vol 58, 1991, p 2354
17. J.E. Sanchez, Jr., J.W. Morris, Jr., and J.R. Lloyd, *J. Met.*, Vol 42, 1990, p 41
18. P.F. Tang, A.G. Milnes, C.L. Bauer, and S. Mahajan, *Proc. MRS Symp.*, Vol 167, 1989
19. P.S. Marcoux, P.P. Merchant, V. Naroditsky, and W.D. Rehder, *Hewlett-Packard J.*, June 1989, p 79
20. P.P. Meng, "Computer Simulation of Electromigration in Thin Films," M.S. thesis, Rensselaer Polytechnic Institute, May 1988

21. J. Cho and C.V. Thompson, *Appl. Phys. Lett.*, Vol 54, 1989, p 2577
22. A. Bobbio, A. Ferro, and O. Saracco, *IEEE Trans. Rel.*, Vol R-23, 1974, p 194
23. R.E. Hummel and H.H. Hoang, *J. Appl. Phys.*, Vol 65, 1989, p 1929

Index